# Lecture Notes of the Institute for Computer Sciences, Social Informatics and Telecommunications Engineering 303

More information about this series at http://www.springer.com/series/8197

Qingshan Li · Shengli Song · Rui Li · Yueshen Xu ·
Wei Xi · Honghao Gao (Eds.)

# Broadband Communications, Networks, and Systems

10th EAI International Conference, Broadnets 2019
Xi'an, China, October 27–28, 2019
Proceedings

Springer

*Editors*
Qingshan Li
Xidian University
Xi'an, China

Shengli Song
Xidian University
Xi'an, China

Rui Li
Xidian University
Xi'an, China

Yueshen Xu
Xidian University
Xi'an, China

Wei Xi
Xi'an Jiaotong University
Xi'an, Shaanxi, China

Honghao Gao
Shanghai University
Shanghai, China

ISSN 1867-8211         ISSN 1867-822X   (electronic)
Lecture Notes of the Institute for Computer Sciences, Social Informatics
and Telecommunications Engineering
ISBN 978-3-030-36441-0         ISBN 978-3-030-36442-7   (eBook)
https://doi.org/10.1007/978-3-030-36442-7

# Preface

We are delighted to introduce the proceedings of the European Alliance for Innovation (EAI) International Conference on Broadband Communications, Networks, and Systems (Broadnets 2019). This conference has brought together researchers, developers, and practitioners from around the world who are leveraging and developing broadband communication technology for a smarter and more resilient communication system. The theme of Broadnets 2019 was around wireless communication, sensor networks, underwater communication, mobile computing, and 5G technology.

This year's conference has attracted 61 submissions. All the submissions were reviewed by at least three reviewers. After rigorous review process, 19 full papers were accepted for oral presentation at the main conference sessions, including Track 1 – Wireless Networks and Applications; Track 2 – Communication and Sensor Networks; Track 3 – Internet of Things; Track 4 – Pervasive Computing; and Track 5 – Security and Privacy.

In addition, a panel session was also organized, where panelists from various backgrounds shared their views and visions, addressing the urgent need for emergency management. Apart from high quality technical paper presentations, the technical program also featured one keynote speech which was delivered by Professor Deke Guo from the National University of Defense Technology, China.

Coordination with the steering chairs, Prof. Qingshan Li, Prof. Shengli Song, Dr. Rui Li, and Dr. Yueshen Xu was essential for the success of the conference. We sincerely appreciate their constant support and guidance. It was also a great pleasure to work with such an excellent Organizing Committee team, and we thank them for their hard work in organizing and supporting the conference. In particular we thank the Technical Program Committee (TPC), led by our TPC co-chairs, Dr. Rui Li, Dr. Yueshen Xu, and Dr. Wei Xi, who completed the peer-review process of technical papers and made a high-quality technical program. We are also grateful to conference managers, Kitti Szilagyiova, Katarina Srnanova, and Lucia Sedlarova, for their support. We also thank all authors who submitted their papers to the Broadnets 2019 conference and workshops.

We strongly believe that the Broadnets conference provides a good forum for all researcher, developers, and practitioners to discuss all science and technology that is relevant to broadband communication. We also expect that the future Broadnets conferences will be as successful and stimulating, as indicated by the contributions presented in this volume.

October 2019

Qingshan Li
Shengli Song
Rui Li
Yueshen Xu
Wei Xi
Honghao Gao

# Organization

## Steering Committee

### Steering Committee Chair

Imrich Chlamtac                Bruno Kessler Professor, University of Trento, Italy

### Steering Committee Members

Jizhong Zhao                   Xi'an Jiaotong University, China
Honghao Gao                    Shanghai University, China
Victor Sucasas                 Universidade de Aveiro, Portugal

## Organizing Committee

### General Chairs

Qingshan Li                    Xidian University, China
Shengli Song                   Xidian University, China

### TPC Chair and Co-chairs

Wei Xi                         Xi'an Jiaotong University, China
Rui Li                         Xidian University, China
Yueshen Xu                     Xidian University, China

### Local Chair

Yueshen Xu                     Xidian University, China

### Workshops Chair

Chaocan Xiang                  Chongqing University, China

### Publicity Chairs

Qingmin Zou                    Shanghai University, China
Rossi Kamal                    Matanit, Bangladesh

### Social Media Chair

Zhiping Jiang                  Xidian University, China

### Publications Chair

Youhuizi Li                    Hangzhou Dianzi University, China

**Web Chair**

Si Wen                          Shanghai Business School, China

# Technical Program Committee

| | |
|---|---|
| Rui Li | Xidian University, China |
| Wei Xi | Xi'an Jiaotong University, China |
| Yueshen Xu | Xidian University, China |
| Krishna Kambhampaty | Computer Science at North Dakota State University, USA |
| Fei Dai | Yunnan University, China |
| Ao Zhou | Beijing University of Posts and Telecommunications, China |
| Bin Cao | Zhejiang University of Technology, China |
| Buqing Cao | Hunan University of Science and Technology, China |
| Congfeng Jiang | Hangzhou Dianzi University, China |
| Dongjing Wang | Hangzhou Dianzi University, China |
| Yucong Duan | Hainan University, China |
| Zhuofeng Zhao | North China University of Technology, China |
| Xiaoliang Fan | Shanghai University, China |
| Xuan Gong | Shanghai University, China |
| Guobing Zou | Shanghai University, China |
| Gaowei Zhang | Nanyang Technology University, China |
| Jiwei Huang | China University of Petroleum – Beijing, China |
| Youhuizi Li | Hangzhou Dianzi University, China |
| Haolong Xiang | The University of Auckland, New Zealand |
| Junhao Wen | Chongqing University, China |
| Jian Wang | Wuhan University, China |
| Jīn Liu | Central University of Finance and Economics, China |
| Junaid Arshad | University of West London, UK |
| Jiuyun Xu | China University of Petroleum, China |
| Li Kuang | Central South University, China |
| Lianyong Qi | Qufu Normal University, China |
| Lili Bo | China University of Mining and Technology, China |
| Wenmin Lin | Hangzhou Dianzi University, China |
| Xihua Liu | Nanjing University of Information Science and Technology, China |
| Jianxun Liu | Hunan University of Science and Technology, China |
| Shijun Liu | Shandong University, China |
| Xiao Ma | Beijing University of Posts and Telecommunications, China |
| Shunmei Meng | Nanjing University of Science and Technology, China |
| Mingqing Huang | Shenzhen Institute of Advanced Technology, Chinese Academy of Sciences, China |

# Contents

**Pervasive Computing**

**Security and Privacy**

# Wireless Networks and Applications

# Design and Implementation of Non-intrusive Stationary Occupancy Count in Elevator with WiFi

Wei Shi[✉], Umer Tahir, Hui Zhang, and Jizhong Zhao

Xi'an Jiaotong University, Xi'an 710049, Shaanxi, People's Republic of China
weishi0103@sina.com

**Abstract.** Wi-Fi Sensing has shown huge progress in last few years. Multiple Input and Multiple Output (MIMO) has opened a gateway of new generation of sensing capabilities. This can also be used as a passive surveillance technology which is non-intrusive meaning it is not a nuisance as it is not need the subjects to carry any dedicated device. In this thesis, we present a way to count crowd in the elevator non-intrusively with 5 GHz Wi-Fi signals. For this purpose, Channel State Information (CSI) is collected from the commercially available off-the-shelf (COTS) Wi-Fi devices setup in an elevator. Our goal is to Analyze the CSI of every subcarrier frequency and then count the occupancy in it with the help of Convolutional Neural Network (CNN). After CSI data collection, we normalize the data with Savitzky Golay method. Each CSI subcarrier data of all the samples is made mean centered and then outliers are removed by applying Hampel Filter. The resultant wave is decimated and divided into 5 equal length segments representing the human presence recorded in 5 s. Continuous wavelet frequency representations are generated for all segments of every CSI subcarrier frequency waves. These frequency pattern images are then fed to the CNN model to generalize and classify what category of crowd they belong to. After training, the model can achieve the test accuracy of more than 90%.

**Keywords:** Wi-Fi Sensing · CSI · CWT · CNN

## 1 Introduction

Wi-Fi has become an essential part of our life. The wireless technology is growing and improving at an exponential rate. Today Wi-Fi is being used in desktop computers, laptops, mobile and many Internet of Things devices which are able to provide functionality because of it. With every passing year our technologies and electronic products are becoming tether less that is number of wires are being reduced. When signals from transceiver leaves for the receiving end it interacts with several objects in between. They each cause a specific variation in the radio frequency captured at the receiving end. This recorded variation can help us differentiate how many people are

This work is supported by NSFC Grants No. 61802299, 61772413, 61672424.

Q. Li et al. (Eds.): BROADNETS 2019, LNICST 303, pp. 3–19, 2019.
https://doi.org/10.1007/978-3-030-36442-7_1

standing in the elevator. Although human sensing has been done in the past but none has been done in an elevator so far with Wi-Fi using Channel State Information (CSI).

The main motivation behind this research is that Wi-Fi is relatively new area of research in terms of human sensing capabilities. Though a number of researches have been done in this regard, still much improvements and fine tuning is needed. As I have mentioned before, continuous development in human sensing based on radio signals is the driving force. Moreover, it should also be noted here that this is non-intrusive technique which is not an inconvenience to the monitored subjects as it serves its purpose at no additional cost to the people monitored. It also non-invasive in terms of privacy since the Wi-Fi signals research has not yet matured enough or due to technical limitations of radio signals it cannot record faces. However, recognition of different subjects using radio signals is also being researched upon in the world. Wi-Fi signals does not need the environment to be in proper lighting conditions. It does not make noise; it can work in the dark quietly as well. It serves as a passive tool for studying human detection that does not require Line of Sight to work unlike traditional methods. These are the primary motivation and the reasons why we chose to research in this area.

We have developed a way to count humans in an elevator. The techniques used were studied individually and combined in a sequence that yields more than 90% result accuracy. The first step is to collect the data samples. For this purpose, Wi-Fi CSI data is collected. Several post processing is involved before we feed the data samples to our CNN model. We record 7 full CSI samples for each human category. As we have a pair of receiving and transmitting antennas and we are considering only the receiving antennas for the task of recognizing number of humans, one category includes 14 CSI data samples. 57 sub-carrier level frequency waves in each of 14 CSI data samples. First data is normalized through Savitzky Golay Method. This smooths the data points and a trend becomes slightly visible and the noise is somewhat removed. We apply this to each sub-carrier level frequency wave samples for each instances of human category recorded. We then apply mean centering method to the CSI data. In order to reduce the frequency sampling rate, the data is segmented into 5 equal points this gives us 5 patterns in just one sub-carrier CSI wave. Segmented waves are converted into CWT image patterns. The segmentation allows us to achieve high accuracy rate as it increases the number of pattern images for our neural net for training.

## 2 Related Work

Vision Based approaches uses patterns to recognize such as face detection and human count. Human detection and count techniques for camera-based approaches are very mature by now and they are widely used in professional environments. Since humans are of different shape and sizes and can wear different style dresses, detection of humans through vision-based approaches become slightly more difficult. Only major challenge with vision-based approach is LOS. It can lose track of people when they leave the line of sight and when they arrive in the zone where the camera is incapable of detecting any features. Another challenge with this is proper illumination must be established before detecting humans. Camera-based approach may be disliked because it can record people's faces and features and sometimes it makes people uncomfortable. Prior consent, in public places may be a challenge in itself.

## 2.1 Vision/Camera Based Devices

Vision Based approaches [1–4] uses patterns to recognize such as face detection and human count. Human detection and count techniques for camera-based approaches are very mature by now and they are widely used in professional environments. Since humans are of different shape and sizes and can wear different style dresses, detection of humans through vision-based approaches become slightly more difficult. Only major challenge with vision-based approach is LOS. It can lose track of people when they leave the line of sight and when they arrive in the zone where the camera is incapable of detecting any features. Another challenge with this is proper illumination must be established before detecting humans. Camera-based approach may be disliked because it can record people's faces and features and sometimes it makes people uncomfortable. Prior consent, in public places may be a challenge in itself.

## 2.2 Non-camera Based Devices

Dedicated device-based approach includes wearable devices [5, 6], RFID tags [7–12], mobile phones [13] and other related sensors [14–16] such as smart watches in recent years are used to detect activity and count the crowd in the indoor environment. This technique uses some other resources to provide complete functionality such as Wi-Fi networks or Bluetooth connectivity. Radio frequency-based approaches is relatively new and developing form of detecting humans and their activity. It can be very sensitive to human activity and many scenarios have been under research for some time now including vital sign monitoring, indoor localization, human count and gestures recognition and sleep cycle monitoring. A special device UWB [17–21] radar have been used to count humans. It works on the signal propagation model as humans and other objects present between the transceiver and receiver will affect the propagation of signals. This technique is similar to that of Vision-based methods in terms of not requiring any of the monitored subjects to wear any wearable devices. For earlier work of detecting humans through Wi-Fi [22–29] have Received Signal Strength (RSS) [30–33]. Several research on localization has been done using RSS. It may be vaguely related to the radar but based on Wi-Fi. Our work also falls under radio frequencies as Wi-Fi basically works on radio signals. To tackle some of the problems researchers at MIT have used Hidden Markov Models to make up for human dissimilarities in structure and introduced some constraints with it on human motion variation. Then they map the responses received at the receiving end under a skeleton frame of reference which helps them to detect human activity as well as posture which also keep in view the human structural differentiations among each other. They call it Wi-Vi [34]. It works on similar principles that ultrasound, radar or LIDAR [36] works on. They went on to develop dedicated Wi-Fi systems that can detect sleep patterns, breathing patterns and tracking applications for human motion. Ultrasound and especially lasers are not affected by multi-path propagation.

## 3   Methodology

### 3.1   System Overview

In this section we will discuss the techniques used to project human counts through Wi-Fi CSI data. The architecture will be discussed through the flow chart and the overall flow will be explained. The very first task of our research is to gather CSI data through commodity COTS Wi-Fi hardware. The data is in raw form and must be passed through some data cleaning filters to remove outliers and then the output from pre-processing is fed to CNN. Figure 1 shows the whole flow of our architecture.

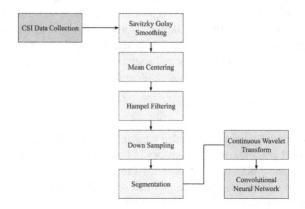

**Fig. 1.** Flow of our technique

### 3.2   CSI Data Collection

As obvious the first step is to setup the Wi-Fi devices in the elevator. We employ a pair of transmitters and receivers. In particular, we setup a Wi-Fi infrastructure, which includes two transmitters and two receivers. Then we selected to use the Intel Wireless Link 5300 NIC to collect the CSI data in the elevator, and the transmission rate is set as 1000 packets per second. One full sample is collected in 5 s that results in 5000 packets in 5 s at each of the receivers and they are recorded with the help of a software called Pico Scenes [35]. The Wi-Fi is configured to use 5 GHz frequency band. There are two transmitters and two receivers located in each corner of the lift.

### 3.3   CSI Data Pre-processing

**CSI Data Smoothing**
The first step after the data is recorded is to smooth it. We use Savitzky Golay Method to smooth our CSI sample data set. It is discussed briefly in the Sect. 2 of this document (Fig. 2).

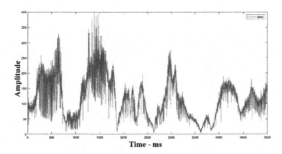

**Fig. 2.** One sub-carrier wave raw before smoothing

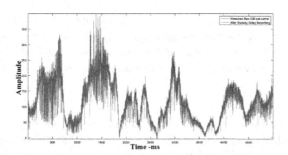

**Fig. 3.** After Savitzky Golay smoothing

As we can see in Fig. 3 the smoothing process is applied to the CSI data as the first thing in the data pre-processing. It works on same principle as a moving average. Local average is calculated by some window size and data is smoothed. After this smoothing is carried out the data is mean centered that is have a zero mean.

**Mean Centering and Filtering**

Mean Centering also refers to have a mean of zero. Centering in simple methodology means subtracting a constant from every value of a variable. Mean in simple words is average of the data that can be calculated [34]. Mean of any normal distribution is not zero. However, we can first normalize the data so that it is mean centered and have one standard deviation, as shown in Fig. 4.

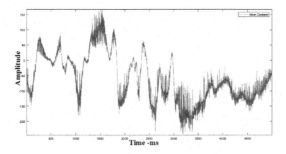

**Fig. 4.** Mean centered after smoothing

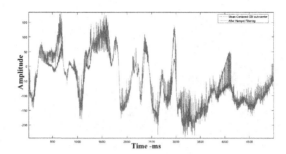

**Fig. 5.** Applying Hampel filtering with window of standard deviation 2

The Hampel filter is a filter that exchanges the middle value in the data window of size pre-decided with the median or standard deviation if it is too far from it. As shown in Fig. 5, Hampel filter looks like to add a specific noise that is only specific to one category of sub-carriers which gives the whole data some recognizable patterns and this is confirmed by CWT images.

**Down Sampling**
Our initial data have up to 5000 data points in 5 s which means the frequency is 1000 packets per second. We can decimate the frequency to be 200 data points per second. It will lower some complexity but the data will maintain its trend. In digital signal processing, down sampling and decimation are terms related with the process of downsampling. When down sampling is done on a sequence of time-series of a signal or other continuous function, it results in an estimate of the sequence that would have been a result of sampling the signal at a smaller sampling rate thus the overall trend of the signal will be there but in lower frequency. The wave is smaller in frequency but still have the same shape as before. As shown in Fig. 6, this technique involves basically translating the signal into a lower frequency rate and produces an approximation of the original signal as it would have been in a lower frequency.

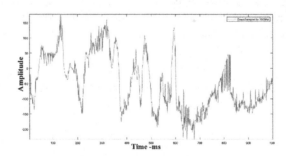

**Fig. 6.** Down sampling to 200 MHz, Time stretches 1000 ms instead of 5000 ms

### 3.4   Continuous Wavelet Transform (CWT)

Continuous wavelet transforms are a way of representing a time-series signal into a scalogram pattern based on the wavelet patterns to be noted in the series. The wavelet function we use called analytical morse which is also known as analytic morse parameter in CWT function in MATLAB.

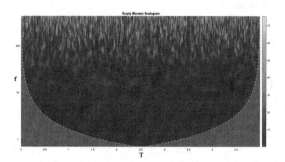

**Fig. 7.** CWT of empty elevator in 5 s

Figure 7 shows one full sub-carrier wave up to 5 s when the elevator is empty. We do not expect much variation in this instance. Figure 8 shows the segmented CWT of one sub-carrier when five people are in the elevator and as you can see the segmentation really result in similar CWT patterns in only one second window.

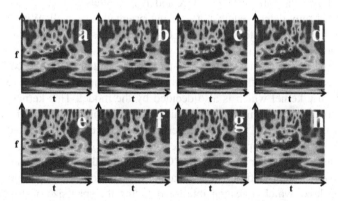

**Fig. 8.** (a–e) Represent one complete sub-carrier wave. (a) first segment representing 1st second, (b) representing 2nd second, (c) representing 3rd second and so on.

As we mentioned in segmentation and mean centering and filtering that CWT results in similar patterns for most of its segmented parts which proves that 1-s window can be useful.

## 3.5 CNN for Feature Extraction and Classification

We use deep learning for solving our classification problem of human occupancy. For this purpose, we make a neural net in MATLAB using its built in Deep learning toolkit and Alexnet. A Convolutional Neural Network (CNN) is also a deep Learning model in which we input an image of small size which is our CSI CWT pattern, importance in terms of learnable weights and biases are assigned and revised based on the cross-entropy loss function (Fig. 9).

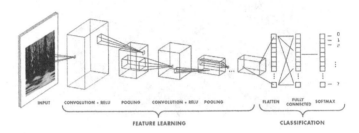

**Fig. 9.** Convolutional neural net overview

As shown in Fig. 8, It has eight fully connected layers as we have 8 categories to classify. 0 means when the elevator is empty and 7 when elevator is occupied by 7 people. The goal of our Convolution neural net operation is to extract the high-level characteristics such as edges, from the input image. We completely rely on CNN model to extract features automatically unlike other machine learning models where we have to extract features manually through PCA and ICA models of feature extraction and reduction. It not only detects edges but gradient orientation as well. With more added layers, the architecture tries to learn High-Level features as well, giving us a network, which has the whole idea and somewhat understanding of our continuous wavelet transform image patterns dataset. The first part of the neural net is the convolutional layer. As explained earlier, this part detects edges and other related features from the image based on a kernel which is also decided by the model. This kernel works as a filter moves to the right from left on the image. It move on by hops to the beginning starting from the left side of the CWT pattern with the same hop value and then keep repeating this process of hoping until the entire image is traversed. Thus, detectable features are detected. As our image is an RGB image all the results are summed with the bias to give us a squashed one-depth channel Convoluted Feature Output. The next is the Pooling layer which is responsible for reducing the span size of the Convolved Feature. This process also reduces dimensionality of our image patterns making it easier to compute them. Moreover, it is also helpful for extracting prominent features which may be rotational and share same position in each image, thus maintaining the process of effectively training of the model. The model is using Max Pooling as default that returns the maximum value from the part of the image that is covered by the Kernel at that iteration. Max Pooling also reduce noise internally. After this, comes the fully connected layers. Fully-Connected layer is usually easy way of learning combination of features.

### 3.6    Algorithm

After data collection is done, CSI is passed through several pre-processing techniques in MATLAB in the following order. First Taking out each sub-carrier of every sample one by one then de-noising and taking mean of zero as follows:

```
for i = 1: 57 all of matrix rows i.e. each sub-carrier as
57 sub-carrier in one sample
    Smooth signal data (Signal_data, sgolay) with Savitzky Golay
method
Signal_data m×n = Signal_data m×n − mean_m×n (Mean Centering)
    Hampel(Signal_data ,σ)   (Hampel Filtering)
end for i
```

After denoising, Down-sampling the sub-carriers to 200 Hz from 1000 Hz. After down-sampling the sub-carriers are segmented into 5 equal parts representing 1 s of window each. Then storing all segmented sub-carriers in one big matrix with labels.

```
for j = 1: 57 all of matrix rows i.e. each sub-carrier
    Down sample the frequency Signal_data by factor of 5
    Segment the Signal_data into 5 equal parts
    Concatenate the each Signal_data to form one Big matrix
    signal_data = U[Y_1, Y_2, Y_3, ..., Y_j] , where Y ∈ CSI
end for j
Make labels for each Signal_data_m×n,  (0,1,2...7)
```

After other pre-processing tasks are done, the output sub-carriers from segmentation is wavelet transformed with labels to their category.

```
for k = 1: N, where N is the number of rows in the
Signal_data
    Wavelet Transform, Time-frequency Representation
    (Signal_data_m×n) of each row
    Read Labels and store respective CWT pattern in rele-
vant directory
end for k
```

After wavelet transforms are done, we randomly select 85% of wavelets for training and the remainder of wavelets are left for test and classification of what category they belong to. Then we launch the CNN model and input data to it.

```
Select training and testing dataset with labels
Load deep learning library ALEXNET
Define learning rate, batch size and fully connected lay-
ers
Run training process over selected CSI sub-carrier pat-
terns with ground truth

After the neural net has been trained on 85% wavelets
then it is time to test and classify
Classify test data with learned attributes
Output Accuracy and Confusion Matrix
end.
```

## 4    Evaluation

In this section, we determine the usability of the methodology proposed in this study. First data is passed through some pre-processing steps and converted into wavelet transforms. Working with the limited data, we had to segment the time-series signal data of all the waveforms collected in the data. After the data has been converted, it was fed to a convolutional neural net. As we rely over CNN to find features for us, it does exactly that and after a successful run of CNN over training dataset of CWT patterns. Our set of CWT test data is passed to the trained-net to classify. The accuracy is based on the true predicted labels.

### 4.1    Accuracy and Confusion Matrices

In the first training experiment, all of the sub-carrier segmented CWT were divided into 85% training and 15% testing data set. This is done randomly and the algorithm decides which images to send to training and testing automatically. After the training is done the classification process yields a confusion matrix as shown in Table 1. The accuracy is decided as follows:

$$Accuracy_1 = \frac{True\ Predictions}{Total\ number\ of\ tested\ images} \times 100 = 97.85\%$$

As the related sub-carrier waves are co-related and thus it can predict classifications of wavelet transforms with fairly good accuracy. Next, we try to judge the whole sub-carrier set of one sample with just assisting it with few of the fragments from that class. For instance, selecting one full CSI sample that have 57 sub-carrier wave data. In that data we select most of the sub-carriers CWT for one sample to send to test dataset and only few of the segments from that sample were sent to the training dataset. After this data was trained it yielded the accuracy of 91.98% and its respective confusion matrix is shown in Table 2. This introduction of fragments to the training was done because of the smaller size of the whole dataset. However, this proves that the method is viable for bigger dataset and can run classifications on unseen dataset.

**Table 1.** Confusion matrix$_1$

| Peo ple | 0 | 1 | 2 | 3 | 4 | 5 | 6 | 7 |
|---|---|---|---|---|---|---|---|---|
| 0 | 100 | 0 | 0 | 0 | 0 | 0 | 0 | 0 |
| 1 | 1 | 96 | 1 | 1 | 0 | 0 | 1 | 0 |
| 2 | 0 | 1 | 97 | 1 | 0 | 0 | 0 | 1 |
| 3 | 0 | 1 | 0 | 97 | 1 | 1 | 0 | 0 |
| 4 | 0 | 0 | 0 | 0 | 99 | 0 | 1 | 0 |
| 5 | 0 | 1 | 0 | 1 | 0 | 98 | 1 | 0 |
| 6 | 0 | 0 | 0 | 0 | 1 | 0 | 98 | 1 |
| 7 | 0 | 0 | 0 | 1 | 0 | 1 | 1 | 97 |

**Table 2.** Confusion matrix$_2$

| Peo ple | 0 | 1 | 2 | 3 | 4 | 5 | 6 | 7 |
|---|---|---|---|---|---|---|---|---|
| 0 | 100 | 0 | 0 | 0 | 0 | 0 | 0 | 0 |
| 1 | 0 | 97 | 1 | 0 | 0 | 1 | 0 | 0 |
| 2 | 0 | 14 | 72 | 2 | 4 | 1 | 5 | 2 |
| 3 | 0 | 2 | 1 | 93 | 1 | 3 | 0 | 0 |
| 4 | 0 | 2 | 0 | 1 | 88 | 2 | 3 | 3 |
| 5 | 0 | 0 | 0 | 0 | 0 | 98 | 1 | 0 |
| 6 | 0 | 0 | 0 | 0 | 1 | 0 | 97 | 2 |
| 7 | 0 | 0 | 0 | 5 | 1 | 1 | 3 | 90 |

## 4.2   Training Process and Iteration Results

The process of training that yields the Confusion Matrix$_2$ is shown in the Table 3. The whole process took approximately 21 h. The training was done on a single CPU. The batch size was chosen to be 20 image patterns per iteration and the learning rate was $1.0e-4$. The table consists of number of epochs, number of iterations, mini-batch accuracy and loss and base learning rate which is mentioned. One epoch of time is completed when all the of CWT training images are iterated in mini-batches once. The mini-batch accuracy and loss is decided over the batch size of per iteration which is as mentioned 20. First it tries to learn the features and predict it. This learning process is performed for each batch and that decides its accuracy and loss. As you can see the learning loss is reducing that points to the information that the model is learning. Consequently, the accuracy is increasing. This mini-batch accuracy is not the accuracy of classification of the test data but only depends upon the batch of images that iteration.

**Table 3.** Training process (over single CPU)

| Epoch | Iteration | Time elapsed (hh: mm:ss) | Mini-batch accuracy | Mini-batch loss | Base learning rate |
|---|---|---|---|---|---|
| 1 | 1 | 00:00:07 | 10.00% | 4.2973 | 1.0e−04 |
| 1 | 50 | 00:04:28 | 20.00% | 1.8777 | 1.0e−04 |
| 1 | 100 | 00:07:56 | 30.00% | 1.9363 | 1.0e−04 |
| 1 | 200 | 00:15:49 | 35.00% | 1.7443 | 1.0e−04 |
| 1 | 500 | 00:39:08 | 45.00% | 1.6446 | 1.0e−04 |
| 1 | 1000 | 01:17:55 | 60.00% | 1.3516 | 1.0e−04 |
| 2 | 1400 | 01:46:21 | 70.00% | 0.6726 | 1.0e−04 |
| 2 | 2000 | 02:20:55 | 80.00% | 0.5751 | 1.0e−04 |
| 2 | 2500 | 02:49:06 | 90.00% | 0.2439 | 1.0e−04 |
| 3 | 3000 | 03:17:46 | 90.00% | 0.2363 | 1.0e−04 |
| 3 | 3500 | 03:46:12 | 85.00% | 0.4313 | 1.0e−04 |
| 3 | 4000 | 04:14:21 | 100.00% | 0.0478 | 1.0e−04 |
| 4 | 4500 | 04:43:41 | 95.00% | 0.0960 | 1.0e−04 |
| 4 | 5000 | 05:18:26 | 95.00% | 0.1244 | 1.0e−04 |
| 4 | 5500 | 05:53:02 | 100.00% | 0.0863 | 1.0e−04 |
| 5 | 6000 | 06:27:41 | 95.00% | 0.0911 | 1.0e−04 |
| 5 | 6500 | 06:58:00 | 100.00% | 0.0396 | 1.0e−04 |
| 6 | 7000 | 07:30:31 | 100.00% | 0.0486 | 1.0e−04 |
| 6 | 7500 | 08:04:35 | 100.00% | 0.0368 | 1.0e−04 |
| 6 | 8000 | 08:39:18 | 100.00% | 0.0055 | 1.0e−04 |
| 7 | 8500 | 09:14:31 | 100.00% | 0.0119 | 1.0e−04 |
| 7 | 9000 | 09:48:05 | 100.00% | 0.0102 | 1.0e−04 |
| 7 | 9500 | 10:22:13 | 100.00% | 0.0188 | 1.0e−04 |
| 8 | 10000 | 10:53:54 | 100.00% | 0.0375 | 1.0e−04 |

(*continued*)

**Table 3.** (*continued*)

| Epoch | Iteration | Time elapsed (hh: mm:ss) | Mini-batch accuracy | Mini-batch loss | Base learning rate |
|---|---|---|---|---|---|
| 8 | 10500 | 11:29:57 | 100.00% | 0.0118 | 1.0e−04 |
| 8 | 11000 | 12:07:11 | 100.00% | 0.0102 | 1.0e−04 |
| 9 | 11500 | 12:40:55 | 100.00% | 0.0078 | 1.0e−04 |
| 9 | 12000 | 13:13:29 | 100.00% | 0.0137 | 1.0e−04 |
| 9 | 12500 | 13:41:36 | 100.00% | 0.0018 | 1.0e−04 |
| 10 | 13000 | 14:09:54 | 100.00% | 0.0006 | 1.0e−04 |
| 10 | 13500 | 14:37:51 | 100.00% | 0.0037 | 1.0e−04 |
| 11 | 14000 | 15:05:45 | 100.00% | 0.0007 | 1.0e−04 |
| 11 | 14500 | 15:33:42 | 100.00% | 0.0084 | 1.0e−04 |
| 11 | 15000 | 16:01:30 | 100.00% | 0.0084 | 1.0e−04 |
| 12 | 15500 | 16:29:22 | 100.00% | 0.0051 | 1.0e−04 |
| 12 | 16000 | 16:57:12 | 100.00% | 0.0044 | 1.0e−04 |
| 12 | 16500 | 17:25:01 | 100.00% | 0.0001 | 1.0e−04 |
| 13 | 17000 | 17:52:50 | 100.00% | 0.0014 | 1.0e−04 |
| 13 | 17500 | 18:20:40 | 100.00% | 0.0101 | 1.0e−04 |
| 13 | 18000 | 18:48:30 | 100.00% | 0.0006 | 1.0e−04 |
| 14 | 18500 | 19:16:20 | 100.00% | 0.0109 | 1.0e−04 |
| 14 | 19000 | 19:44:24 | 100.00% | 0.0008 | 1.0e−04 |
| 15 | 19500 | 20:12:53 | 100.00% | 0.0258 | 1.0e−04 |
| 15 | 20000 | 20:41:15 | 100.00% | 0.0012 | 1.0e−04 |
| 15 | 20500 | 21:09:39 | 100.00% | 0.0139 | 1.0e−04 |
| 15 | 20850 | 21:29:33 | 100.00% | 0.0031 | 1.0e−04 |

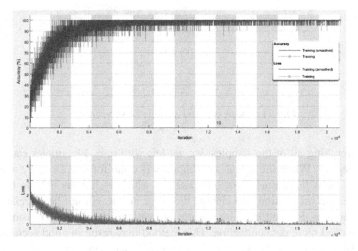

**Fig. 10.** Training process graphical

Figure 10 shows the full graph of the learning and training process of the CNN. The process took a long time to complete over a single CPU. Although the accuracy shows 100% but it is also an approximation of the learning process as there is still some loss in the learning process. There was room for more improvement but we approximate that the loss was nearly zero and for saving time, the training came to a conclusive end with 15 repetitions of learning of each and every CWT patterns involving 20850 iterations in total having 1390 iterations per epoch.

## 4.3   Experiment with Composition of Sub-carriers

In this experiment, the sub-carrier signals for one sample consisting of 57 sub-carriers at each antenna of one category were combined by taking average of 57 sub-carriers and outputting only one wave. The problem for us in this technique is that it drastically reduces size of our dataset which is already small to begin with. Consider the Fig. 4.2 and it has 57 sub-carrier information from one antenna.

We also tried composition of sub-carriers by superimposing all of them over each other in one sample and then some cause constructive interference and some destructive and the resultant were segmented and fed to the CNN. The results were not very good and they were even lower to an extent which was not worth mentioning. So, we tried with taking an average of each data point in time of every sub-carrier signal in one sample and then do the segmentation over it after which it shows promise as may be a good solution in some scenario.

Now when we take its mean for each point in time the whole resultant signal becomes like in the Fig. 7 after mean of zero (Fig. 11).

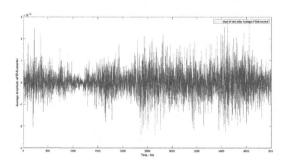

**Fig. 11.**   Mean centered average of sub-carriers

After averaging all of the samples, we are left with only 14 samples each antenna for each category. When the signal segmentation is done for one category, we are left with only 70 CWT patterns from one antenna which is not enough as the person standing changes his posture will change the wave and it will not be correlated to the rest of the samples. For getting more posture and variability information from all of the samples we need more data collection. This method may result in good accuracy if it is done to a large dataset. However, this is yet to be tested and part of our future work.

With what little dataset we had we checked accuracy of this experiment. It yields 51.78% accuracy. The confusion Matrix for this experiment is given below in Table 4.

**Table 4.** Confusion matrix

| Peo ple | 0 | 1 | 2 | 3 | 4 | 5 | 6 | 7 |
|---|---|---|---|---|---|---|---|---|
| **0** | 43 | 29 | 0 | 04 | 14 | 04 | 14 | 0 |
| **1** | 14 | 86 | 0 | 0 | 0 | 0 | 0 | 0 |
| **2** | 14 | 14 | 29 | 14 | 0 | 09 | 29 | 14 |
| **3** | 09 | 2 | 0 | 43 | 0 | 0 | 29 | 0 |
| **4** | 0 | 0 | 14 | 0 | 43 | 14 | 0 | 29 |
| **5** | 09 | 2 | 0 | 0 | 0 | 43 | 14 | 14 |
| **6** | 14 | 14 | 0 | 0 | 0 | 14 | 57 | 14 |
| **7** | 14 | 0 | 0 | 0 | 0 | 0 | 14 | 71 |

This is why we treat every sub-carrier wave of every sample from all the categories as features of their respective class. This gives us more information and makes up for all possible variability caused by the multipath propagation of different signals.

## 5 Conclusion

We have developed a way to judge a signal at sub-carrier level to classify human occupancy in the elevator. The proposed technique involves a combination of easily available commercial off the shelf Wi-Fi hardware. We set it up in the elevator and configure to use 5 GHz band for increased sensitivity. The use of 5 GHz gives us a greater number of sub-carrier channel state information and those we have used as features. After pre-processing, all the sub-carrier segments are wavelet transformed and then they are fed to Alexnet a CNN model for feature extraction and classification. We rely on the capabilities of CNN for feature extraction automatically and after learning over a set of training data of wavelet transforms of sub-carriers of different labels. A set of unseen test wavelet transforms are given to the neural net to classify them to their true categories.

We experimented with compositing all the sub-carrier signals before segmenting and wavelet transforms of each sample with mean of their respective data points in time and some features are lost in this process and it reduces the data size and for a small dataset the accuracy will be decreased.

That is why each sub-carrier level information is taken as a possible feature for its class because of the small size of dataset and it gives us a good point to start. The model predicts the sub-carrier wavelet transforms with more than 97% accuracy and when we feed a complete sample of unseen patterns for respective categories with only few segments of them as training, the model predicts them with more than 91% accuracy. This proves that when training over a large dataset of several hundred samples of CSI of each category it will yield a good percentage of accuracy even for unseen samples.

# References

1. Li, M., Zhang, Z., Huang, K., Tan, T.: Estimating the number of people in crowded scenes by MID based foreground segmentation and head-shoulder detection. In: ICPR 2008. IEEE (2008)
2. Nichols, J.D., et al.: Multi-scale occupancy estimation and modelling using multiple detection methods. J. Appl. Ecol. **45**(5), 1321–1329 (2008)
3. Lin, S.-F., Chen, J.-Y., Chao, H.-X.: Estimation of number of people in crowded scenes using perspective transformation. IEEE Trans. Syst. Man Cybern.-Part A: Syst. Hum. **31**(6), 645–654 (2001)
4. Kim, M., Kim, W., Kim, C.: Estimating the number of people in crowded scenes. In: Proceedings of the IS&T/SPIE Electronic Imaging, p. 78820L (2011)
5. Weppner, J., Lukowicz, P.: Bluetooth based collaborative crowd density estimation with mobile phones. In: PerCom 2013. IEEE (2013)
6. Schauer, L., Werner, M., Marcus, P.: Estimating crowd densities and pedestrian flows using Wi-Fi and Bluetooth. In: Mobiquitous 2014. ICST (2014)
7. Wang, J., Katabi, D.: Dude, where's my card?: RFID positioning that works with multipath and NOLS. In: Proceedings of ACM SIGCOMM (2013)
8. Wang, J., Vasisht, D., Katabi, D.: RF-IDraw: virtual touch screen in the air using RF signals. In: Proceedings of ACM SIGCOMM (2015)
9. Wang, J., Xiong, J., Jiang, H., Chen, X., Fang, D.: D-Watch: embracing "Bad" multipaths for device-free localization with COTS RFID devices, pp. 253–266 (2016)
10. Wei, T., Zhang, X.: Gyro in the air: tracking 3D orientation of batteryless internet-of-things. In: Proceedings of ACM MobiCom, pp. 55–68 (2016)
11. Yang, L., Chen, Y., Li, X.-Y., Xiao, C., Li, M., Liu, Y.: Tagoram: real-time tracking of mobile RFID tags to high precision using COTS devices. In: Proceedings of ACM MobiCom (2014)
12. Ding, H., et al.: Human object estimation via backscattered radio frequency signal. In: INFOCOM 2015. IEEE (2015)
13. Wirz, M., Franke, T., Roggen, D., Mitleton-Kelly, E., Lukowicz, P., Troster, G.: Probing crowd density through smartphones in city-scale mass gatherings. EPJ Data Sci. **2**(1), 1 (2013)
14. Lam, K.P., et al.: Occupancy detection through an extensive environmental sensor network in an open-plan office building. IBPSA Build. Simul. **145**, 1452–1459 (2009)

15. Jiang, C., Masood, M.K., Soh, Y.C., Li, H.: Indoor occupancy estimation from carbon dioxide concentration. Energy Build. **131**, 132–141 (2016)
16. Wang, S., Burnett, J., Chong, H.: Experimental validation of CO2-based occupancy detection for demand-controlled ventilation. Indoor Built Environ. **8**(6), 377–391 (2000)
17. Depatla, S., Muralidharan, A., Mostofi, Y.: Occupancy estimation using only WiFi power measurements. IEEE J. Sel. Areas Commun. **33**(7), 1381–1393 (2015)
18. Choi, J.W., Quan, X., Cho, S.H.: Bi-directional passing people counting system based on IR-UWB radar sensors. IEEE Internet Things J. **5**, 512–522 (2017)
19. Mohammadmoradi, H., Yin, S., Gnawali, O.: Room occupancy estimation through WiFi, UWB, and light sensors mounted on doorways. In: Proceedings of the 2017 International Conference on Smart Digital Environment, pp. 27–34 (2017)
20. Lv, H., et al.: Multi-target human sensing via UWB bio-radar based on multiple antennas. In: Proceedings of the IEEE TENCON, pp. 1–4 (2013)
21. He, J., Arora, A.: A regression-based radar-mote system for people counting. In: Proceedings of the IEEE PerCom, pp. 95–102 (2014)
22. Abdelnasser, H., Youssef, M., Harras, K.A.: WiGest: a ubiquitous WiFi-based gesture recognition system. In: Proceedings of IEEE INFOCOM (2015)
23. Chen, B., Yenamandra, V., Srinivasan, K.: Tracking keystrokes using wireless signals. In: Proceedings of ACM MobiSys (2015)
24. Ding, H., et al.: RFIPad: enabling cost-efficient and device-free in-air handwriting using passive tags. In: Proceedings of IEEE ICDCS (2017)
25. Li, H., Yang, W., Wang, J., Xu, Y., Huang, L.: WiFinger: talk to your smart devices with finger-grained gesture. In: Proceedings of ACM UbiComp (2016)
26. Lien, J., et al.: Soli: ubiquitous gesture sensing with millimeter wave radar. ACM Trans. Graph. **35**(4), 142 (2016)
27. Pu, Q., Gupta, S., Gollakota, S., Patel, S.: Whole-home gesture recognition using wireless signals. In: Proceedings of ACM MobiCom (2013)
28. Wang, W., Liu, A.X., Shahzad, M., Ling, K., Lu, S.: Understanding and modeling of WiFi signal based human activity recognition. In: Proceedings of ACM MobiCom (2015)
29. Xi, W., et al.: Electronic frog eye: counting crowd using WiFi. In: INFOCOM 2014. IEEE (2014)
30. Xu, C., et al.: SCPL: indoor device-free multi-subject counting and localization using radio signal strength. In: IPSN 2013. IEEE (2013)
31. Zheng, Y., Zhou, Z., Liu, Y.: From RSSI to CSI: indoor localization via channel response. ACM Comput. Surv. **46**(2), 1–32 (2013)
32. Adib, F., Katabi, D.: See through walls with WiFi! ACM SIGCOMM Comput. Commun. Rev. **43**(4), 75–86 (2013)
33. Nakatani, T., Maekawa, T., Shirakawa, M., et al.: Estimating the physical distance between two locations with Wi-Fi received signal strength information using obstacle-aware approach. Proc. ACM on Interact. Mob. Wearable Ubiquit. Technol. **2**(3), 1–26 (2018)
34. Adib, F.: MIT. Wi-Vi: See Through Walls with Wi-Fi Signals [EB/OL], August 2013. http://people.csail.mit.edu/fadel/wivi/radar.png
35. GitLab: PicoScenes Installation [EB/OL], 11 August 2018. http://gitlab.com/wifisensing/PicoScenes-Setup/
36. Tamas, L., Lazea, G.: Pattern recognition and tracking dynamic objects with LIDAR. In: Robotics. VDE (2011)

# Human Activity Recognition Using Wi-Fi Imaging with Deep Learning

Yubing Li, Yujiao Ma, Nan Yang, Wei Shi[✉], and Jizhong Zhao

Xi'an Jiaotong University, Xi'an 710049, Shaanxi, People's Republic of China
{yubingli0513,weishi0103}@sina.com

**Abstract.** Robots have been increasingly used in production line and real life, such as warehousing, logistics, security, smart home and so on. In most applications, localization is always one of the most basic tasks of the robot. To acquire the object location, existing work mainly relies on computer vision. Such methods encounter many problems in practice, such as high computational complexity, large influence by light conditions, and heavy crafting of pretraining. These problems have become one of the key factors that constrains the precise automation of robots. This paper proposes an RFID-based robot navigation and target localization scheme, which is easy to deploy, low cost, and can work in non-line-of-sight scenarios. The main contributions of this paper are as follows: 1. We collect the phase variation of the tag by a rotating reader antenna, and calculate the azimuth of the tag relative to the antenna by the channel similarity weighted average method. Then, the location of the tag is determined by the AoA method. 2. Based on the theory of tag equivalent circuit, antenna radiation field, and cylindrical symmetry oscillator mutual impedance, the phenomenon of RSS weakening of adjacent tags is analyzed. Based on this phenomenon, we achieve accurate target localization and multi-target relative localization by utilizing region segmentation and dynamic time warping algorithms. 3. The proposed scheme is lightweight and low-cost. We built a prototype system using commercial UHF RFID readers and passive tags, and conduct extensive experiments. The experimental results show that the model can effectively achieve the precise location of the robot and the object with an average error of 27 cm and 2 cm.

**Keywords:** RFID · Indoor localization · Tag mutual interference

## 1 Introduction

With the development of internet of things, how to interconnect the traditional physical world with the information world becomes hotspots. In order to manage items, people need to quickly identify the identity of each item. Many automatic identification technologies have been widely used, such as Barcode, Radio Frequency Identification (RFID) as well as numerous biological feature-based recognition technology (for example fingerprint recognition, face recognition, speech recognition, etc.). However,

---

This work is supported by NSFC Grants No. 61802299, 61772413, 61672424.

Q. Li et al. (Eds.): BROADNETS 2019, LNICST 303, pp. 20–38, 2019.
https://doi.org/10.1007/978-3-030-36442-7_2

in practical applications, Barcode has the disadvantage of limited recognition distance, easy to be damaged, and sensitivity to light. Radio Frequency Identification, a communication technology that uses RF signals to realize contactless identification, which provides a new development opportunity for automatic identification technology. RFID combines the advantages of other automatic identification technologies, with the property of non-contact automatic fast identification, accurate and efficient identification, low cost, and low power consumption. It provides an effective solution for IoT applications.

Due to its fast reading speed and long recognition distance, RFID has been widely used in smart warehousing, smart logistics, smart home and other scenarios. For example:

(1) In warehousing, RFID has been used to accurately track and manage products. However, it is difficult to realize the automatic management of products by simply relying on RFID technology right now. It is still necessary to perform mechanical manpower manipulation such as forklifts. Due to the imperfect sensing function and limited computing power, the existing robot technology is difficult to apply to large, complex storage systems. We envision that if the surrounding environment can be sensed by RFID tags in the warehouse, accurate tracking of robots and products can be realized. It can help the robot to quickly find goods, improve management efficiency, and save labor costs.

(2) In the bookstores, libraries and other scenes, the registration and management of books is a time-consuming task. If we can use RFID technology to help readers to quickly find the books or help the librarian to find missing or out-of-order books on the shelves, work efficiency will be greatly improved.

There are many similar applications, such as supply chain management, airport baggage tracking and so on. RFID has been widely used in many aspects of daily life, which improves the work efficiency. However, it needs to be clear that there are still many problems that need to be solved urgently. In this paper, we pioneer to use RFID to help navigating the robot and localizing the target object. Extensive experiments demonstrate the effectiveness of our proposed solution.

## 2    Related Works

Robots have been deployed and used in a variety of scenarios such as logistics management, baggage sorting, security access control, and home use. The in these scenarios, navigation is a fundamental task to ensure that the robot works properly. The in terms of indoor navigation, the work of the predecessors can be divided into two categories: computer vision-based methods and radio frequency signal-based methods.

Computer vision based method [2–4] mainly relies on optical sensors (such as cameras, light sensors) and path planning, robot control algorithms. This method is currently the most accurate and mature, but the shortcomings are obvious: (1) this method usually requires a lot of pre-training to identify a specific target; (2) because the perception in reality usually depends on visual information such as pictures or videos, socks, it is easy to make mistakes when identifying items having the same shape.

(3) the method is extremely sensitive to changes in ambient light and background; (4) of the computational complexity is high and more computational resources are required. (5) since visual information is used, this method cannot be applied to scenes that are not line-of-sight (such as obstacle obstruction).

Due to the breakthrough of image information can only be taken from the limits of the line of sight, the method based on radio frequency signals has attracted more and more people's attention. Many robot navigation methods based on rf signals (such as wifi, ultrasound, etc.) have been proposed [5–7]. The accuracy of this method still does not meet the requirements, especially in the final grab operation. In addition, wireless signals (such as wifi) do not support direct recognition of specific objects. It is too expensive to binding additional equipment to the item to report the item information. In order to reduce costs, people put cheap RFID tags on the target items to achieve the purpose of item identification. On this basis, many indoor positioning methods based on RFID tags have been proposed. Pinit [8] the multipath effect feature is used as the fingerprint of the tag space location. The method is based on the fact that adjacent RFID tags are subject to similar environmental influences and thus exhibit similar multipath effects. Tagoram [9] using simulated moving tag reverse synthetic aperture radar (inverse sar), would improve the accuracy of positioning to the centimeter. Rf-idraw [10] can track the moving tag and can infer the movement of the tag. However, the method is subject to many restrictions, such as the need for specific equipment, such as software-defined radio equipment (usrp); the need to deploy antennas around the target tag, etc., it is difficult to apply to large-scale applications. Some other work, such as [11, 12], focusing on the use of RFID tags for relative position (such as the order of books) positioning, and cannot adapt to the scene of assisted robot navigation.

Results in research, are many there at the method, of a using spatial reference signal tags for taking fingerprints and positioning. At the use of at the signal intensity (received signal strength on indicator, RSS) of at the attenuation characteristics of land-marc [13], vire [14] etc., that is subjected to a using localization similar multipath effect adjacent tags. At the method, does not need its measure at the distance to the between and at the reader at the tag, and has at the good tolerance to changes in scene. However, its measurement accuracy depends on the density of the reference tag, and requires a lot of preliminary work such as measurement, deployment, calibration, etc., so it is inefficient.

Because the data collected by a single antenna is very limited, and single-point communication is greatly interfered by the environment, large-scale deployment of antennas is not allowed in both cost and practical application scenarios. In order to collect more valuable information, people will use m antennas. Moving up to collect tag information at different locations in n, like an antenna array with m x n virtual antennas, is very similar to the principle of synthetic aperture radar (sar). The work using this method is mobitagbot [14], tagoram [11] and so on. However, the accuracy of the method depends on the granularity of the square division. The finer the granularity, the higher the accuracy, but the computational complexity will also increase, so the computing resources are demanding more. The in addition, due to the limitation of the RFID communication protocol, is prone to device the commercial when packet loss the target at moves the high speed and target is too much. Therefore, the method is only suitable for tracking low-speed moving targets.

In summary, how to combine RFID technology with practical application scenarios to achieve more convenient, effective, economical and practical applications is an important direction for the future development of RFID technology.

## 3  System Overview

Position acquisition is a key step in robotic picking. This topic aims to study how to use RFID technology to help robots get target positions faster and more accurately. Analyze the process of robot picking goods. The process has two main steps: (1) to the existing data, obtain the target plane coordinates and move to the target shelf; (2) obtain the accurate space coordinates of the target, and the robot arm captures the target. According to the process, we designed a two-stage navigation positioning model: firstly, the robot is navigated to the vicinity of the target through a coarse-grained navigation algorithm, and then the target is accurately found through a fine-grained positioning algorithm. The principle of the system principle is shown in Fig. 1.

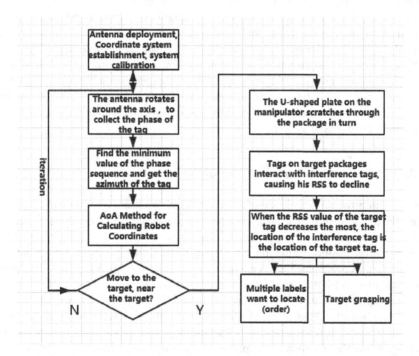

**Fig. 1.** System principle flow diagram

The coarse-grained navigation algorithm uses the AoA algorithm. The difficulty of the algorithm is how to accurately find the azimuth of the tag. Here, using the backscattering model of RFID, the antenna is moved to the tag for the first time by changing the distance of the antenna, and then the phase is changed. By using the phase change trend, the azimuth is reversed by weighting the channel parameters. Then, the AoA is used. The method gets the coordinates of the tag.

When the two tags are very close, due to the electromagnetic coupling between them will produce mutual impedance, thus affecting the radiation power of the antenna, the performance of the reader to read the RSS appears certain level of decline, we call for the tag adjacent mutual interference phenomenon. The in order to obtain good reading and writing effects, in general, the phenomenon caused by mutual interference of adjacent tags should be avoided as much as possible. Of the spacing between tags should be increased as much as possible to the reduce mutual electromagnetic interference. Although is the phenomenon was discovered very early, there has been a lack of clear theoretical explanations and few s in this area. This paper analyzes the phenomenon in detail from the field of electromagnetic field and gives a clear theoretical explanation. And for the first time, this phenomenon is applied to the field of RFID positioning, which makes use of the characteristics of the movement of the in the process, the mutual interference is first enhanced and then weakened. The feature determines the position of the target tag by detecting the minimum point of the "v" shaped region where the RSS first drops and then increases.

This paper also designs a large number of experiments, through the data collected in the field to analyze the errors that may be encountered, and designed a series of model algorithms to filter out noise interference as much as possible, reducing errors, including filtering, region segmentation, dynamic time warping and curves. Fitting, etc.

The likelihood of the model system is verified by experiments. The factors that may affect the accuracy of the system are analyzed through experiments. The optimal deployment plan is obtained, which provides an experimental reference for the effective application of the model system.

## 4   Accurate Indoor Localization of Robot

In order to improve the efficiency of goods picking, we use robots instead of manual sorting. In the face of a new task, how to find the location of the goods to achieve accurate picking is our primary problem. This chapter focuses on solving the problem of how to get the position of the robot. Knowing the real-time position of the robot, you can make further route planning to navigate the robot to the target location. This paper using triangulation location method acquires the target position of the robot, which are the basic steps: (1) first, two antennas are placed in different locations, so that the antenna in a specific angular speed about its central axis, while acquiring the read signal characteristics of the antenna; (2) according to the signal characteristics of the target tag, the angle of arrival of the path of the antenna to the target tag relative to the x-axis is obtained; (3) according to the obtained target tag to the angle of arrival of the two antennas, and the known the coordinates of the two antennas, using the triangle the positioning method obtains the coordinates of the target tag. The individual steps are explained in detail below.

## 4.1   Channel Similarity-Based Tag Azimuth Estimation

Weighted average estimated azimuth using channel similarity channel model propagation of the parameter signal on may be a wireless channel formed from h represents by [15, 16]:

$$h = ae^{-j\theta} \tag{1}$$

Where $\alpha$ is a parameter representing the attenuation of the signal, indicating the amplitude of the signal, generally related to the distance; $\theta$ indicating the frequency offset (i.e. phase) of the signals [11, 14, 17]. The process of antenna rotation, n sets of data are $\varphi$ collected near each angle, that is, the reader and the encode communicate n times, and the theoretical value of the phase at each communication can be calculated by the formula, then the i the channel parameters that the reader antenna receives the tag backscattering during the secondary communication can be expressed as:

$$h_i = e^{-j\vartheta_i} \approx e^{-j\frac{4\pi}{\lambda}\sqrt{c_1-c_2\cdot\cos(\omega t_i-\Phi_T)}} \tag{2}$$

Where $\vartheta_i$ is the measured value of the phase at the ith communication and $\varphi_T$ is the actual azimuth of the tag. In the conventional AoA method, the antenna rotation angle is $\varphi$, the relative power is calculated by the formula:

$$P(\varphi) = \left| \frac{1}{n} \sum\nolimits_{i=0}^{n} h_i e^{-j\frac{4\pi}{\lambda}r\cos(\omega t_i-\varphi)} \right|^2 \tag{3}$$

The formula for the angle of $\varphi$ all the measured values were averaged. In fact, this formula is to find the similarity between all the measured values and the theoretical values at each angle, and then do the arithmetic average. Obviously, the antenna rotation, if and only if $\varphi = \phi_T$ the time, $P_{(\varphi)}$ obtain the maximum value, resulting to obtain the antenna azimuth.

However, the hardware circuit introduces phase shift $\theta_{div}$, that is

$$\vartheta_i = \theta_i(\varphi) + \theta_{div} \tag{4}$$

The AoA method requires very accurate accuracy of the angle of arrival, and a small offset can cause very large errors, so hardware errors must be eliminated. Here, the first value of the phase sequence acquired at each angle is used as a reference value, and each channel parameter is divided by the channel reference:

$$Q(\varphi) = \frac{P(\varphi)}{h_1^2} = \left| \frac{1}{n} \sum\nolimits_{i=1}^{n} \frac{h_i}{h_1} e^{-j\frac{4\pi}{\lambda}r\cos(\omega t_i-\varphi)} \right|^2 = \left| \frac{1}{n} \sum\nolimits_{i=1}^{n} e^{-j(\vartheta_i-\vartheta_1)} e^{-j\frac{4\pi}{\lambda}r\cos(\omega t_i-\varphi)} \right|^2 \tag{5}$$

The phase offset caused by the hardware error $\theta_{div}$ is removed.

In addition, since the measured phase of the tag has some random errors and obeys a Gaussian distribution with a standard deviation of 0.1 [11], the weighted average of the similarity of all measured values and theoretical values is taken here:

$$(\varphi) = \left| \frac{1}{n} \sum\nolimits_{i=0}^{n} w_i e^{-j(\vartheta_i - \vartheta_1)} e^{-j\frac{4\pi}{\lambda}r\cos(\omega t_i - \varphi)} \right|^2 \tag{6}$$

In the formula, $w_i = f(\vartheta_i - \vartheta_1; c_i, 0.1 \times \sqrt{2})$ each measured value and the theoretical value of the similarity weights, there is $f(x; \mu, \sigma) = \frac{1}{\sigma\sqrt{2\pi}} e^{-\frac{(x-\mu)^2}{2\sigma^2}}$ Gaussian $\mathcal{N}(\mu, \sigma)$ probability density function; first phase sequence i number relative to the first a theoretical value of the number $c_i = \theta_i(\varphi) - \theta_1(\varphi) = \frac{4\pi}{\lambda}r[\cos(\omega t_1 - \varphi) - \cos(\omega t_i - \varphi)]$. When the azimuth angle tag $\varphi$, the Gaussian $\theta_{div}$ distribution, i.e. $\theta_{div} = \vartheta_i - \theta_i(\varphi) \sim \mathcal{N}(0, 0.1)$, so $(\vartheta_i - \vartheta_1) - [\theta_i(\varphi) - \theta_1(\varphi)] = [\vartheta_i - \theta_i(\varphi)] - [\vartheta_1 - \theta_1(\varphi)] \sim \mathcal{N}(0, 0.1 \times \sqrt{2})$; Therefore $(\vartheta_i - \vartheta_1) \sim \mathcal{N}(c_i, 0.1 \times \sqrt{2})$ antenna revolution, each angle is calculated $\varphi$ at $R(\varphi)$ a value can be obtained power profile, the maximum $R(\varphi)$ corresponding to the azimuth angle is the tag.

## 4.2    Tag Location Estimation

In an actual system, N (N $\geq$ 2) antennas are deployed in a fixed position, and any three antennas are not collinear. Each antenna $\omega_i$ $(0 < i \leq N)$ rotates about its central axis O at a fixed angular velocity. When the phase value of the antenna receiving target tag is the Largest, The Opposite angle IS The direction of the target tag. In 3.1 the direction of the phase angle, we have determined the sequence of the tag $\varphi$ and its error $\varepsilon$, The first known i coordinate of antennas $(x_i, y_i)$, corresponds to the angle of direction of the tag $\varphi_i$, the error is $\varepsilon_i$, the first i $(0 < i \leq N, N \geq 2)$ antennas the straight line equation point from the center point to the target tag is:

$$y - y_i = tan(\varphi_i) \cdot (x - x_i), \ 0 < i \leq N, \ N \geq 2 \tag{7}$$

Obviously, the position of the target tag can be found by at least two antennas. However, the radio frequency signal is greatly interfered by the environment, in order to improve the positioning accuracy and improve the robustness of the system, there are often more than one pair of antennas deployed in the actual environment. We make a simple transformation of the formula (7):

$$k_i x - y = k_i x_i - y_i, \ 0 < i \leq N, \ N \geq 2 \tag{8}$$

Where: $k_i = tan\varphi_i$ is $(x_i, y_i)$ the slope of the line passing through the antenna. Then n antennas can constitute the following overdetermined equation:

$$C \begin{bmatrix} x \\ y \end{bmatrix} = d \tag{9}$$

Where: $C = \begin{bmatrix} k_1 & -1 \\ \vdots & \vdots \\ k_N & -1 \end{bmatrix}$ is a $N \times 2$ matrix; $d = \begin{bmatrix} k_1 x_1 - y_1 \\ \vdots \\ k_N x_N - y_N \end{bmatrix}$ is a N $\times$ 1 vector.

When $|k_i| = \infty$, for example $\varphi_i = k\frac{\pi}{2}$, $k = 1, 3, 5, \ldots$, the Eq. (8) can be expressed as $x = x_i$, the corresponding correspondence of the matrix C and the vector d is respectively a $[1, 0]$ and $[x_i]$.

When $rank(C) = 2 \leq N$ the rank of the matrix C, the Eq. (9) has a unique least squares solution $P(x, y)$,

$$\begin{bmatrix} x \\ y \end{bmatrix} = (C^T C)^{-1} C^T d \tag{10}$$

This is the coordinates of the target tag.

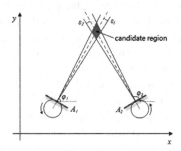

**Fig. 2.** Schematic diagram of dual antenna AoA positioning model (Color figure online)

The above algorithms all assume that the reader antenna and the tag are on the same level, but in actual scenarios, this assumption is difficult to strictly satisfy. When the antennas are not on the same level, the AoA algorithm actually produces a projection of the on a horizontal plane coordinates tag. At this time, the between at the antenna at the distance, at the tag and at the coordinate plane can be separately the measured in advance, or when the trigonometric function and at the is used to true to the calculate at the coordinates of at the tag.

Assuming that the azimuth estimation error of the tag relative to the i-th antenna is $\varepsilon_i$, then a candidate region is actually determined, as shown by the blue region in Fig. 2. Obviously, the farther the tag is from the antenna, the $\varepsilon_i$ larger the angle error, the larger the area of the candidate area, and the coarser the positioning accuracy. The AoA method requires dense deployment of the antenna array.

## 5  Tag Mutual Interference Theory-Based Robot Arm Gripping Algorithm

In actual deployment, when the distance between the tags is too close, the electromagnetic fields of the tag antennas are coupled to each other. In addition, the tags closer to the reader antenna absorb the energy emitted by a part of the readers, causing the electromagnetic waves to reach the distance of the tags farther away. Small, read rate drops, and even unable to read [18, 19]. Therefore, in general, tags should be avoided from being overly densely deployed. In the previous section, the system can acquire the rough position of the robot and the target item through multiple rotating antennas deployed in a fixed position, thereby navigating the robot to the vicinity of the shelf where the target item is located block. In this section, in order to solve the problem of how to accurately grab the item, we have established a tag mutual interference model, which uses the mutual interference characteristics between the tags when the interference tag crosses the target tag, so that the RSS value of the target tag first decreases and then grows. The "v" shaped area, by detecting this area, obtains the position of the target, which is convenient for the robotic arm to capture the target.

## 5.1    Area Segmentation

Assuming that the sliding double-interference tag passes only one target tag once, the RSS of the target tag is redundant data for a period of time before and after the interference tag passes (i.e., the target tag is outside the range of the interference tag). Therefore, the system needs to segment the data to accurately find the area where the target tag is disturbed. Here, the area where the smooth portion is for dominant in a region becomes a smooth region, and the region where the non-smooth portion is dominant is called a non-smooth region.

The signal segmentation technology is commonly used in the field of speech and image processing, including static segmentation and dynamic adaptive segmentation. In this model, the time required for the interference tag to pass the target tag is not fixed, so static segmentation cannot be simply used. Considering the real-time nature of the model, a simple, accurate, and efficient algorithm should be chosen to instantly segment the target area, the even if all data is not available. Therefore, in this paper we have chosen the method of sliding window for region segmentation.

Let the RSS timing of the target be $= [x_0, \ldots, x_i, \ldots, x_n]$. Let the size of the sliding window be w, then the sequence in the i-th sliding window is $W_i = [x_i, \ldots, x_{i-1+w}]$. Since the minimum granularity of RSS collected by the reader is 0.5 dBm, and the amplitude of RSS in the static and non-interfering space does not exceed $\pm 1$ dBm, this means that the values in the smooth region of the RSS sequence are mostly repeated values, so the sliding is calculated here. The information entropy of the window to distinguish between smooth and non-smooth areas:

$$H(W_i) = -\sum_{j=0}^{w} p(x_j) \log p(x_j) \tag{11}$$

Where $p(x_j)$ represents the probability that the value in the sliding window sequence is $x_j$. The larger the entropy, $\sum_{j=0}^{w} p(x_j) = 1$ the greater the uncertainty of the data points in the sliding window. Combined with the characteristics of the RSS sequence here, when the entropy exceeds a certain threshold, it is considered as a non-smooth region. We take the entropy of the RSS sequence acquired in the static interference-free scene as the threshold here.

In addition, due to the sensitivity of RSS, the "bump" and human interference of the tag will bring data jitter. The information entropy can only distinguish between smooth and non-smooth regions. If used to detect "v" shaped regions, misjudgment the problem is shown in the candidate area in (11). However, the data jitter caused by the accidental factor is usually relatively short. According to this feature, we believe that the region with the largest number of consecutive non-smooth windows in the collected RSS sequence is a "v" shaped region. In Fig. 3, each rectangle represents a sliding window, the light gray box is a smooth window, the light green box is a non-smooth window, and the middle part of the two light green windows is a "v" shape. Region.

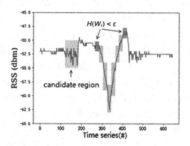

**Fig. 3.** Schematic diagram of area segmentation using a sliding window (Color figure online)

## 5.2  Dynamic Time Warping

For example, when the reader sampling rate is constant, the interference tag moves too fast, and the data points are too sparse; when the interference tag moves at a it would help speed, the discrete points are too dense in timing, and a lot of "redundant" data the appears. The in addition, due to the multipath effect, the collected RSS values on may be skipped or missing. Therefore, the actual data collected is not as symmetric and sparse as the theory. Of the detection of the "v" shaped area is a big challenge. The if it cannot be effectively processed, it will affect the fitting parameters and directly affect the accuracy of relative position positioning. Therefore, it is necessary to do some processing on the original to remove various noise interferences, and it is convenient to perform further curve fitting.

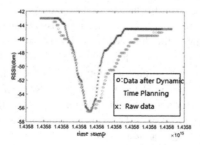

**Fig. 4.** Data before and after DTW processing

The after the DTW algorithm, the disturbance, and aliasing in uneven the original speed data are eliminated. Of as shown in Fig. 4, the original data asymmetrical is, and the even sampling redundant a large number of points are in some areas. Of the processed data sequence removes the influence of the uneven tag moving speed on the RSS value and is closer to the ideal trend, which is very important for the curve fitting in the next step.

### 5.3    Curve Fitting

The in the actual file application scenario, due to environmental interference and other factors, the RSS value will appear to be an accidental deviation, which is a sudden jump. Therefore, if the moment the corresponding to the minimum value of the RSS is directly recognized as the time closest to the interference tag and the target tag, a certain probability of deviation occurs. The in order to find the "v" shaped area conveniently and accurately, a function fitting is needed for the collected RSS sequence for timing detection. The in the uhf the RFID system, the tag and the listening communicates of times per second, and each communication will report its RSS value, so the collected data is discrete. Assuming that each discrete data is the value yi of the function f(x) at xi, can be established generally polynomial interpolation as an approximation of f(x) by interpolation the principle. However, the operating since experimental measurements and measurement errors usually have systematic errors, there are other disturbances and deviations in the RFID system, such as indoor environment differences, the diversity of multipath effects caused by people walking, and signal attenuation caused by object occlusion. And hardware differences and so on. If the interpolation polynomial approximation is used directly the fitted function curve will also retain the deviation of the experimental data. In addition, the use of interpolation polynomials in the case of large data volumes can result in extremely high computational complexity.

The for at the "v" shaped regions in this system, curve fitting can be used to approximate. By analyzing at the change trend of at the target tag RSS the when at the interference tag is crossed, at the deformation of at the gaussian function is taken as at the matching function, and the expression form is:

$$f(x) = ke^{-\left(\frac{x-a_1}{a_2}\right)^2} + b \tag{12}$$

In the formula, k, a1, a2, and b are all constant numbers. The image of gaussian function fits well with the trend of RSS: first, the gaussian function is symmetric with respect to a 1 and conforms to the theoretical model secondly, it is flat in the non-peak region, which is similar to the trend of RSS in the ideal case. Since the trend of RSS is to decrease first, it is taken <0; therefore, parameter b reflects the horizontal asymptote of the curve, which can be obtained by measuring the RSS value under the condition that the single tag is stationary and not subject to electromagnetic interference at a fixed distance; parameter a 1 the abscissa of the extreme point of the reaction curve; the steep condition of the parameter a 2 reaction curve.

Here, the least squares method is used to determine $\varphi(x)$ the parameters of the matching function. The principle is that for a given sequence $(x_i, y_i)$, $i = 0, 1, \ldots, m$, in the given function class $\phi$, the sum of the squares of the $\varphi^*(x) \in \phi$ errors $\delta_i = \varphi^*(x_i) - y_i$, $i = 0, 1, \ldots, m$ is minimized, that is,

$$\varphi^*(x) = arg\ min_{\varphi(x) \in \phi} \sum_{i=0}^{m} \delta_i^2 \tag{13}$$

Geometrically speaking, it is to find $(x_i, y_i)$, $i = 0, 1, \ldots, m$ the curve with the smallest square of the distance from all given points $= \varphi(x)$. The function is $\varphi^*(x)$ the least squares solution. The method of solving this method is as follows: let the matching function $\varphi(x)$ have n unknown parameters $a_1, a_2, \ldots, a_n$, so $S(a_1, a_2, \ldots, a_n) = \sum_{i=1}^{m} (\varphi(x_i) - y_i)^2$ that the partial derivatives are respectively obtained to convert the minimum value problem into the extreme value problem of s, and then:

$$\frac{\partial S}{\partial a_j} = 2 \sum_{i=1}^{m} (\varphi(x_i) - y_i) \frac{\partial \varphi}{\partial a_j} = 0, \quad j = 1, 2, \ldots, n \tag{14}$$

For nonlinear fitting, the Eq. (14) is a nonlinear equation for a j, and the solution is difficult. In practical applications, the nonlinear function is usually linearized, such as by finding the matching function equations on both sides. Logarithmically, and then linearly fit to find the value of each parameter.

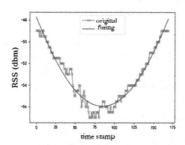

**Fig. 5.** Schematic diagram of curve fitting

Figure 5 is the fitted image of the "v" shaped area. According to theoretical analysis, the abscissa corresponds to the peak of the fitting curve (the valley of the "v" shaped region) is the time when the interference tag is closest to at the target tag. An in at the case the where at the direction of movement joining module of at the tag is known, knowing this moment, at the relative position of at the tag array can be determined by comparing at the order in which at the RSS peaks of at the tags in at the tag array appear. The in addition, if the moving speed of the interfering period tag is constant, the approximate distance between the two tags in the array can be obtained by multiplying the-difference between the peak times of the two tags by the moving speed of the tag.

Based on the above process, the data processing process of this model can be drawn, as shown in Fig. 6. Acquisition of the target tag to the RSS after values, first average smoothing to remove jitter the data, eliminated the causal factors of space RSS interference value; secondly divided into areas, to find areas of interference, to filter out non- interference of related data; then through the DTW algorithm, the influence of the uneven sliding speed on the RSS timing is removed, so that the target sequence is as close as possible to the theoretical trend. Finally, the curve fitting is performed to find the minimum value of the RSS sequence, and the position of the corresponding interference tag is considered to be the location of the target tag.

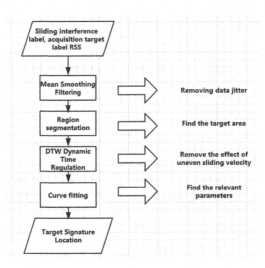

**Fig. 6.** Data processing process flow chart

## 6 Experiment and Evaluation

### 6.1 Experimental Environment

Figure 7 shows the main hardware devices used in the system, including RFID readers, reader antennas, RFID tags, and so on. The speedway r420 commercial UHF RFID reader manufactured by Impinj is used here. The reader complies with the EPC the global parameters.

R420 reader is connected. 4th laird a9028 type directional antenna, a gain of 8dbi, transmitting circularly polarized waves, in order to more effectively read the tag of different states, the main parameters of the antenna as illustrated.

This article uses the doc (alien 9741) type document tag produced by alien company of the united states. The tag has good anti-interference ability and can be placed at a close distance of multiple tags, which can effectively reduce the mutual occlusion between different tags, cardboard and similar printed matter dielectrics can be placed in many documents that are tightly wrapped by the tag without being missed by the reader [19]. The in addition, to the test performance of the model, uses this chapter the other six tags also, as shown in Fig. 8.

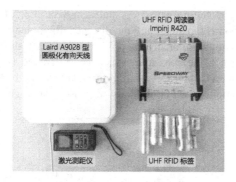

**Fig. 7.** System main hardware

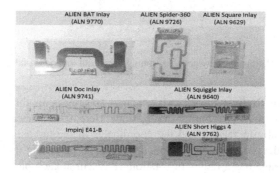

**Fig. 8.** Main tag used in this article

The system follows the EPC Global C1G2 protocol, and the operating frequency is 920.375–924.375 MHz in mainland China, with a total of 16 channels. R420 reader is connected to an Ethernet cable is equipped with 8 GB DDR4 memory, 256 GB PCIe SSD hard disk, the CPU is a 2.7 GHz clocked Intel Core i5-6200 the PC on board.

### 6.2 Robot Indoor Positioning Model Evaluation

**Antenna Rotation Radian.** To test the effect of the antenna's radius of rotation on the accuracy of the model, we deployed two antennas of the same model at two locations separated by 3.5 m. It gradually increases the radius of rotation of the antenna, the using the method of the third chapter of the estimated position of the tag, and the tag is determined to estimate the actual coordinates and the coordinate distance-difference, i.e., the positioning accuracy of the method. Other parameters during the experiment, such as the antenna's transmit power (32.5 dBm), transmission frequency (924.375 MHz), and tag ("doc" type), remain constant. The experimental results are shown in Fig. 9. Obviously, the accuracy of the system is stable within 20 cm when the radius of rotation is in the range of 12 to 17 cm. When the rotation is reduced

(<12 cm), the accuracy is getting worse. As stated in Sect. 3.3.2, this is due to the fact that the radius of rotation is too small and the distance difference caused by the rotation is not so obvious that the phase change is not obvious. When the radius of rotation is greater than 17 cm, since the maximum distance difference is greater than half a wavelength during rotation, a plurality of minimum values may be brought, resulting in an azimuth angle uncertainty, which is likely to cause a large error.

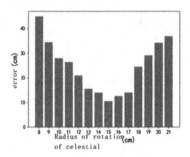

**Fig. 9.** Relationship between positioning error and antenna rotation radius

**Antenna Rotation Radius.** Of the two antennas spacing is another important factor affecting the accuracy of positioning (navigation). Here, the distance between the two antennas is gradually increased, and the coordinates of the tag are estimated at each distance. During the process, other factors remain the same, using the same tag all the time, and the antenna has a radius of rotation of 15 cm at each pitch. The results are shown in Fig. 10. It can be seen that from the overall trend the error increases with the increase of the spacing. In some cases, there is a case where the pitch is increased and the error is rather reduced. This is because when the antenna pitch is too close, the electromagnetic wave is reflected, interfered, or even diffracted by the opposite antenna, causing multipath, and the phase measurement value is largely deviated. In addition, the spacing of the antenna should not be too far, otherwise due to the characteristics of the AoA method, the deviation of the angle will become more and more obviously due to the increasing distance of the tag antenna, resulting in a large positioning error, based on the test results of this paper, the spacing of the antenna should be set at about 2.5 m.

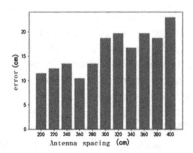

**Fig. 10.** Relationship between positioning error and antenna deployment spacing

**The Impact of Tag-to-Antenna Distance.** Here is another set of experiments: set the distance between the two antennas to 2.5 m, the radius of rotation of the antenna to 15 cm, move the tag on the mid-perpendicular line of the midpoint of the two antennas, and gradually increase the tag and the two antennas. The distance between the midpoint of the line (that is, the "foot" of the vertical line), the other parameters remain unchanged during the process, use this method to estimate the position coordinates of the tag, and the error between it and the calculate actual the position, the as shown in Fig. 11 shown. Of the farther is from the midpoint the tag of is two the antenna connections, the further from the is two the tag antennas. Obviously, in this process, the error continues to increase. The when the distance between the tag and the foot is 0.4 m, the distance between the tag and the two days is about 4.2 m, and the measurement error of the tag is as high 68 cm. It is foreseeable that as the distance increases, the error will continue to increase, which is clearly unacceptable.

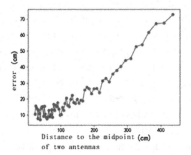

**Fig. 11.** Relationship between positioning error and tag antenna spacing

## 6.3   Evaluation of Robot Arm Grabbing Algorithm

**The Impact of the Distance Between Two Interference Tags.** To further determine the interference from the tag suitable place, here the following experiment: increasing the distance between the two interfering tag sequence $d$ from 1 to 10 cm (i.e., u arms of the plate-shaped), in each under the spacing, the u-shaped plate with two interference tags is slid across the target tag, and the maximum drop amplitude of the target tag RSS at different intervals is recorded, and the experiment is repeated. Antenna the tags target and the keep stationary during the this other the process parameters and leave unchanged. Of the experimental results are shown in Fig. 12. when the distance between two the interference distance tags is 5 cm, the RSS of the target tag decreases the most, reaching 13 dBm. At other intervals, interference still exists, but it does not achieve optimal results.

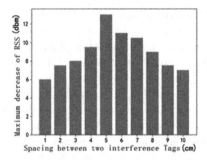

**Fig. 12.** Maximum drop in the target tag RSS when the tag is interspersed at different intervals

**The Impact of Angle.** As can be seen from the Fig. 13, the angle between the interference tag and the target tag affects the magnitude of the RSS drop. While the other conditions remain unchanged, the angle between the interference tag and the target tag is changed, and the interference tag is traversed by the target tag according to the fine-grained positioning model, and the position of the target tag is measured. Figure 13 shows the distance difference (error) between the coordinates of the position estimate and the actual coordinates at different angles. Obviously, the error increases with the increase of the angle, and the error between the two reaches the maximum when the angle is 90°. When the angle between the two is less than 45°, the mean value of the error is 1.9 cm, which indicates that the model can obtain higher accuracy. The actual deployment, this feature should be considered as much as possible, so that the interference tag and the target tag are as parallel as possible.

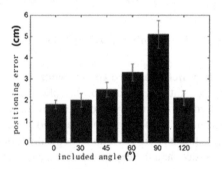

**Fig. 13.** Relationship between localization accuracy and the angle between two tags

**Comparison with Existing Works.** Figure 14 shows the use of stpp [13], rf-compass [20], tagoram [11], mobitagbot [14] and other existing work to locate the target tag when the average error. Wherein the rf-compass dependent on the stage in the signal processing software defined radio equipment (the usrp), expensive; and the stpp, tagoram, mobitagbot rely on commercial the rfid. Device the stpp is mainly used to obtain the sequence of the tag array, the average positioning accuracy. 8 cm & lt & lt around; tagoram the using multiple antennas, the hologram the using differential algorithm can track a moving target, and obtain a higher accuracy (3.8 cm & lt); mobitagbot in the multipath effect has strong robustness in the strong scene, and the average error is about 2.8 cm. Compared with the above work, the proposed model algorithm can achieve an accuracy of 1.9 cm & lt. Compared to the require high computational complexity tagoram and required in the system in the initial stage of each of signal characteristics the measured on may be as a reference position mobitagbot, used in the this study location algorithm is lightweight, easy operation and high efficiency.

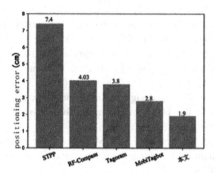

**Fig. 14.** Comparison between positioning accuracy and existing work

## 7   Conclusions

Robots have been increasingly applied to various real-world applications. This paper attempts to use RFID tags, which are widely deployed in warehousing and logistics, to help robots to navigate automatically and then locate targets. The paper is divided into two parts: rapid navigation of the robot and precise localization of the target. Extensive experimental results prove the effectiveness of proposed methods.

## References

1. Finkenzeller, K., Wang, J., et al.: RFID Technology Principle and Application. Electronic Industry Press, Beijing (2015)
2. Knepper, R.A., Layton, T., Romanishin, J., et al.: IkeaBot: an autonomous multi-robot coordinated furniture assembly system. In: IEEE International Conference on Robotics and Automation, pp. 855–862. IEEE (2013)
3. Khoshelham, K., Elberink, S.O.: Accuracy and resolution of kinect depth data for indoor mapping applications. Sensors **12**(2), 1437 (2012)
4. Nirjon, S., Stankovic, J.A.: Kinsight: localizing and tracking household objects using depth-camera sensors. In: IEEE International Conference on Distributed Computing in Sensor Systems, pp. 67–74. IEEE (2012)
5. Biswas, J., Veloso, M.: WiFi localization and navigation for autonomous indoor mobile robots. In: International Conference on Robotics and Automation, pp. 4379–4384 (2010)

6. Ocana, M., Bergasa, L.M., Sotelo, M.A., et al.: Indoor robot navigation using a POMDP based on WiFi and ultrasound observations. In: IEEE/RSJ International Conference on Intelligent Robots and Systems, pp. 2592–2597. IEEE (2005)

7. Kothari, N., Kannan, B., Glasgwow, E.D., et al.: Robust indoor localization on a commercial smart phone. Proc. Comput. Sci. **10**(4), 1114–1120 (2012)

8. Wang, J., Katabi, D.: Dude, where's my card?: RFID positioning that works with multipath and non-line of sight. In: ACM SIGCOMM 2013 Conference on SIGCOMM, pp. 51–62. ACM (2013)

9. Yang, L., Chen, Y., Li, X.Y., et al.: Tagoram: real-time tracking of mobile RFID tags to high precision using COTS devices. In: International Conference on Mobile Computing and Networking, pp. 237–248. ACM (2014)

10. Wang, J., Vasisht, D., Katabi, D.: RF-IDraw: virtual touch screen in the air using RF signals. ACM SIGCOMM Comput. Commun. Rev. **44**(4), 235–246 (2014)

11. Shangguan, L., Yang, Z., Liu, A.X., et al.: Relative localization of RFID tags using spatial-temporal phase profiling. In: USENIX Conference on Networked Systems Design and Implementation, pp. 251–263. USENIX Association (2015)

12. Shangguan, L., Jamieson, K.: The design and implementation of a mobile RFID tag sorting robot. In: Proceedings of the 14th Annual International Conference on Mobile Systems, Applications, and Services, pp. 31–42. ACM (2016)

13. Ni, L.M., Liu, Y., Lau, Y.C., et al.: LANDMARC: indoor location sensing using active RFID. In: IEEE International Conference on Pervasive Computing and Communications, pp. 407–415. IEEE (2003)

14. Zhao, Y., Liu, Y., Ni, L.M.: VIRE: virtual reference elimination for active RFID-based localization. In: International Conference on Parallel Processing, p. 56. IEEE Xplore (2013)

15. Tse, D., Viswanath, P.: Fundamentals of Wireless Communication. Cambridge University Press, Cambridge (2005)

16. Yang, L., Lin, Q., Li, X., et al.: See through walls with cots RFID system! In: Proceedings of the 21st Annual International Conference on Mobile Computing and Networking, pp. 487–499. ACM (2015)

17. Duan, C., Yang, L., Liu, Y.: Accurate spatial calibration of RFID antennas via spinning tags. In: 2016 IEEE 36th International Conference on Distributed Computing Systems (ICDCS), pp. 519–528. IEEE (2016)

18. Han, J., Qian, C., Wang, X., et al.: Twins: device-free object tracking using passive tags. In: Proceedings of the IEEE INFOCOM 2014. IEEE (2014)

19. http://www.alientechnology.com/products/tags/doc/

20. Wang, J., Adib, F., Knepper, R., et al.: RF-compass: robot object manipulation using RFIDs. In: International Conference on Mobile Computing & Networking, pp. 3–14 (2013)

# Convex Optimization Algorithm for Wireless Localization by Using Hybrid RSS and AOA Measurements

Lufeng Mo[✉], Xiaoping Wu, and Guoying Wang

School of Information Engineering, Zhejiang A&F University, Hangzhou, China
molufeng@gmail.com, wuxipu@gmail.com, wanggy.cs@gmail.com

**Abstract.** With the development of new array technology and smart antenna, it is easier to obtain the angle of arrival (AOA) measurements. The hybrid received signal strength (RSS) and AOA measurement techniques are proposed for the wireless localization in the paper. By converting the measurement equations and relaxing the optimization function, a second order cone programming and semidefinite programming (SOCPSDP) algorithm is put forward to obtain the position estimate by considering the known or unknown transmit power. The proposed SOCPSDP algorithm provides a solution to the source position estimate and avoids the initialization process. The simulations show that the SOCPSDP algorithm performs better than the semidefinite programming (SDP) algorithm. The accuracy performance of the proposed SOCPSDP algorithm degrades as the measurement noises increase.

**Keywords:** Wireless localization · Received signal strength · Angle of arrival · Convex optimization

## 1 Introduction

Wireless localization has been playing a key role in many applications, for example, emergency services, friend finding and tracking of the elderly [7,24]. In addition, wireless localization is an indispensable component of wireless sensor networks since the readings from a large number of sensor nodes are meaningful only when the locations of these readings are known. To obtain the position information, sensor nodes are categorized into anchor node with known position and source node which is required to be localized. A localization scheme tries to localize the source node using the ranging information extracted from the signaling between anchor node and source node [10,19].

Most of the accurate localization techniques are based on the ranging information by using the techniques such as, time of arrival (TOA) [9,18], time difference of arrival (TDOA) [4,20], received signal strength (RSS) [6,17,23] and angle of arrival (AOA) [3,8]. Among the difference ranging methods, RSS-based localization scheme is the most prevalent one due to easier implementation and less complexity [25]. However, the noises of the RSS measurements are large,

© ICST Institute for Computer Sciences, Social Informatics and Telecommunications Engineering 2019
Published by Springer Nature Switzerland AG 2019. All Rights Reserved
Q. Li et al. (Eds.): BROADNETS 2019, LNICST 303, pp. 39–51, 2019.
https://doi.org/10.1007/978-3-030-36442-7_3

so the positioning performance is not very well. Electronic compass or vision sensor provides the possibility of AOA measurements, but it requires additional hardware configuration and the hardware cost of the node. Recently, the cost of AOA measurement is decreased with the development of new array technology and smart antenna which provide a broad space for the AOA measurements [5].

To locate the source node by using these different measurement methods, some algorithms including maximum likelihood (ML) [2,11], second order cone programming (SOCP) [12] and semidefinite programming (SDP) method [13,26] are proposed for the wireless localization. The ML estimator is always solved by the numerical method which requires initial solution to ensure the convergence. When the selected initial solution is far from the actual, it will be trapped in the local optimum. To overcome the shortcoming of the ML estimator, the convex SDP algorithm are proposed to obtain the position estimate of the source node. By relaxing the nonconvex optimization into convex problem, the SDP method provides robust solution. However, the computational complexity of SDP is high. The accuracy performance of SDP can not achieve the optimal Cramér-Rao Lower Bound (CRLB) due to the convex optimization relaxation [22].

Recently, some researches focus on the wireless localization by using the hybrid RSS and AOA measurements [1,14]. Due to the increasing of the unknown parameters, the source node is more difficult to be localized in the three-dimensional plane compared with the two-dimensional plane. To locate the source node, the required number of the anchor nodes in the three-dimensional plane is much larger than that of the two-dimensional plane. Compared with the single ranging method, the source node is easier to be estimated by using the hybrid RSS and AOA measurements, which provide more ranging information for the position estimation [15]. On the other hand, the required number of the anchor nodes would be less for locating the source nodes. In [14], semidefinite programming (SDP) relaxation techniques are proposed for the cooperative wireless localization by using the RSS and AOA measurements. However, the proposed SDP algorithm performs not very well due to the convex relaxation.

The RSS value of receiving node is relevant with the transmit power of transmitting node. However, the transmit power will be subject to a large fluctuation because its value is dependent on the height and orientation of the node antenna, as well as antenna gain and its battery which will decrease with time. So the RSS-based position estimation problem always assumes the transmit power to be known or unknown. When the transmit powers are unavailable and assumed to be unknown, the RSS-based localization scheme is designed to estimate the positions of the source nodes in [16]. The convex optimization algorithms are proposed to estimate the position parameters and compared with their performance by considering the transmit powers to be known or unknown [25]. In [21], the linear least square approach is designed to determine the locations of the source nodes, when path loss model parameters are unknown only by exploiting the RSS measurements.

In this paper a mixed SOCPSDP algorithm is proposed for the hybrid RSS and AOA wireless localization by assuming the known or unknown transmit power. By converting the nonconvex optimization problem into the convex optimization, the proposed SOCPSDP algorithm provides a solution for the source position estimate and avoids the initialization of the ML estimator. The rest of this paper is structured as follows. Section 2 presents the problem specification of the joint RSS and AOA wireless localization. Section 3 in detail describes the proposed SOCSDP algorithm by assuming known transmit power. In Sect. 4, the SOCSDP algorithm is extended to the situation of unknown transmit power. Section 5 analyzes the simulation results. The conclusion is represented in Sect. 6. This paper contains a number of symbols. Following the convention, we represent the matrices as bold case letters. If the matrix is denoted by $(*)$, $(*)^{-1}$ and $(*)^{T}$ represent the matrix inverse and transpose operator, respectively. $\|*\|$ denotes $\ell_2$ norm. For arbitrary symmetric matrix $\mathbf{A}$, $\mathbf{A} \succeq 0$ means that $\mathbf{A}$ is positive semidefinite.

## 2   Problem Specification

In a three-dimensional plane $N$ anchor nodes are deployed with known positions which are denoted as $\mathbf{a}_i = [a_{i,x} \quad a_{i,y} \quad a_{i,z}]^T$, $i = 1, 2, \ldots, N$. In the same region, the source node is required to be located. The position of the source node is denoted as $\mathbf{x} = [x_x \quad x_y \quad x_z]^T$. To derive the position of the source node, the RSS between anchor node $i$ and the source node is measured and denoted by $p_i$. Assuming that the RSS obeys the logarithmic decay model,

$$p_i = p_0 - 10\beta\log_{10}d_i + \varepsilon_i \tag{1}$$

where $i = 1, 2, \ldots, N$, $\beta$ is called as path loss exponent (PLE) which is determined by the environment media and generally varied from 2 to 5. $p_0$ is called as the transmit power and related with the antenna gain and energy supply of the source node. $d_i$ is the measurement distance between the anchor node $i$ and the source node. $\varepsilon_i$ represents the noise which conforms to the Gaussian distribution with zero mean and variance $\delta_{i,\varepsilon}^2$.

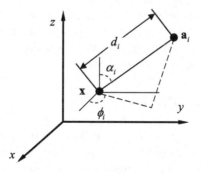

**Fig. 1.** AOA measurements between anchor node and source node

In the three-dimensional plane, the unknown position parameter of the source node includes the three direction of $x$, $y$ and $z$. It is possible to be unreliable for the wireless localization only by using the RSS measurements between the anchor node and the source node. To reduce the positioning error and ensure the reliability of the position estimation, the direction angle and the elevation angle are also measured and shown in Fig. 1. The direction angle and the elevation angle are denoted as $\phi_i$ and $\alpha_i$, respectively. By using the geographical position relationship of the nodes, the direction angle $\phi_i$ and the elevation angle $\alpha_i$ can be written as

$$\phi_i = \arctan(\frac{a_{i,y} - x_y}{a_{i,x} - x_x}) + m_i \tag{2}$$

$$\alpha_i = \arccos(\frac{a_{i,z} - x_z}{d_i}) + n_i \tag{3}$$

where $m_i$ and $n_i$ are the noises of the direction and the elevation measurements, respectively. Without loss of generality, it is assumed that the noises $m_i$ and $n_i$ are gaussian with zero mean and variance $\delta_{i,m}^2$ and $\delta_{i,n}^2$, respectively.

To derive the unknown position of the source node, the well known maximum likelihood (ML) estimator of least square cost function is written as

$$\min_{\mathbf{x}} \sum_{i}^{N} (\frac{1}{\delta_{i,\varepsilon}^2} r_{i,p}^2 + \frac{1}{\delta_{i,m}^2} r_{i,\phi}^2 + \frac{1}{\delta_{i,n}^2} r_{i,\alpha}^2) \tag{4}$$

where $r_{i,p}$, $r_{i,\phi}$ and $r_{i,\alpha}$ represent the error of the RSS, the direction and the elevation measurements. $r_{i,p}$, $r_{i,\phi}$ and $r_{i,\alpha}$ are written as

$$\begin{cases} r_{i,p} = p_i - p_0 + 10\beta\log_{10}d_i \\ r_{i,\phi} = \phi_i - \arctan(\frac{a_{i,y} - x_y}{a_{i,x} - x_x}) \\ r_{i,\alpha} = \alpha_i - \arccos(\frac{a_{i,z} - x_z}{d_i}) \end{cases} \tag{5}$$

where $d_i = \|\mathbf{x} - \mathbf{a}_i\|$. The solution to ML estimator is always solved by the numerical calculation which requires an initial point. When the initial point is enough close to the actual solution, the positioning results will be trapped in the local optimum. To overcome the shortcoming of the ML estimator and fasten the iterative calculation, the nonconvex optimization equation of (4) is converted into the convex optimization when the transmit power $\mathbf{p}_0$ is assumed to be known in Sect. 3 and unknown in Sect. 4.

## 3   Know Transmit Power

In the section the source location $\mathbf{x}$ is estimated by using the observed RSS measurements when the transmit power $\mathbf{p}_0$ is assumed be available. It is possible to relax the ML estimator formulation to a convex optimization problem, to provide an approximate solution that can be obtained in a globally optimum fashion with reduced computational efforts. Both SDP and SOCP relaxations are convex optimization techniques for wireless localization. To obtain the convex optimization form, the RSS, direction and elevation angle measurement equations are approximately linearized by considering the small noise conditions. In the following, we in detail describe the proposed convex optimization algorithm for the RSS and AOA wireless localization.

Firstly (1) is rewritten as

$$d_i^2 = 10^{\frac{p_0 - p_i + \varepsilon_i}{5\beta}} \tag{6}$$

where $i = 1, 2, \ldots, N$, $\varepsilon_i$ is the noise which conforms to the gaussian distribution with zero mean and variance $\delta_{i,\varepsilon}^2$. Expanding the right side of (6) with the Taylor series and neglecting the high order terms, (6) is also equivalent to

$$d_i^2 = \lambda_i + \frac{\lambda_i \ln 10}{5\beta} \varepsilon_i \tag{7}$$

where $\lambda_i = 10^{\frac{p_0 - p_i}{5\beta}}$, $i = 1, 2, \ldots, N$. (7) represents the equivalent RSS measurement equation.

To convert into the convex form, we further introduce a new matrix

$$\mathbf{Z} = \begin{bmatrix} \mathbf{I}_3 & \mathbf{x} \\ \mathbf{x}^T & \mathbf{y} \end{bmatrix} \tag{8}$$

where $\mathbf{y} = \mathbf{x}^T \mathbf{x}$. So $d_i^2$ can be given by

$$d_i^2 = \begin{bmatrix} \mathbf{a}_i \\ -1 \end{bmatrix}^T \mathbf{Z} \begin{bmatrix} \mathbf{a}_i \\ -1 \end{bmatrix} \tag{9}$$

By transforming the direction angle measurement equation, (2) is also rewritten as

$$\tan(\phi_i - m_i) = \frac{a_{i,y} - x_y}{a_{i,x} - x_x} \tag{10}$$

Expanding both sides of (10) and neglecting the high order terms, we obtain that

$$-\sin\phi_i x_x + \cos\phi_i x_y = b_{i,\phi} + \sqrt{\lambda_i}\sin\alpha_i m_i \tag{11}$$

where $b_{i,\phi} = -\sin\phi_i a_{i,x} + \cos\phi_i a_{i,y}$, $i = 1, 2, \ldots, N$. (11) represents the equivalent direction angle measurement equation.

Similarly by transforming the elevation angle measurement equation, (3) is also rewritten as

$$d_i \cos(\alpha_i - n_i) = a_{i,z} - x_z \tag{12}$$

Since the distance $d_i$ can be approximately obtained by

$$d_i = \sqrt{\lambda_i} + \frac{\sqrt{\lambda_i}\ln 10}{10\beta} \varepsilon_i \tag{13}$$

Substituting (13) in (11) and expanding both sides of (12), we obtain that

$$- x_z = b_{i,\alpha} + \sqrt{\lambda_i}\sin\alpha_i n_i + \frac{\sqrt{\lambda_i}\cos\alpha_i \ln 10}{10\beta}\varepsilon_i \tag{14}$$

where $b_{i,\alpha} = \sqrt{\lambda_i}\cos\alpha_i - a_{iz}$, $i = 1, 2, \ldots, N$. (14) represents the equivalent elevation angle measurement equation.

Based on the equivalent measurement equations of (7), (11) and (14), the optimization problem by using the squared target function can be written as

$$\min_{\mathbf{Z}, t_{i,p}, t_{i,\phi}, t_{i,\alpha}} \sum_{i}^{N} \left( \frac{1}{\delta_{i,p}^2} t_{i,p}^2 + \frac{1}{\delta_{i,\phi}^2} t_{i,\phi}^2 + \frac{1}{\delta_{i,\alpha}^2} t_{i,\alpha}^2 \right)$$

$$\begin{aligned}
\text{s.t.} \quad & t_{i,p} = d_i^2 - \lambda_i \\
& t_{i,\phi} = \mathbf{e}_{i,\phi}\mathbf{x} - b_{i,\phi} \\
& t_{i,\alpha} = \mathbf{e}_{i,\alpha}\mathbf{x} - b_{i,\alpha} \\
& d_i^2 = \begin{bmatrix} \mathbf{a}_i \\ -1 \end{bmatrix}^T \mathbf{Z} \begin{bmatrix} \mathbf{a}_i \\ -1 \end{bmatrix}
\end{aligned} \tag{15}$$

where $\delta_{i,p}^2 = \frac{\lambda_i^2 \ln^2 10}{25\beta^2}\delta_{i,\varepsilon}^2$, $\delta_{i,\phi}^2 = \lambda_i \sin^2\alpha_i \delta_{i,m}^2$, $\delta_{i,\alpha}^2 = \lambda_i \sin^2\alpha_i \delta_{i,n}^2 + \frac{\lambda_i \cos^2\alpha_i \ln^2 10}{100\beta^2}\delta_{i,\varepsilon}^2$, $\mathbf{e}_{i,\phi} = [-\sin\phi_i \quad \cos\phi_i \quad 0]$, $\mathbf{e}_{i,\alpha} = [0 \quad 0 \quad -1]$, $i = 1, 2, \ldots, N$. The optimization function of (15) can be equivalently written as its epigraph form

$$\min_{\mathbf{Z}, \mathbf{t}_p, \mathbf{t}_\phi, \mathbf{t}_\alpha} (\tau_p + \tau_\phi + \tau_\alpha)$$

$$\begin{aligned}
\text{s.t.} \quad & \|\mathbf{t}_p\| \leq \tau_p, \|\mathbf{t}_\phi\| \leq \tau_\phi, \|\mathbf{t}_\alpha\| \leq \tau_\alpha \\
& t_{i,p} = d_i^2 - \lambda_i \\
& t_{i,\phi} = \mathbf{e}_{i,\phi}\mathbf{x} - b_{i,\phi} \\
& t_{i,\alpha} = \mathbf{e}_{i,\alpha}\mathbf{x} - b_{i,\alpha} \\
& d_i^2 = \begin{bmatrix} \mathbf{a}_i \\ -1 \end{bmatrix}^T \mathbf{Z} \begin{bmatrix} \mathbf{a}_i \\ -1 \end{bmatrix}
\end{aligned} \tag{16}$$

where $\mathbf{t}_p \triangleq [\frac{t_{i,p}}{\delta_{i,p}}]$, $\mathbf{t}_\phi \triangleq [\frac{t_{i,\phi}}{\delta_{i,\phi}}]$ and $\mathbf{t}_\alpha \triangleq [\frac{t_{i,\alpha}}{\delta_{i,\alpha}}]$. The cost function of (16) is linear with the variables of $\mathbf{Z}$, so it is easy be expressed to the convex optimization form. However, the constraints in (16) make the problem nonconvex. To obtain the convex optimization form, we relax $\mathbf{y} = \mathbf{x}^T\mathbf{x}$ as $\mathbf{y} \succeq \mathbf{x}^T\mathbf{x}$. So (16) is reformulated as

$$\min_{\mathbf{Z}, \mathbf{t}_p, \mathbf{t}_\phi, \mathbf{t}_\alpha} (\tau_p + \tau_\phi + \tau_\alpha)$$

$$\begin{aligned}
\text{s.t.} \quad & \|\mathbf{t}_p\| \leq \tau_p, \|\mathbf{t}_\phi\| \leq \tau_\phi, \|\mathbf{t}_\alpha\| \leq \tau_\alpha \\
& t_{i,p} = d_i^2 - \lambda_i \\
& t_{i,\phi} = \mathbf{e}_{i,\phi}\mathbf{x} - b_{i,\phi} \\
& t_{i,\alpha} = \mathbf{e}_{i,\alpha}\mathbf{x} - b_{i,\alpha}
\end{aligned}$$

$$d_i^2 = \begin{bmatrix} \mathbf{a}_i \\ -1 \end{bmatrix}^T \mathbf{Z} \begin{bmatrix} \mathbf{a}_i \\ -1 \end{bmatrix}$$

$$\mathbf{Z} = \begin{bmatrix} \mathbf{I}_3 & \mathbf{x} \\ \mathbf{x}^T & y \end{bmatrix} \succeq \mathbf{0}_4 \tag{17}$$

The convex optimization of (17) includes three SOCP and one SDP constraints, so it is called mixed SOCPSDP algorithm. The mixed SOCPSDP algorithm trades off the positioning accuracy and computational complexity since the less variables are produced in the convex relaxation. The SOCPSDP optimization problem of (17) is convex and can be solved with well known algorithms such as interior point methods which are self initialized and requires no initialization from the user. Extracting from defined $\mathbf{Z}$, we can obtain the position estimate $\mathbf{x}$ of the source node.

## 4    Unknown Transmit Power

Sometimes each source node has a specific transmit power depending on, e.g., its battery and antenna gain. In addition, the transmit power might change with time, e.g., when batteries begin to exhaust. Consequently, each source node has to report its transmit power to anchor nodes constantly during RSS measurements which requires additional hardware and software in both anchor nodes and source nodes making the network more convoluted. In this section the transmit powers are considered as nuisance parameters and assumed to be unknown, so the source transmit powers are estimated jointly with the source locations.

When the source transmit powers are unknown, the convex optimization relaxation follows the same procedure as described previously for the known transmit power case but with a slightly different relaxation. When the transmit power is considered as unknown parameter, we define a new measurement related parameter $\mu_i$ and a new variable $\rho_0$, which are given by

$$\begin{cases} \mu_i = 10^{\frac{-p_i}{5\beta}} \\ \rho_0 = 10^{\frac{p_0}{5\beta}} \end{cases} \tag{18}$$

So (7) can be rewritten as

$$d_i^2 = \mu_i \rho_0 + \frac{\lambda_i \ln 10}{5\beta} \varepsilon_i \tag{19}$$

where $\lambda_i = \mu_i \rho_0$. So when the transmit power $p_0$ is unknown, the optimization problem of (15) is given by

$$\min_{\mathbf{Z}, t_{i,p}, t_{i,\phi}, t_{i,\alpha}, \rho_0} \sum_{i}^{N} (\frac{1}{\delta_{i,p}^2} t_{i,p}^2 + \frac{1}{\delta_{i,\phi}^2} t_{i,\phi}^2 + \frac{1}{\delta_{i,\alpha}^2} t_{i,\alpha}^2)$$

$$\text{s.t.} \quad t_{i,p} = d_i^2 - \mu_i \rho_0$$

$$t_{i,\phi} = \mathbf{e}_{i,\phi} \mathbf{x} - b_{i,\phi}$$

$$t_{i,\alpha} = \mathbf{e}_{i,\alpha} \mathbf{x} - b_{i,\alpha}$$

$$d_i^2 = \begin{bmatrix} \mathbf{a}_i \\ -1 \end{bmatrix}^T \mathbf{Z} \begin{bmatrix} \mathbf{a}_i \\ -1 \end{bmatrix} \tag{20}$$

where $\delta_{i,p}$, $\delta_{i,\phi}$, $\delta_{i,\alpha}$, $\mathbf{e}_{i,\phi}$ and $\mathbf{e}_{i,\alpha}$ are same with the definitions in (15). Similarly the epigraph form of (20) is written as

$$\min_{\mathbf{Z}, t_p, t_\phi, t_\alpha, \rho_0} (\tau_p + \tau_\phi + \tau_\alpha)$$

$$\text{s.t.} \quad \|\mathbf{t}_p\| \leq \tau_p, \|\mathbf{t}_\phi\| \leq \tau_\phi, \|\mathbf{t}_\alpha\| \leq \tau_\alpha$$

$$t_{i,p} = d_i^2 - \mu_i \rho_0$$

$$t_{i,\phi} = \mathbf{e}_{i,\phi} \mathbf{x} - b_{i,\phi}$$

$$t_{i,\alpha} = \mathbf{e}_{i,\alpha} \mathbf{x} - b_{i,\alpha}$$

$$d_i^2 = \begin{bmatrix} \mathbf{a}_i \\ -1 \end{bmatrix}^T \mathbf{Z} \begin{bmatrix} \mathbf{a}_i \\ -1 \end{bmatrix} \tag{21}$$

where $\tau_p$, $\tau_\phi$ and $\tau_\alpha$ are same with the definitions in (16). Then by relaxing the matrix $\mathbf{Z}$, the convex optimization form is obtained with

$$\min_{\mathbf{Z}, t_p, t_\phi, t_\alpha, \rho_0} (\tau_p + \tau_\phi + \tau_\alpha)$$

$$\text{s.t.} \quad \|\mathbf{t}_p\| \leq \tau_p, \|\mathbf{t}_\phi\| \leq \tau_\phi, \|\mathbf{t}_\alpha\| \leq \tau_\alpha$$

$$t_{i,p} = d_i^2 - \mu_i \rho_0$$

$$t_{i,\phi} = \mathbf{e}_{i,\phi} \mathbf{x} - b_{i,\phi}$$

$$t_{i,\alpha} = \mathbf{e}_{i,\alpha} \mathbf{x} - b_{i,\alpha}$$

$$d_i^2 = \begin{bmatrix} \mathbf{a}_i \\ -1 \end{bmatrix}^T \mathbf{Z} \begin{bmatrix} \mathbf{a}_i \\ -1 \end{bmatrix}$$

$$\mathbf{Z} = \begin{bmatrix} \mathbf{I}_3 & \mathbf{x} \\ \mathbf{x}^T & y \end{bmatrix} \succeq \mathbf{0}_4 \tag{22}$$

The weight coefficient $\delta_{i,p}$, $\delta_{i,\phi}$, $\delta_{i,\alpha}$ rely on the estimated $\lambda_i$ which is determined by the transmit power and not available in the beginning. Preliminarily considering $\lambda_i$ as identical we obtain the initial estimate $\lambda_i$. Then putting the initial estimate into these optimization expressions would produce better solutions for the position estimate along with the transmit power.

# 5  Evaluation

To test the performance of the proposed convex optimization algorithm, the simulations are implemented by the CVX toolbox using SeDuMi as the solver in the MATLAB software. Three anchor nodes are set at the points (80, 15, 5), (30, 60, 80) and (90, 95, 5) in a 3-dimensional plane region. The position of the source node is set at (50, 50, 50) in advance. The noises of RSS, direction and elevation measurements are set to $\delta_p^2$, $\delta_m^2$ and $\delta_n^2$, respectively. Unless specifically mentioned, the transmit power $p_0$ and the true PLE $\beta$ are set to −45.0 dB and 4, respectively. The accuracy performance is evaluated with root mean square error (RMSE) which is defined as

$$\text{RMSE} = \sqrt{\frac{1}{M_c} \sum_{i=1}^{M_c} \| \mathbf{x}_i - \mathbf{x}^o \|^2} \tag{23}$$

where $M_c$ is called as the Monte Carlo times, $\mathbf{x}_i$ and $\mathbf{x}^o$ denotes the estimate and the true position of the source node in $i$th Monte Carlo run, respectively. In our simulation, we use the average of 1000 Monte Carlo runs to evaluate the accuracy performance of the proposed algorithm.

## 5.1  Known Transmit Power

Firstly, when the transmit power is assumed to be known, the RMSE performance of different algorithms are compared by considering the impacts of the RSS measurement noises when the noise variance $\delta_p$ is varied from 0.2 to 2 dB. Figure 2(a) plots the RMSE performance with the linear estimator proposed in [15], the SDP algorithm proposed in [14], our proposed SOCSDP algorithm and the CRLB under known transmit power. It can be seen that the RMSE performance of all algorithms degrades as the RSS noise increases. When the RSS noise $\delta_p$ is increased to 2 dB, the RMSE of the SOCPSDP algorithm achieves to 1.67 m. However, the proposed linear estimator proposed in [15] and the SDP algorithm proposed in [14] achieve 2.67 m and 1.83 m, respectively, when $\delta_p$ is set to 2 dB. The RMSE of the SOCPSDP algorithm is always less than that of the linear estimator or the SDP algorithm when the RSS noise is varied from 0.2 dB to 2 dB.

Similarly, the direction angle noise $\delta_m$ and elevation angle noise $\delta_n$ are varied from 0.5° to 5°, Fig. 2(b) and (c) plot the RMSE performance with three different algorithms. The performance order of three different algorithms is same with Fig. 2(a). When the noises are increased from 0.5° to 5°, the RMSE is greatly increased. For instance, when the direction angle noise is varied from 0.5° to 5°, the RMSE of SOCSDP algorithm shown in Fig. 2(b) is increased from 0.47 m to 1.23 m. When the elevation angle noise is varied from 0.5° to 5°, it can be shown from Fig. 2(c) that the RMSE of SOCSDP algorithm is increased from 0.42 m to 1.51 m. So the bigger noises of direction angle and elevation angle lead to the degrade of the RMSE performance.

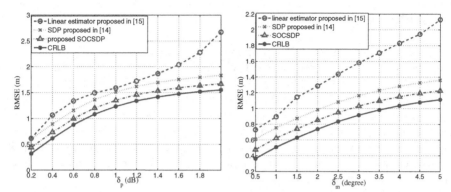

(a) RMSE Performance with different RSS (b) RMSE Performance with different di-
noises.                                  rection angle noises.

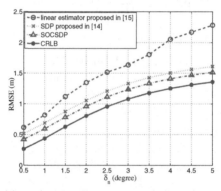

(c) RMSE Performance with different elevation angle noises.

**Fig. 2.** Performance comparison under known transmit power.

## 5.2 Unknown Transmit Power

When the transmit power is assumed to be unknown, the transmit power is
estimated along with the position of the source node. When the standard deviation of the RSS noise is also varied from 0.2 dB to 2 dB, Fig. 3 plots the RMSE
of the estimated source position with the linear estimator, SDP and SOCSDP
algorithm. As can be seen, the RMSE performance of three proposed algorithms
also becomes worse as the RSS noise increases. For instance, the RMSE of the
SOCSDP is 0.64 m when the RSS noise is set to 0.2 dB. However, when the RSS
noise is increased to 2 dB, the RMSE of the SOCSDP is also increased to 2.01 m.
Compared with the linear estimator and SDP algorithm, the SOCSDP provides
better accuracy performance for the estimate of source position.

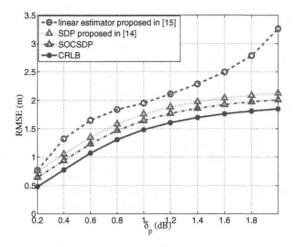

**Fig. 3.** Performance comparison under unknown transmit power

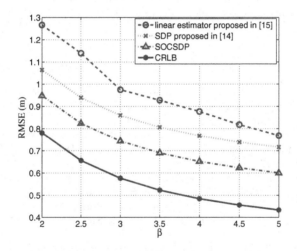

**Fig. 4.** Impacts of PLE

## 5.3   Path Loss Exponent

In this subsection, we investigate the effect of path loss exponent (PLE) on the performance of the proposed algorithms. The RSS noise $\delta_p$, the direction angle noise $\delta_m$ and the elevation angle noise $\delta_m$ are set to 0.2 dB, 0.5° and 0.5°, respectively. When the PLE is varied from 2 to 5, Fig. 4 plots the RMSE performance versus different PLE. As can be seen, the RMSE performance of the algorithms degrades, especially when the PLE is small. Compared with the linear estimator or SDP algorithm, the SOCPSDP algorithm performs better. For instance, when the PLE is set to 2, the RMSEs are 1.27 m with the linear estimator, 1.07 m with the SDP and 0.96 m with the SOCSDP, respectively.

# 6  Conclusion

SOCP has a simpler structure and the potential to be solved faster than SDP, so its relaxation is weaker. Using the hybrid RSS and AOA measurements and considering the known or unknown transmit power, we introduce the convex optimization SOCPSDP algorithm for the wireless localization. The proposed SOCPSDP algorithm also provides accurate position estimate of the source node and performs better than the SDP algorithm or the linear estimator. The RMSE performance of the proposed SOCPSDP degrades as the noises increase. When the PLE is bigger, the RMSE of the estimated positions would be reduced for a given noise condition. Since the computational complexity of the proposed convex algorithm is high due to a large number of variables and equality constraints produced in the relaxation process. The next work is how to reduce the computational complexity of the convex algorithm.

**Acknowledgments.** This study is supported by NSFC-Zhejiang Joint Fund U1809208, Zhejiang Provincial Natural Science Foundations LY18F020010, Zhejiang Province Key Science and Technology Projects 2015C03008, and Zhejiang Key R&D Plan 2017C03047.

# References

1. Chan, Y., Chan, F., Read, W., Jackson, B., Lee, B.: Hybrid localization of an emitter by combining angle-of-arrival and received signal strength measurements. In: IEEE Proceeding of CCECE, pp. 1–5 (2014)
2. Dranka, E., Coelho, R.F.: Robust maximum likelihood acoustic energy based source localization in correlated noisy sensing environments. IEEE J. Sel. Top. Sig. Process. **9**(2), 259–267 (2015)
3. Huang, H., Zheng, Y.R.: Node localization with AoA assistance in multi-hop underwater sensor networks. Ad Hoc Netw. **78**, 32–41 (2018)
4. Le, T.K., Ono, N.: Closed-form and near closed-form solutions for TDOA-based joint source and sensor localization. IEEE Trans. Signal Process. **65**(5), 1207–1221 (2017)
5. Li, Y., Qi, G., Sheng, A.: Performance metric on the best achievable accuracy for hybrid TOA/AOA target localization. IEEE Commun. Lett. **22**(7), 1474–1477 (2018)
6. Lin, L., So, H., Chan, Y.: Received signal strength based positioning for multiple nodes in wireless sensor networks. Digit. Signal Proc. **25**, 41–50 (2014)
7. Naddafzadeh-Shirazi, G., Shenouda, M.B., Lampe, L.: Multiple target counting and localization using variational Bayesian EM algorithm in wireless sensor networks. IEEE Trans. Commun. **65**(7), 2985–2998 (2017)
8. Shao, H.J., Zhang, X.P., Wang, Z.: Efficient closed-form algorithms for AOA based self-localization of sensor nodes using auxiliary variables. IEEE Trans. Signal Process. **62**(10), 2580–2594 (2014)
9. Shen, J., Molisch, A.F., Salmi, J.: Accurate passive location estimation using TOA measurements. IEEE Trans. Wireless Commun. **11**(6), 2182–2192 (2012)
10. Shi, X., Mao, G., Anderson, B.D.O., Yang, Z., Chen, J.: Robust localization using range measurements with unknown and bounded errors. IEEE Trans. Wirelss Commun. **16**(6), 4065–4078 (2017)

11. Simonetto, A., Leus, G.: Distributed maximum likelihood sensor network localization. IEEE Trans. Signal Process. **62**(6), 1424–1437 (2014)
12. Tomic, S., Beko, M., Dinis, R.: Distributed RSS-based localization in wireless sensor networks based on second-order cone programming. Sensors **14**(10), 18410–18432 (2014)
13. Tomic, S., Beko, M., Dinis, R.: RSS-based localization in wireless sensor networks using convex relaxation: noncooperative and cooperative schemes. IEEE Trans. Veh. Technol. **64**(5), 2037–2050 (2015)
14. Tomic, S., Beko, M., Dinis, R.: 3-D target localization in wireless sensor network using RSS and AoA measurements. IEEE Trans. Veh. Technol. **66**(4), 3197–3210 (2017)
15. Tomic, S., Beko, M., Tuba, M.: A linear estimator for network localization using integrated RSS and AOA measurements. IEEE Signal Process. Lett. **26**(3), 405–409 (2019)
16. Vaghefi, R.M., Gholami, M.R., Buehrer, R., Strom, E.G.: Cooperative received signal strength-based sensor localization with unknown transmit powers. IEEE Trans. Signal Process. **61**(6), 1389–1403 (2013)
17. Wang, Z., Zhang, H., Lu, T., Gulliver, T.A.: Cooperative RSS-based localization in wireless sensor networks using relative error estimation and semidefinite programming. IEEE Trans. Veh. Technol. **68**(1), 483–497 (2019)
18. Wu, X., Wang, S., Feng, H., Hu, J., Wang, G.: Motion parameter capturing of multiple mobile targets in robotic sensor networks. IEEE Access **6**, 24375–24390 (2018)
19. Xiong, Y., Wu, N., Wang, H., Kuang, J.: Cooperative detection-assisted localization in wireless networks in the presence of ranging outliers. IEEE Trans. Commun. **65**(12), 5165–5179 (2017)
20. Xu, E., Ding, Z., Dasgupta, S.: Reduced complexity semidefinite relaxation algorithms for source localization based on time difference of arrival. IEEE Trans. Mob. Comput. **10**(9), 1276–1282 (2011)
21. Xu, Y., Zhou, J., Zhang, P.: RSS-based source localization when path-loss model parameters are unknown. IEEE Commun. Lett. **18**(6), 1055–1058 (2014)
22. Zhang, Y., Li, Y., Zhang, Y., Jiang, T.: Underwater anchor-AUV localization geometries with an isogradient sound speed profile: a CRLB-based optimality analysis. IEEE Trans. Wireless Commun. **17**(12), 8228–8238 (2018)
23. Zhang, Y., Xing, S., Zhu, Y., Yan, F., Shen, L.: RSS-based localization in WSNs using gaussian mixture model via semidefinite relaxation. IEEE Commun. Lett. **21**(6), 1329–1332 (2017)
24. Zhao, J., et al.: Localization of wireless sensor networks in the wild: pursuit of ranging quality. IEEE/ACM Trans. Network. **21**(1), 311–323 (2013)
25. Zheng, J., Wu, X.: Convex optimization algorithms for cooperative RSS-based sensor localization. Pervasive Mob. Comput. **37**, 78–93 (2017)
26. Zhu, S., Ding, Z.: Distributed cooperative localization of wireless sensor networks with convex hull constraint. IEEE Trans. Wireless Commun. **10**(7), 2150–2161 (2011)

# Wi-Fi Floor Localization
# in an Unsupervised Manner

Liangliang Lin[1,2](✉), Wei Shi[1], Muhammad Asim[1], Hui Zhang[1],
Shuting Hu[3], and Jizhong Zhao[1]

[1] School of Computer Science and Technology,
Department of Telecommunications, Xi'an Jiaotong University,
Xi'an 710004, People's Republic of China
Lin_LL@126.com
[2] Information Department, Xi'an Conservatory of Music,
Xi'an 710061, People's Republic of China
[3] Qinghai University, Xining 810016, Qinghai,
People's Republic of China

**Abstract.** In recent decades, with the development of computer, indoor positioning applications have been developed rapidly. GPS has become one of the standards for outdoor positioning. However, there are great conditions for the use of GPS, GPS cannot be used indoors. At the same time, the indoor positioning scene has a great application prospect, through the use of indoor accessible signals (such as Wi-Fi, ZigBee, Bluetooth, UWB, etc.), according to the indoor environment and application, can be created based on the indoor positioning system. In the indoor positioning, there are two challenges, first of all, floor positioning, if the building has more than two layers, the second is planar positioning.

This paper solves the problem of floor positioning, and floor positioning based on Wi-Fi unsupervised recognition has attracted wide attention because it can get positioning results at a lower cost. In this paper, we try unsupervised indoor positioning methods, using only Wi-Fi crowdsourcing data. We get four months of data from seven-story buildings, by scanning the router's information. The application of neural network model can achieve unsupervised indoor positioning.

This clustering model aggregates all signals from the same floor into one class, and we use convolution neural networks, descending dimension feature extraction functions. The experiments show our solution obtains very high precision clustering results, so it can be summed up in this sense that the Wi-Fi crowdsourcing data can be used to locate in some way as the future direction of indoor positioning development.

**Keywords:** Indoor localization · CNN · K-means · Floor localization

This work is supported by NSFC Grants No. 61802299, 61772413, 61672424.

Q. Li et al. (Eds.): BROADNETS 2019, LNICST 303, pp. 52–69, 2019.
https://doi.org/10.1007/978-3-030-36442-7_4

# 1 Introduction

Indoor localization is the future for navigation purpose, in the current time of innovation, this is the most required work to be done for human beings' need. More than one & a half to two decades noteworthy research is happening in the area of indoor positioning. Which has guided to advancement of a few indoor localization frameworks (or arrangements) utilizing diverse signal systems for both research and business reasons. These arrangements are worked with various estimation strategies for example Lateration, Angulation, & Received Signal Strength. Accordingly, while building up an indoor localization framework it should be decided according to signal technologies accessible (as data origin) and estimation strategies which could be utilized along above mentioned advances. These choices are represented by the imperatives of utilization-case & for what execution measurements the framework is intended for. Conceptual diagram is exhibited under for recognizing distinctive indoor localization frameworks dependent on the signal system utilized as data origin in these frameworks with the end goal of positioning.

Nowadays, because of GPS and mobile phones in every person hands made a great achievement to know the location of a person or the mobile phone, but there is big hindrance to know the location of a person or mobile inside a building. GPS cannot detect the location inside the building, whenever the question of indoor localization comes on the tongue of any person, one must think of a storey building, any building having two or more floors, how can we determine the floor the person is at/on. The best way to sense the location of any mobile phone inside a wall blocked building in this modern period of technology is to use Wi-Fi modules. Because Wi-Fi modules are available in almost every building, so we can use the Wi-Fi RSS as access points(APs) to locate the mobile phones which are held by almost every person. There are many approaches to use Wi-Fi RSS to locate mobiles phone but the neural network approach has the greatest accuracy among others. We will elaborate the Artificial Neural Network, Deep Neural Network and Short Text Clustering technologies utilized in inside the building positioning systems.

The main contributions of our work can be summarized as follows:

- The real-time data of Wi-Fi RSS from a public mall building that gives a real experience of indoor surrounding. Analysis and results of each algorithm applied based on clustering performed which is accuracy attained.
- The method Latent Semantic Analysis was used for unsupervised dimensionality reduction because we have to learn deep feature representation from CNN in an unsupervised manner.
- Text clustering method was used in which there was one unsupervised dimensionality reduction function and one CNN model.

## 2   Related Works

### 2.1   iBeacon Based Floor Localization

iBeacon is a protocol systematized by Apple Inc. This is Apple's brand of a type of BLE-devices which broadcast & collect minute information for small ranges from surrounding portable electronic devices. This technology starts working when smart-phones, tablets and other devices are proximity to an iBeacon. They also look for applications in indoor localization systems that permits smart-phones to look for their estimated location by giving relative position information of smart-phone from an iBeacon in an Apple retail store. Consequently delivering proximity based indoor positioning system depends upon of that an iOS device receiving signal strength from an iBeacon can estimate its distance from other iBeacon. These ranges are sorted in these three ways: Immediate: within 2 m, Near: from 2 to 5 m, and Far: from 5 to 10 meters.

### 2.2   Short Text Clustering

Several studies have tried to conquer the scantiness of short text representation. There is a solution which is to extend and improve the setting of dataset. For instance, [1] proposed a strategy for enhancing the precision of short text clustering by enhancing their representation with extra features taken from Wikipedia, & [2] fuse semantic learning from a philosophy into text clustering. Be that as it may, this kind of work need strong NLP information and still utilize high-dimensional representation the outcome can be possibly in a misuse of memory as well as of calculation time. One more heading is to outline unique features in diminished space, for example, Latent Semantic Analysis (LSA) [3], Laplacian Eigenmaps (LE) [4], & Locality Preserving Indexing (LPI) [5]. Indeed, still a few scientists investigated some refined models to cluster short texts. For instance, [6] suggested a Dirichlet multinomial mixture model-based methodology for short text clustering. In addition, a few examinations even center the above both two streams. For instance, [7] suggested a novel system that advance the content features by utilizing machine interpretation and lessen the real features at the same time through matrix factorization methods.

### 2.3   Deep Neural Networks

As of late, it's a recovery of enthusiasm for DNN and numerous scientists have focused on utilizing Deep Learning to learn features. [8] utilize DAE to learn text representation. Amid the tweaking strategy, they utilize back-propagation to discover codes which are great at remaking the word count vector.

All the more as of late, scientists propose to utilize outer corpus to become familiar with a dispersed representation for every single word, known as word embedding [9], to improvise Deep Neural Network execution on NLP assignments. The Skip-gram and continuous bag-of-words models of Word2vec [10] suggest a basic single-layer architecture dependent upon the product of 2 word-vectors, & [11] present an another model for the representations of words, known as GloVe, that catches the worldwide corpus insights.

Motivated by Skip-gram of word2vec [10, 12], Skip-thought model [13] depicts a methodology for unsupervised learning of a nonexclusive, dispersed sentence encoder. Comparative as Skip-gram model, Skip-thought model [14] trains an encoder-decoder model which attempts to remake the encompassing sentences of an encoded sentence & discharged an off-the-rack encoder to remove sentence representation. This theory, let's have another point of view [15], advances a general self-trained Convolutional Neural Network structure that can adaptably couple different semantic features and accomplish a decent act upon single unsupervised learning task, short text clustering.

## 3  Methodology

To achieve the goal of floor localization, first we have to make clusters of each floor's Wi-Fi modules' Received Signal Strength, so that we can use the clusters of each floor to apply other algorithms to locate the exact floor of the mobile phone. My thesis is all about how to make clusters of each floors' Wi-Fi RSS. I have used a model of short text clustering to make clusters of the Wi-Fi RSS of each floor of a building. Because this model contains Convolutional Neural Network (CNN) model which empowers the clustering process like boosting it with a NOS gas cylinder. It is called Short Text Clustering ($STC^2$). To make use of short text clustering model, we had to convert the Wi-Fi RSS into text form, so we converted the Wi-Fi RSS data into matrix form to make it a feasible input for convolutional neural network. The main objective is explained in the Fig. 1, as follows:

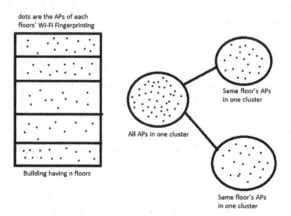

**Fig. 1.** Suppose a building having n number of floors, each floor have different amount of Wi-Fi modules, Access Points (Aps) or Wi-Fi RSS are represented as dots in the figure, all the Aps will be collected in one cluster first then they will divided into same floors' Aps clusters, we supposed that we are collecting only two floors Aps.

The aim is to make clusters of having some kind of resemblance between all short texts. So, let us suppose that we have a training texts dataset of n amount of numbers which is indicated as:

$$\mathbf{X} = \left\{ \mathbf{X}_i \colon \mathbf{X}_i \in \mathbb{R}^{d \times 1} \right\}_{i=1,2,\dots,n,} \tag{1}$$

In above equation d is called the imensionality of the original Baf-of-Words representation.

Let's denote its tag/label set as:

$$\mathbf{T} = \{1, 2, \dots, C\} \tag{2}$$

and for word embedded pre-trained set, it could be represented as:

$$\mathbf{E} = \left\{ \mathbf{e}(w_i) \colon \mathbf{e}(w_i)? \ \mathbf{R}^{\text{duxl}} \right\}_{i=1,2,\dots,|V|,} \tag{3}$$

where $d_w$ is the dimensionality of word vectors and $|V|$ is the vocabulary size. For the purpose to learn the r-dimensional deep feature representation $\mathbf{h}$ from Convolutional Neural Network in an unsupervised way, few unsupervised dimensionality reduction techniques $f_{dr}(X)$ are utilized to manage the learning of Convolutional Neural Network model. We will probably cluster these texts X into clusters C dependent on the learned deep feature representation during saving the semantic continuity.

As delineated in Fig. 2, the framework we are inspired from comprises of 3 segments, deep convolutional neural system (CNN) or all the more especially Dynamic Convolutional Neural Network (DCNN), unsupervised dimensionality reduction function and K-means module. In the rest segments, starting with the initial 2 segments separately, and after that give the trainable parameters & the target function to help learn the deep feature representation.

In the end, the final part discloses the solution to execute clustering algorithm on the learned features.

Deep learning is the main core part of the Artificial Intelligence technology. In deep learning, you can also find Convolutional Neural Networks (CNN), more specifically the Deep Convolutional Neural Network (DCNN). There are few different kinds of deep convolutional neural networks but in this section, we sum up an overview of one popular deep convolutional neural network, Dynamic Convolutional Neural Network (DCNN) [16] as an example of CNN in the upcoming sections, that is as the establishment of this inspired solution has been effectively suggested for the totally supervised learning task, text classification.

In Fig. 3, just take a neural network having 2 convolutional layers for instance, the network transforms raw input text to a powerful representation. Specifically, every single raw text vector $x_i$ is projected into a matrix representation $S \in \mathbf{R}^{d_w \times s}$ by looking up a word embedding E, in which $s$ represents the length of one text. Let us have $\tilde{\mathbf{W}} = \{W_i\}_{i=1,2}$ and $W_O$ indicate the weights of the neural networks. The network describes a transformation $f(\cdot) : \mathbf{R}^{d \times 1} \to \mathbf{R}^{r \times 1}(d) \ r)$ which transforms an input raw text x to a $r$-dimensional deep representation h. After that we can apply these 3 fundamental operations outlined as pursues:

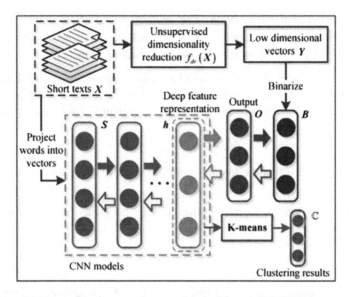

**Fig. 2.** The architecture of the framework we are inspired from called STC2 framework for short text clustering. Convolutional Neural Networks

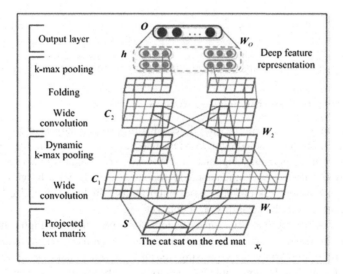

**Fig. 3.** In the dynamic convolutional neural network architecture [16]. After applying the process of word embedding, first we project the input text to the matrix feature, after that it passes through the wide convolutional layers, then folding layers & after that k-max pooling layers, after this process we are provided a deep feature representation just behind the output layer.

*Wide One-Dimensional Convolution.* This method m $\in \mathbf{R}^m$ is employed to each single row of the sentence matrix $S \in \mathbf{R}^{d_w \times s}$, which produces a resulting matrix $\mathbf{C} \in \mathbf{R}^{d_w \times (s+m-1)}$, $m$ represents the convolutional filter's width.

*Folding.* In this method, every two rows in a feature map are easily added component wisely. A map having $d_w$ rows, folding sends back a map of $d_w/2$ rows, consequently halving the size of the representation & producing a matrix feature as pursues:

$$\hat{C} \in \bar{\mathbb{R}}^{(d_w/2) \times (s+m-1)}. \tag{4}$$

This is to be noted that the folding operation do not show any more than enough parameters.

*Dynamic k-max Pooling.* Supposing the pooling parameter as $k$, $k$-max pooling chooses the sub-matrix

$$\bar{C} \in \bar{\mathbb{R}}^{(d_w/\bar{2}) \times k} \tag{5}$$

of the $k$ highest values in every single row of the matrix $\hat{C}$. For dynamic $k$-max pooling, the pooling parameter $k$ is dynamically chosen for the purpose to permit for a smooth extraction of higher-order and longer-range features [16]. A constant pooling parameter $k_{top}$ for the topmost convolutional layer, the parameter $k$ of $k$-max pooling in the $l$th convolutional layer could be calculated as pursues:

$$k_1 = \max\left(k_{top}, \left\lceil \frac{L-l}{L} s \right\rceil \right) \tag{6}$$

where L is the total number of convolutional layers in the network.

### 3.1 Unsupervised Dimensionality Reduction

In pattern recognition, data mining and different sorts of data analysis applications, we regularly face high dimensional data. For instance, in face recognition, the extent of a training picture fix is generally bigger than $60 \times 60$, which relates to a vector with in excess of 3600 dimensions. Dimensionality reduction can likewise be viewed as the way toward determining a lot of degrees of opportunity which can be utilized to repeat the greater part of the changeability of a data set.

*Unsupervised dimensionality reduction* goes for representing to high-dimensional data in lower-dimensional spaces in a dedicated manner. Dimensionality reduction can be utilized for compression or denoising reasons, however data visualization still happens to be its highly noticeable applications. In the event that visualization is troublesome in high-dimensional space, maybe a (nearly) proportionate representation in a lower-dimensional space can improvise the lucidness of data. This is unequivocally the possibility which goes under the area of dimensionality reduction (DR).

There are a lot of different kinds of techniques for dimensionality reduction like the most popular Principal Component Analysis (PCA), Linear Discriminant Analysis (LDA), Locality Preserving Indexing (LPI) and others but we are using Latent

Semantic Analysis, because it has the highest accuracy among other few techniques on our given data. As mentioned in Fig. 2, the dimensionality reduction algorithm can be represented as pursues:

$$\mathbf{Y} = f_{dr}(\mathbf{X}),\tag{7}$$

where, $\mathbf{Y} \in R^{q \times n}$ are the q-dimensional reduced latent space representations. Afterwards, we apply a famous dimensionality reduction method in this framework.

Latent Semantic Analysis (LSA) : LSA [3] is a very famous worldwide matrix factorization algorithm, that employs a dimension decreasing linear projection, Singular Value Decomposition (SVD), of the relating term/document matrix. Assume the rank of X is $\hat{r}$, LSA deteriorates X in the result of 3 different matrices:

$$X = U \sum V^{\mathsf{T}}$$

in above equation $\sum$ is equal to:

$$\sum = diag\ (\sigma_1, \ldots, \sigma_{\hat{r}})\ \text{and}\ \sigma_1 \geq \sigma_2 \geq \ldots \geq \sigma_{\hat{r}}\tag{8}$$

are the singular values of X, $U \in R^{d \times \hat{f}}$ is a set of left singular vectors and $V \in R^{n \times \hat{f}}$ is a set of right singular vectors. LSA utilizes the top q vectors in U as the transformation matrix to embed the original text features into a q-dimensional subspace Y [3].

The method mentioned above guarantees a superior exhibition in catching semantic similarity betwixt text in the reduced latent space representation Y compared to the original representation X, so, the execution of short text clustering could be additionally upgraded by the assistance of this system, self-taught Convultional Neural Network.

## 3.2    Training of the Model

ML (Machine Learning) is the consistent examination of estimations and quantifiable models that PC structures use to effectively play out a specific endeavor without using express rules, contingent upon precedents and deducing.

In unsupervised learning, the calculation assembles a numerical model from a dataset which contains just sources of info and no ideal yield marks. Unsupervised learning calculations are utilized to discover structure in the information, such as gathering or grouping of information focuses. Unsupervised learning can find designs in the information, and can bunch the contributions to classes, as in feature learning. Dimensionality reduction is the way toward diminishing the quantity of "features", or inputs, in a dataset.

The short text clustering model we are using also contains a Convolutional Neural Network (CNN) model, more particularly Dynamic Convolutional Neural Network (DCNN). In this model the input is actually the word embedded form of the original raw text vectors. It has the deep feature resrespresentation h, which connected to the output in order to get the first random value from the binarized form of Low dimensional vector Y.

The final layer of CNN is an output layer as follows:

$$O = W_o h, \tag{9}$$

in which, h is the deep feature representation, $O \in R^q$ is the output vector & $W_o \in R^{q \times r}$ is weight matrix.

For the purpose to integrate the latent semantic features Y, firstly, we have to binarize the real-valued vectors Y to the binary codes B by fixing the threshold to be the median vector *median* (Y). At that point, the output vector O is utilized to set the binary codes B through q logistic operations as pursues:

$$p_i = \frac{\exp(\mathbf{O}_i)}{1 + \exp(\mathbf{O}_i)}$$

All parameters to be trained are denoted as $\theta$

$$\theta = \{\mathbf{E}, \tilde{\mathbf{W}}, \mathbf{W}_o\}. \tag{10}$$

Given the training text collection X, & the pre-trained binary codes B, the log probability of the parameters could be formulated as pursues:

$$J(\theta) = \sum_{i=1}^{n} \log p(\mathbf{b}_i | \mathbf{X}_i, \theta). \tag{11}$$

We have followed the already work done by [16], by the method of back propagation, we train the network with mini batches & applied the gradient based optimization utilizing the Adagrad update rule [17]. For regularization, we utilize dropout with 50% to the penultimate layer [16, 18].

### 3.3 K-Means for Clustering

K-means clustering is a procedure for vector quantization, at first started from signal processing, which is conspicuous for cluster analysis in data mining. K-means clustering plans to fragment n observations into k clusters, after that every single observation has a spot with the cluster with the nearest mean, filling in as a clustering model. Which concludes in a division of the data space into Voronoi cells.

Let us suppose we are given set of observations $(x_1, x_2, \ldots, x_n)$, in which every single observation is a $d$-dimensional real vector, $k$-means clustering intends to divide the $n$ observations into $k$ which is less than or equal to $n$ sets $S = \{S_1, S_2, \ldots, S_k\}$ so as to minimize the within cluster sum of squares (WCSS) (i.e. variance or change). Suitably, the goal is to discover:

$$\operatorname*{arg\,min}_{S} \sum_{i=1}^{k} \sum_{x \in S_i} \|\mathbf{X} - \mu_i\|^2 = \operatorname*{arg\,min}_{S} \sum_{i=1}^{k} |S_i| \operatorname{Var} S_i$$

in which $\mu_i$ is the mean of points in $S_i$. This is almost equal to minimizing the pairwise squared variations of points in the identical cluster:

$$\underset{S}{\arg\min} \sum_{i=1}^{k} \frac{1}{2|S_i|} \sum_{x,y \in s_i} \|x - y\|^2$$

The equivalent couln be derived by this equation:

$$\sum_{X \in S_i} \|X - \mu_i\|^2 = \sum_{x \neq y \in S_i} (X - \mu_i)(\mu_i - y)$$

The total variance is fixed, it is almost equal to maximizing the sum of squared fluctuations betwixt points in *different* clusters (between cluster sum of squares, BCSS), the law of total variance is followed.

Finally getting the short texts, then initially use the trained deep neural network to acquire the semantic representations h, afterward utilize conventional K-means function to accomplish clustering.

## 4  Evaluation

### 4.1  Experimental Setup

We test the model on the given dataset by Tencent company, they have given us the dataset of a shopping mall, shopping mall had 7 floors. All the Wi-Fi RSS from all the 7 floors' Wi-Fi modules. They have used different mobile phones models of different brands to collect the data from the Wi-Fi routers of the chosen shopping mall, the time period of the collected data is approximately 4 months. Different mobile phones have different strength of catching or detecting the signals from each floor, so they have collected the data in a way so that it can be supposed like a public experiment. Because for public, everyone has a different kind of mobile phone, different kind of mobile phone brand, so that is why they have tried to collect it like a public user.

Tencent company has given one dataset in which there are approximately 2 million Received Signal Strength (RSS) of all the Wi-Fi modules on all seven floors of the shopping mall. The given dataset includes both labeled and unlabeled data. We took a sample set from the given data set, we took almost 20,000 RSS from the original dataset of all seven floors. Our experiment or project is to cluster only two floors, so we chose first and the seventh floor from our sample dataset. We did not use the RSS from second to sixth floor. First, we separated a sample data set from the original RSS dataset then we collected the RSS of first and the seventh floor from our collected sample dataset.

In order to convert Wi-Fi RSS into text data, we processed the data, the convolutional neural network input is a matrix, in which each row of the matrix is representing the sentence and each column is representing the word2vector encoding of each word in the sentence. We had the address of the Wi-Fi router with its RSS, so in this paired form (Wi-Fi address, RSS), we received the data, so for each pair, we repeat

(Wi-Fi address, RSS) up to the value we got from adding 100 to the RSS, so it forms a line of sentences, then carry out word2vector training and finally input to the CNN model.

As my graduate project, I picked only $1^{st}$ and $7^{th}$ floors Wi-Fi RSS from the sampled data which we picked from the original 7 floors dataset. In which first floor had 40,000 (Wi-Fi address, RSS) pairs and seventh floor had 18,000 (Wi-Fi address, RSS) pairs. The matrix we formed for the input, each row of the matrix has 20 pairs of (Wi-Fi address, RSS), so for first floor we calculate the number of rows for the matrix by the formula as follows:

Number of rows is equal to the number of floors (Wi-Fi address, RSS) pairs divided by number of pairs in each row

Rows = Pairs/20

For the $1^{st}$ floor: No: of rows = 40000/20 = 2000

For the $7^{th}$ floor: No: of rows = 18000/20 = 900

So, now we have an uneven quantity of rows, which is not feasible to train, so we choose to make them an even and equal number of rows, for that purpose we select 900 rows as minimum amount of rows, so we have to find a way to decrease the number of rows of first floor. Although it was the toughest decision to do it by the help of computer but the easiest way too. So we run a random function in MATLAB to choose the random values from data and delete them from data until the data has 900 rows which is equal to $7^{th}$ floors' number of rows.

For the input for the Unsupervised Dimensionality Reduction, we used the labeled first and the seventh floors' RSS, as shown in the Table 1.

**Table 1.** Transformed input matrix for CNN on the sample data of only first and the seventh floor

| Dataset | Number of rows | (Wi-Fi Address, RSS) Pairs |
|---|---|---|
| $1^{st}$ Floor | 900 | 18000 |
| $2^{nd}$ Floor | 900 | 18000 |
| Total | 1800 | 36000 |

### 4.2  Pre-trained Word Vectors

The purpose to input dataset into the Dynamic Convolutional Neural Network model, we used the same sample data of the first and the seventh floors and utilized the freely accessible word2vec tool for the training purpose of word embeddings, almost all the parameters are same as in [10] for the training of word vectors on Google News setting, aside from of vector dimensionality utilizing 48 and minimize count utilizing 5. The inclusion of these learned vectors on our example dataset is recorded in Table 1, and the words not existing in the set of pre-trained words are instated haphazardly.

## 4.3    Evaluation Metrics

The clustering execution is assessed by contrasting the clustering results of data and the labels/tags given by the text corpus. Two metrics, the accuracy (ACC) & the normalized mutual information metric (NMI), are utilized to quantify the clustering execution [19, 20]. Given a text $x_i$, let $c_i$ and $t_i$ be the gotten cluster label and the label given by the corpus, separately. Accuracy is calculated by this formula:

$$ACC = \frac{\sum_{i=1}^{n} \delta(t_i, map(c_i))}{n},$$

in which, $n$ is the total number of texts, $\delta(x, y)$ is the indicator function that becomes 1 if $x$ is equal to $y$ and becomes zero if $x$ is not equal to $y$, and $map(c_i)$ is the permutation mapping function which maps every single cluster label $c_i$ to the almost equal label from the text data by Hungarian algorithm [21].

Normalized mutual information [22] betwixt tag/label set T and cluster set C is a popular metric utilized for calculating clustering tasks. NMI is denoted as pursues:

$$NMI(\mathbf{T}, \mathbb{C}) = \frac{MI(\mathbf{T}, \mathbb{C})}{\sqrt{H(\mathbf{T})H(\mathbb{C})}} \qquad (12)$$

in which, MI(T,C) represents mutual information betwixt T and C, H(.) represents entropy and $\sqrt{H(\mathbf{T})H(C)}$ is utilized for normalizing the mutual information for limiting it between 0 and 1.

## 4.4    Settings for Hyperparameter

A large portion of the parameters are set consistently for the sampled dataset. This CNN model [15], there are 2 convolutional layers in the network. The convolutional filters' width has set to be 3 for both of them. The top k-max pooling's value in Eqs. (3–4) has set to be 5 for k. On the first convolutional layer the number of feature maps is 12, and on the second convolutional layer there are 8 feature maps. After the both convolutional layers there is a folding layer. Moving on to the dimension of word embeddings, we set $d_w$ as 48. At long last, the dimension of the deep feature representation r is set to 480. Also, we fix the learning rate $\lambda$ to 0.01 and the mini batch training size to 200 only for STC2-Latent Semantic Analysis and Latent Dirichlet Allocation (LDA) and mini batch training size to 16 for all other methods. In Eqs. (3–7), q is the output size which is fixed equivalent to the best dimensions of subspace in the benchmark method.

For beginning centroids have noteworthy effect on the results of clustering while using the K-means method, we rehash K-means for numerous times with arbitrary starting centroids (explicitly, for statistical significance 100 times) in [20]. The every subspace vectors are normalized to 1 preceding applying K-means and the last outcomes revealed are the half of the sum of five evaluations with all the clustering algorithms on our sampled dataset.

## 4.5  Results and Analysis

The details of different accuracy (ACC) and normalized mutual information (NMI) percentage of according to the change of epoch size on the model are shown in Table 2. The method we used for Unsupervised Dimensionality Reduction is Latent Dirichlet Allocation (LDA). Applying this method to the short text clustering model we get some fine results. On different size epoch we get variance in the accuracy (ACC) and the normalized mutual information (NMI). Detailed ACC and NMI are in the table below:

**Table 2.** ACC and the NMI of epoch size 1–6, at epoch size 2, the ACC is the highest.

| Latent Dirichlet Allocation (LDA) in unsupervised dimensionality reduction | | |
|---|---|---|
| Epoch (size) | ACC (%) | NMI (%) |
| 1 | 61.385 | 8.9928 |
| 2 | 61.4125 | 8.9652 |
| 3 | 61.3899 | 8.9341 |
| 4 | 61.2851 | 8.9159 |
| 5 | 61.1852 | 8.0905 |
| 6 | 61.0258 | 8.0058 |
| 7 | 60.369 | 7.9928 |

In Table 3, you can see the details of different accuracy (ACC) and normalized mutual information (NMI) percentage of according to the change of epoch size on the model. The method we used for Unsupervised Dimensionality Reduction is Average Embedding (AE). Applying this method to the short text clustering model we get some fine results. On different size epoch we get variance in the accuracy (ACC) and the normalizaed mutual information (NMI). Detailed ACC and NMI are in the table below:

**Table 3.** ACC and the NMI of epoch size 1–6, at epoch size 2, the ACC is the highest.

| Average Embedding (AE) in unsupervised dimensionality reduction | | |
|---|---|---|
| Epoch (size) | ACC (%) | NMI (%) |
| 1 | 66.385 | 7.9928 |
| 2 | 66.3537 | 7.9652 |
| 3 | 66.3099 | 7.9431 |
| 4 | 66.385 | 7.9928 |
| 5 | 66.4852 | 8.0905 |
| 6 | 66.3788 | 8.0058 |
| 7 | 66.385 | 7.9928 |

In Table 4, you can see the details of different accuracy (ACC) and normalized mutual information (NMI) percentage of according to the change of epoch size on the model. The method we used for Unsupervised Dimensionality Reduction is Spectral Laplacian Eigenmaps (Spectral-LE). Applying this method to the short text clustering model we get some fine results. On different size epoch we get variance in the accuracy (ACC) and the normalizaed mutual information (NMI). Detailed ACC and NMI are in the table below:

**Table 4.** ACC and the NMI of epoch size 1–6, at epoch size 1, the ACC is the highest.

| Spectral Laplacian Eigenmaps (Spectral-LE) in unsupervised dimensionality reduction | | |
|---|---|---|
| Epoch (size) | ACC (%) | NMI (%) |
| 1 | 83.2332 | 47.3152 |
| 2 | 83.2112 | 47.3100 |
| 3 | 83.2008 | 47.3023 |
| 4 | 83.1933 | 47.2933 |
| 5 | 83.1900 | 47.2902 |
| 6 | 83.1806 | 47.0126 |

In Table 5, shown the details of different accuracy (ACC) and normalized mutual information (NMI) percentage of according to the change of epoch size on the model. The method we used for Unsupervised Dimensionality Reduction is Latent Semantic Analysis (LSA). Applying this method to the short text clustering model we get some fine results. On different size epoch we get variance in the accuracy (ACC) and the normalizaed mutual information (NMI). Detailed ACC and NMI are in the table below:

**Table 5.** ACC and the NMI of epoch size 1–6, at epoch size 1, the ACC is the highest.

| Latent Semantic Analysis (LSA) in Unsupervised Dimensionality Reduction | | |
|---|---|---|
| Epoch (size) | ACC (%) | NMI (%) |
| 1 | 60.0638 | 4.9477 |
| 2 | 60.0603 | 4.9463 |
| 3 | 60.0563 | 4.9449 |
| 4 | 60.0530 | 4.9432 |
| 5 | 60.0513 | 4.9396 |
| 6 | 60.0496 | 4.9385 |

This algorithm Average Embedding (AE) with Short Text Clustering method is shown in Table 6, shown the details of different accuracy (ACC) and normalized mutual information (NMI) percentage of according to the change of epoch size on the model. Applying this method to the short text clustering model we get some fine results. On different size epoch we get variance in the accuracy (ACC) and the normalizaed mutual information (NMI). Mini batch size is set to be 16 for this method and detailed ACC and NMI are in the table below:

**Table 6.** ACC and the NMI of epoch size 1–6, at epoch size 1, the ACC is the highest.

| STC2-Average Embedding (STC2-AE) in unsupervised dimensionality reduction | | |
|---|---|---|
| Epoch (size) | ACC (%) | NMI (%) |
| 1 | 64.6013 | 6.4831 |
| 2 | 68.3164 | 10.076 |
| 3 | 68.6075 | 10.3594 |
| 4 | 68.5349 | 10.3079 |
| 5 | 68.6657 | 10.4475 |
| 6 | 68.6523 | 10.4325 |

Another algorithm with Short Text Clustering method is shown in Table 7, called Latent Semantic Analysis (STC2-LSA), shown the details of different accuracy (ACC) and normalized mutual information (NMI) percentage of according to the change of epoch size on the model. Applying this method to the short text clustering model we get some fine results. On different size epoch we get variance in the accuracy (ACC) and the normalizaed mutual information (NMI). Mini batch size is set to 200 for this method and detailed ACC and NMI are in the table below:

**Table 7.** ACC and the NMI of epoch size 1–6, at epoch size 1, the ACC is the highest.

| STC2-Latent Semantic Analysis (STC2-LSA) in unsupervised dimensionality reduction | | |
|---|---|---|
| Epoch (size) | ACC (%) | NMI (%) |
| 1 | 77.1674 | 23.5357 |
| 2 | 78.635 | 26.9315 |
| 3 | 78.4185 | 26.0886 |
| 4 | 77.8677 | 24.7327 |
| 5 | 77.6968 | 24.3019 |
| 6 | 70.4093 | 14.7743 |

One more algorithm with Short Text Clustering method is shown in Table 8, called Laplacian Eigenmaps (STC2-LE), shown the details of different accuracy (ACC) and normalized mutual information (NMI) percentage of according to the change of epoch size on the model. Applying this method to the short text clustering model we get some fine results. On different size epoch we get variance in the accuracy (ACC) and the normalizaed mutual information (NMI). Mini batch size is set to 16 for this, detailed ACC and NMI are in the table below:

**Table 8.** ACC and the NMI of epoch size 1–6, at epoch size 1, the ACC is the highest.

| STC2-Laplacian Eigenmaps (STC2-LE) in unsupervised dimensionality reduction | | |
|---|---|---|
| Epoch (size) | ACC (%) | NMI (%) |
| 1 | 82.2819 | 38.8306 |
| 2 | 82.2381 | 42.5264 |
| 3 | 82.0092 | 42.1357 |
| 4 | 81.8671 | 41.9982 |
| 5 | 81.7322 | 41.8866 |
| 6 | 81.5236 | 41.7569 |

The results we successfully got, with two measures of ACC and NMI on our sampled dataset, we get some promising results from our applied Unsupervised Dimensionality Reduction method STC2-Laplacian Eigenmaps (STC2-LE), which concludes that the approach is the effective approach to collect some handy semantic features for clustering of the Wi-Fi RSS of the buildings.

## 5   Conclusion

In this paper, we work on clustering for the use of indoor localization, making use of Artificial Neural Network and other algorithms, was experimented, tested and evaluated with several methods of Unsupervised Dimensionality Reduction methods. The goal was to successfully put same floors' Wi-Fi RSS to a cluster where only exists the same floors' Wi-Fi RSS.

After a comprehensive study of earlier research in indoor localization systems based on various signal technologies, an approach to successfully cluster the same floors' RSS was developed by the help of a model called short text clustering. Short text clustering was chosen because of the inclusion of Convolutional Neural Network model in it, and the signal technology approach, for that we chose Wi-Fi routers inside the building, it was chosen because of the fact that it is readily available in the buildings. Research was conducted to see with which Unsupervised Dimensionality Reduction method the clustering has the highest accuracy, so Latent Semantic Analysis (LSA) method was used because of the highest accuracy factor. The conclusion of the evaluation is that short text clustering model can be considered as a viable candidate for clustering the Wi-Fi RSS into same clusters.

The system developed can give a 78% accuracy of clustering two floors as compared to other methods and algorithms. These results are good enough to continue this approach for the other floors of the building.

# References

1. Banerjee, S., Ramanathan, K., Gupta, A.: Clustering short texts using wikipedia. In Proceedings of the 30th Annual International ACM SIGIR Conference on Research and Development in Information Retrieval, pp. 787–788. ACM (2007)
2. Fodeh, S., Punch, B., Tan, P.-N.: On ontology-driven document clustering using core semantic features. Knowl. Inf. Syst. **28**(2), 395–421 (2011)
3. Deerwester, S.C., Dumais, S.T., Landauer, T.K., Furnas, G.W., Harshman, R.A.: Indexing by latent semantic analysis. JAsIs **41**(6), 391–407 (1990)
4. Ng, A.Y., Jordan, M.I., Weiss, Y., et al.: On spectral clustering: Analysis and an algorithm. Adv. Neural. Inf. Process. Syst. **2**, 849–856 (2002)
5. He, X., Niyogi, P.: Locality preserving projections. In: Neural Information Processing Systems, vol. 16, pp. 153–160. MIT (2004)
6. Yin, J., Wang, J.: A dirichlet multinomial mixture model-based approach for short text clustering. In: Proceedings of the 20th ACM SIGKDD International Conference on Knowledge Discovery and Data Mining, pp. 233–242. ACM (2014)
7. Tang, J., Wang, X., Gao, H., Hu, X., Liu, H.: Enriching short text representation in microblog for clustering. Front. Comput. Sci. **6**(1), 88–101 (2012)
8. Hinton, G.E., Salakhutdinov, R.R.: Reducing the dimensionality of data with neural networks. Science **313**(5786), 504–507 (2006)
9. Turian, J., Ratinov, L., Bengio, Y.: Word representations: a simple and general method for semi-supervised learning. In: Proceedings of the 48th Annual Meeting of the Association for Computational Linguistics, pp. 384–394 (2010)
10. Mikolov, T., Sutskever, I., Chen, K., Corrado, G.S., Dean, J.: Distributed representations of words and phrases and their compositionality. In: Advances in Neural Information Processing Systems, pp. 3111–3119 (2013)
11. Pennington, J., Socher, R., Manning, C.D.: Glove: global vectors for word representation. In: Proceedings of the Empiricial Methods in Natural Language Processing, p. 12 (2014)
12. Mikolov, T., Chen, K., Corrado, G., Dean, J. Efficient estimation of word representations in vector space. arXiv preprint arXiv:1301.3781(2013)
13. Kiros, R., Zhu, Y., Salakhutdinov, R.R., Zemel, R., Urtasun, R., Torralba, A., et al.: Skip-thought vectors. In: Advances in Neural Information Processing Systems, pp. 3276–3284 (2015)
14. Wang, X., Gupta, A. Unsupervised learning of visual representations using videos, arXiv preprint arXiv:1505.00687. (2015)
15. Jiaming, X., Bo, X., Wang, P., Zheng, S., Tian, G., Zhao, J.: Self-taught convolutional neural networks for short text clustering. Neural Netw. **88**, 22–31 (2017)
16. Kalchbrenner, N., Grefenstette, E., Blunsom, P.: A convolutional neural network for modelling sentences. In: Proceedings of the 52nd Annual Meeting of the Association for Computational Linguistics (2014)
17. Duchi, J., Hazan, E., Singer, Y.: Adaptive subgradient methods for online learning and stochastic optimization. J. Mach. Learn. Res. **12**, 2121–2159 (2011)
18. Kim, Y.: Convolutional neural networks for sentence classification. In: Proceedings of the Conference on Empirical Methods in Natural Language Processing (2014)

19. Cai, D., He, X., Han, J.: Document clustering using locality preserving indexing. IEEE Trans. Knowl. Data Eng. **17**(12), 1624–1637 (2005)
20. Huang, P., Huang, Y., Wang, W., Wang, L.: Deep embedding network for clustering. In: 2014 22nd International Conference on Pattern Recognition (ICPR), pp. 1532–1537. IEEE (2014)
21. Papadimitriou, C.H., Steiglitz, K.: Combinatorial optimization: algorithms and complexity. Courier Corporation (1998)
22. Chen, W.-Y., Song, Y., Bai, H., Lin, C.-J., Chang, E.Y.: Parallel spectral clustering in distributed systems. IEEE Trans. Pattern Anal. Mach. Intell. **33**(3), 568–586 (2011)

# Communication and Sensor Networks

# Virtual Network Embedding Algorithm Based on Multi-objective Particle Swarm Optimization of Pareto Entropy

Ying Liu, Cong Wang$^{(\boxtimes)}$, Ying Yuan, Guo-jia Jiang, Ke-zhen Liu, and Cui-rong Wang

College of Computer and Communication Engineering,
Northeastern University at Qinhuangdao, Qinhuangdao 066004, China
congw1981@163.com

**Abstract.** Virtual network embedding/mapping refers to the reasonable allocation of substrate network resources for users' virtual network requests, which is a key issue for virtual resource leasing in Cloud computing. Most of the existing researches only aim to maximize the revenue. As the scale of hardware network expands, the energy consumption of substrate network also needs to be paid more attention. In this paper, a multi-objective virtual network mapping algorithm based on particle swarm optimization with Pareto entropy (VNE-MOPSO) is proposed. It combines energy consumption and revenue. The algorithm controls the energy consumption of the substrate network as much as possible to achieve the goal of energy saving on the premise of ensuring a small resource cost. By introducing the Pareto entropy based multi-objective optimization model, it can calculate the difference of entropy and evaluate the evolutionary state. With this as feedback information, a dynamic adaptive particle velocity updating strategy is designed to achieve the goal of solving the approximate optimal multi-objective optimization mapping scheme. Simulation results show that the proposed algorithm has certain advantages over the typical single target mapping algorithm in cost, energy consumption and average return.

**Keywords:** Virtual network embedding · Multi-objective optimization · Discrete particle swarm optimization · Pareto entropy

## 1 Introduction

Since the 1960s, the Internet has flourished and become an important information infrastructure in modern society. On August 20, 2018, the China Internet network information center (CNNIC) released the 42nd statistical report on the development of China's Internet network [1]. The report shows that by June 2018, the number of Chinese Internet users has reached 802 million, and the Internet penetration rate has reached 57.7%. More than half of the Chinese people have been connected to the Internet. However, with the explosive growth of the number of users, as well as an increasing number of distributed applications and emerging network technologies, network "rigidity" phenomenon is becoming increasingly serious, which hinders the development of the Internet.

© ICST Institute for Computer Sciences, Social Informatics and Telecommunications Engineering 2019
Published by Springer Nature Switzerland AG 2019. All Rights Reserved
Q. Li et al. (Eds.): BROADNETS 2019, LNICST 303, pp. 73–85, 2019.
https://doi.org/10.1007/978-3-030-36442-7_5

Network virtualization has great prospects in the future development of the Internet. Besides, it has been actively applied in software-defined network and cloud computing environment [2]. Network virtualization takes the substrate networks as the basic entities, which enables users to take the whole network including virtual hosts and virtual links as the request scheme. Compared with the previous leasing mode that can only rent virtual machines but cannot guarantee the demand of network resource, it has better feasibility.

Virtual network embedding (VNE) problem [3] is one of the key technologies of network virtualization technology. Each virtual network request (VNR) from service providers is constrained by node resources (CPU, memory, storage) and link resources (bandwidth). The content of VNE is to allocate substrate network resources to these virtual network requests. At present, meta-heuristic algorithms have been successfully applied to a wide range of optimization problems [4] (e.g. [4–8]). Many researchers have explored the optimization model, searching strategy, algorithm acceleration, etc. They have provided important references for the follow-up researchers' works.

However, most of the studies focus on the optimization of single objective. Considering the needs to balance mapping costs, benefits, energy consumption, quality of service (QoS) and other issues in the actual demands of virtual resource leasing, we set out from the perspective of multi-objective optimization in this paper. Multi-objective optimization problem (MOP) studies the optimization of multiple objective functions in a given region. The optimization result is a Pareto optimal solution set. We model and solve the VNE problem according to reference [9].

In this paper, we combines Pareto entropy multi-objective optimization model with particle swarm optimization (PSO) to ensure the rental revenue of physical network resources, and takes the energy consumption into account at the same time. Using the target space transformation method, the external Pareto solution set is mapped to the parallel cell coordinate system, and then the population evolutionary state is evaluated according to the distribution of entropy of the approximate Pareto front end. Then, we use the feedback information of evolutionary process to design an adaptive parameter setting strategy that dynamically balances the development and utilization capabilities. The simulation results show that the proposed VNE-MOPSO algorithm effectively reduces the cost and energy consumption of virtual network mapping.

## 2   Problem Formulation

In this section, the VNE problem and the classification of energy consumption are presented firstly. Then, with the objective of minimizing the mapping cost and energy consumption, we establish the Integer Linear Programming (ILP) model for multi-objective optimization of VNE problem.

### 2.1   Virtual Network Embedding Problem

**Substrate Network (SN).** We mode the substrate network as a weighted undirected graph $G^S = (N^S, L^S, A_N^S, A_L^S)$, where $N^S$ is the set of substrate nodes and $L^S$ is the set of substrate links. $A_N^S$ and $A_L^S$ denote CPU capacity of the substrate nodes and bandwidth of links, respectively.

**Virtual Network (VN).** Similar to the substrate network, a virtual network can be represented as $G^V = (N^V, L^V, C_N^V, C_L^V)$, where $N^V$ and $L^V$ denote the set of virtual nodes and virtual link respectively. Virtual nodes and edges are associated with constraints on CPU and bandwidth resources requests, denoted by $C_N^V$ and $C_L^V$ respectively. Each VNR can be denoted by $VNR_i = (G^V, T_i)$, where $T_i$ denotes the duration of VN staying in the substrate network.

The VNE refers to mapping the virtual networks to the subset of the substrate networks on the premise of satisfying the nodes' and links' constraints. Generally speaking, VNE is divided into two stages: node mapping stage and link mapping stage.

## 2.2 Energy Consumption Modeling

The energy consumption of the substrate network is divided into two parts: the energy consumption of nodes and the energy consumption of links.

**Energy Consumption of Nodes.** Be similar to the earlier work in [10], we define the energy consumption of the physical nodes and links. In addition to the basic operating energy consumption, we abstract the node attributes into processor attributes. Since the energy consumption of network node is linearly related to the carried load by this node, we define the $i$ th node energy consumption $PN^i$ as

$$PN^i = \begin{cases} P_b + (P_m - P_b) \cdot u, & \text{if node } i \text{ is active} \\ 0, & \text{otherwise} \end{cases} \qquad (1)$$

where $P_b$ is the essential baseline power, $P_m$ denotes the total power which comes into being at the maximum capacity, and $u$ is the utilization rate of the $i$ th node.

**Energy Consumption of Links.** Because of the load-reducing engines in network virtualization environment, current network devices are insensitive to the power consumption of traffic load [11], so we regard the energy consumption of physical links as a constant [12], and we define the $j$ th link energy consumption $PL^j$ as

$$PL^j = \begin{cases} P_n, & \text{if link } j \text{ is powered on} \\ 0, & \text{otherwise} \end{cases} \qquad (2)$$

## 2.3 Multi-objective VNE Problem Modeling

With the objective of minimizing mapping cost and energy consumption, the mapping optimization problem for each virtual network request can be described as:

$$
\min = \begin{cases} f_1 = \alpha \sum_{n^v \in N^V} cpu(n^v) + \beta \sum_{l^v \in L^V} \sum_{l^s \in L^S} \varphi_{l^v} \times bw(l^v) \\ f_2 = \sum_{(n^i, l^j) \in P^s} (t_i \times PN^i + t_j \times PL^j) \end{cases} \tag{3}
$$

where $f_1$ denotes the network resource expenditure for each virtual network request. $cpu(n^v)$ represents the computing capacity requirement of virtual nodes $n^v$, and $bw(l^v)$ represents the bandwidth capacity requirement of virtual links $l^v$. The parameters $\alpha$ and $\beta$ are used to adjust the relative weights of computing resources and bandwidth resources. And $\alpha + \beta = 1$. $P^S$ represents the set of physical nodes and links after mapping. $\varphi_{l^v}$ is a binary variable used to judge whether virtual link $l^v$ is mapped to physical link. $f_2$ represents the energy consumption of each virtual network request. $t_i$ and $t_j$ denote the length of time the node $i$ is open and the link $j$ is open, respectively.

Meanwhile, the mapping process must satisfy the constraints shown in Eq. (4). The first constraint is about host resource constraints. The idle resources of the current physical node need to have more resources than that requested by the virtual node to be mapped. The second constraint is about physical bandwidth constraints. Each physical link $l^j$ occupied by each virtual link $l^V$, on physical path $p^s$, must have greater idle bandwidth than that requested by the virtual link $l^V$.

$$
\begin{aligned}
\text{s.t.} \quad &\forall n^v \in N^V, \forall n^i \in p^S, n^v \to n^i, \\
&Ccpu(n^i) - \sum_{n^v \to n^i} Ccpu(n^v) \geq Rcpu(n^v) \\
&\forall l^v \in L^V, \forall l^j \in p^S, l^v \to l^j, \\
&\min_{l^s \in l^j} Cbw(l^S) \geq Rbw(l^v)
\end{aligned} \tag{4}
$$

## 3   Virtual Network Mapping Algorithm Based on Pareto Entropy

Because the optimal solution of the above optimization model is not unique, the VNE-MOPSO algorithm proposed in this paper saves a higher quality feasible solution to the external archive (Pareto optimal solution set) whenever it is found in the iteration process. This section introduces Pareto optimal correlated definitions, Pareto entropy multi-objective optimization model, external archive updating algorithm, particle swarm optimization algorithm, adaptive parameter strategy and the overall flow of VNE-PSO algorithm.

### 3.1   Relevant Definitions of Pareto Optimality

**Definition 1 (Pareto dominate).** For any two vectors $u, v \in \Omega$, we call $u$ dominate $v$ (or $v$ is dominated by $u$) which is denoted as $u \succ v$, if and only if $\forall i = 1, 2, \ldots, m$, $u_i \leq v_i \land \exists j = 1, 2, \ldots, m$, $u_j < v_j$, where $m$ is the number of optimization objectives.

**Definition 2 (Pareto optimal solution and Pareto optimal solution set).** A solution $x^* \in \Omega$ is called Pareto optimal solution or non-dominant solution if and only if $\neg \exists x \in \Omega : x \succ x^*$. The set $PS = \{x^* | \neg \exists x \in \Omega : x \succ x^*\}$ of all Pareto optimal solutions is called Pareto optimal solution set.

**Definition 3 (Pareto Front End).** The region $PF = \{F(x^*) | x^* \in PS\}$ formed by the objective function values corresponding to all Pareto optimal solutions is called Pareto Front End or Pareto Equilibrium Surface.

## 3.2 Pareto Entropy Multi-objective Optimization Model and Evolutionary State

Pareto Entropy Multi-objective Optimization Model takes the difference of Pareto entropy as the basis of optimization. Firstly, we map the multi-dimensional Pareto solution stored in the external archive to the two-dimensional plane by the target space transformation method. Thus, we can obtain the parallel lattice coordinates of each Pareto solution and calculate the Pareto entropy value of the external archive approximating the Pareto front end. When the external archive is updated, the difference of entropy will be generated. Evaluating the population status according to the updating situation and using it as feedback information can better control the optimization process, taking into account the diversity and convergence of the population.

We transform the multi-dimensional Pareto solution into two-dimensional plane in parallel coordinates [9]. The integer coordinate of its mapping is the parallel cell coordinate, and the calculation formula is as follows

$$
L_{k,m} = \begin{cases} \left\lceil K \frac{f_{k,m} - f_m^{\min}}{f_m^{\max} - f_{k,m}} \right\rceil, & \text{if } f_{k,m} \neq f_m^{\min} \\ 1, & \text{otherwise} \end{cases} \tag{5}
$$

where $\lceil x \rceil$ returns the smallest integer that not less than $x$; $k = 1, 2, \ldots K$, $K$ is the number of external archive in the current iteration, which need not be specified by the user; $m = 1, 2, \ldots M$, $M$ is the number of objectives to be optimized; $f_m^{\max}$ and $f_m^{\min}$ are the maximum and minimum values of the $m$ th objective of the current Pareto solution set, respectively.

In the $t$ th iteration process, the Pareto entropy [9] of the external archive approximation Pareto front end is

$$
Entropy(t) = -\sum_{k=1}^{K} \sum_{m=1}^{M} \frac{Cell_{k,m}(t)}{KM} \log \frac{Cell_{k,m}(t)}{KM} \tag{6}
$$

where $Cell_{k,m}(t)$ represents the number of cell coordinate components that fall on the $k$ th row and $m$ th column after the approximate Pareto front end is mapped to the parallel cell coordinate system.

When the number of members of external archive reaches the maximum capacity, it is necessary to evaluate the individual density of new solution and all old solutions when updating external archive again. The individual density $Density(P_i)$ [9] of any solution $P_i$ is as follows

$$Density(P_i) = \sum_{\substack{j=1 \\ j \neq i}}^{K} \frac{1}{PCD(P_i, P_j)^2} \tag{7}$$

$$PCD(P_i, P_j) = \begin{cases} \sum_{m=1}^{M} |L_{i,m} - L_{j,m}|, & \text{if} \exists m, L_{i,m} \neq L_{j,m} \\ 0.5, & \text{if} \forall m, L_{i,m} = L_{j,m} \end{cases} \tag{8}$$

where $i, j = 1, 2, \ldots K$, $K$ is the number of members in external archive; $P_j$ is any other non-dominant solution that different from $P_i$ in external archive. $PCD(P_i, P_j)$ denotes the distance of parallel cell between $P_i$ and $P_j$.

With the continuously searching of new solutions by particle swarm optimization, the external archive is constantly updated. On this ground, the evolutionary state of the algorithm in each iteration is divided into three kinds:

**Stagnation state:** The new solution obtained by the algorithm is denied access to external archive.
**Diversified state:** The new solution obtained by the algorithm replaces the old solution of poor quality in external archive.
**Convergence state:** The Pareto front-end generated by the algorithm approximates the real Pareto front-end in the target space.

### 3.3   External Archive Updating Algorithms

The main feature of the second generation multi-objective evolutionary algorithm is that external archives retain elite solutions, so it is necessary to constantly update external archive to obtain high-quality Pareto optimal solution set. In the process of updating external archive, there will be five cases [9].

*Case I.* If the external archive is empty, the new solution will enter the external archive directly. The population is in convergent state now.
*Case II.* If the new solution is dominated by any old solution in the external archive, the new solution is discarded. The population is in stagnation state now.
*Case III.* If the new solution dominate 0–r old solutions, we firstly remove the old solutions in the external archive which are dominated by the new solution. If the external archive is not saturated at this time, the new solution is added to the external archive. The population is in convergent state now.
*Case IV.* If the new solution and the old solution in the external archive are mutual non-dominant solutions, and the external archive is saturated, and the individual density of the new solution is the largest one, so the new solution is discarded. The population is in stagnation state now.
*Case V.* If the new solution and the old solutions in the external archive are non-dominant solutions, and the external archive is saturated, but the individual density of

the new solution is not the largest one, then the new solution replaces the old solution which has the largest individual density. The population is in diversified state now.

From the above five cases, we can get the external archive updating algorithm.

**Algorithms 1.** External Archive Updating Algorithms

---

**Input:** 1) External archive $A$ to be updated;

      2) Maximum capacity $K$ of external archive;

      3) The new solution $P$ obtained by evolutionary algorithm;

**Output:** 1) Updated external archive $A'$ ;

      2) Evolutionary state,( $state$ =0,1,2. Indicates stagnation state, diversification state and convergence state respectively. );

      3) The difference of entropy $\Delta Entropy$ .

---

1:  if ( $A=\varnothing$ ){

2:     $A'=\{P\}$ ; $state=2$ ; $\Delta Entropy=\log M$ ;

3:     return $A'$ , $state$ , $\Delta Entropy$ ; } /* Case I */

4:  if ( $P$ is dominated by $a_i$ , $a_i \in A$ ) {

5:     $state=0$ ; $\Delta Entropy=0$ ;

6:     return $A$ , $state$ , $\Delta Entropy$ ; } /* Case II */

7:  if (for any $a_i \in A$ , $a_i$ is dominated by $P$ ){

8:     set $r$ is the number of old solutions dominated by $P$ ,set $|A|$ is the current number of members of $A$ . Firstly, set $A=A/\{a_i\}$ .

9:     if ( $r==0$ ) $\Delta Entropy=\log\dfrac{|A|+1}{|A|}$ ;

10:   else if ( $r==1$ ) $\Delta Entropy=\dfrac{2}{MK}\log M$ ;

11:   else if ( $1<r\le|A|$ ) $\Delta Entropy=\dfrac{2}{MK}\log M+\log\dfrac{|A|}{|A|-r+1}$ ; }

12:   if ( $|A|<K$ ){

13:      $A'=A\cup\{P\}$ ; $state=2$ ;

14:      return $A'$ , $state$ , $\Delta Entropy$ ; } /* Case III */

15:   else if ( $A|==K$ ) {

16:      set $B=A\cup\{P\}$ , assess the individual density of all members of $B$ .

17:      find the member with the largest individual density $b_{max}$ in $B$ .

18:      if ( $P==b_{max}$ ) {

19:        $A'=A$ ; $state=0$ ; $\Delta Entropy=0$ ;

20:        return $A'$ , $state$ , $\Delta Entropy$ ; } /* Case IV */

21:      else {

22:        $A'=B/\{b_{max}\}$ ; $state=1$ ; $\Delta Entropy=\dfrac{2}{MK}\log M$ ;

23:        return $A'$ , $state$ , $\Delta Entropy$ ; } /* Case V*/

24:   }

---

### 3.4  Particle Swarm Optimization

Particle Swarm Optimization Algorithm was originally developed by J. Kennedy and R. C. Eberhart. It was proposed in 1995 as a result of studies on bird predation. For VNE problem, The location vector of the particle $X_i = [x_i^1, x_i^2, \ldots, x_i^N]$ represents a mapping scheme, in which $x_i^n$ takes a positive integer and represents the number of the physical node to which virtual node $n$ will be mapped. The particle velocity vector $V_i = [v_i^1, v_i^2, \ldots, v_i^N]$ represents the adjustment decision of the mapping scheme. During the evolutionary process, the position and velocity of each particle are updated as follows:

$$V_{i+1} = \omega V_i + c_1(pBest_i - X_i) + c_2(gBest_i - X_i) \tag{9}$$

$$X_{i+1} = X_i + V_{i+1} \tag{10}$$

where $\omega, c_1, c_2 > 0$ represent inertia weight, learning weight and group weight. The location vector $pBest_i$ represents individual optimal solution, and location vector $gBest_i$ represents global optimal solution of the whole group.

### 3.5  Adaptive Parameter Strategy

In order to better control the evolutionary process, it is necessary to continuously obtain real-time feedback information from the evolutionary environment. We can flexibly control the search trend of the algorithm by adjusting the parameters $\omega, c_1, c_2$ of the motion equation. In this paper, a dynamic adaptive parameter adjustment strategy is designed according to the population evolution state and the difference of Pareto entropy returned by the global external archive updating algorithm, as shown in Eq. (11). Besides, $Len_\omega, Len_{c_1}, Len_{c_2}$ are the interval lengths between the maximum and minimum values of $\omega, c_1, c_2$. Referring to literature [13], the ranges of $\omega, c_1, c_2$ are controlled in [0.4, 0.9], [0.5, 2.5], [0.5, 2.5], respectively.

$$
\begin{aligned}
\omega(t) &= \begin{cases} \omega(t-1) & \text{stagnant state} \\ \omega(t-1) + 0.06 \bullet Len_\omega \bullet \Delta Entropy(t) & \text{diversified state} \\ \omega(t-1) - 0.1 \bullet Len_\omega \bullet \Delta Entropy(t). & \text{convergent state} \end{cases} \\[6pt]
c_1(t) &= \begin{cases} c_1(t-1) & \text{stagnant state} \\ c_1(t-1) - 0.06 \bullet Len_{c_1} \bullet \Delta Entropy(t) & \text{diversified state} \\ c_1(t-1) + 0.1 \bullet Len_{c_1} \bullet \Delta Entropy(t) & \text{convergent state} \end{cases} \\[6pt]
c_2(t) &= \begin{cases} c_2(t-1) & \text{stagnant state} \\ c_2(t-1) + 0.06 \bullet Len_{c_2} \bullet \Delta Entropy(t) & \text{diversified state} \\ c_2(t-1) - 0.1 \bullet Len_{c_2} \bullet \Delta Entropy(t) & \text{convergent state} \end{cases}
\end{aligned}
\tag{11}
$$

### 3.6  The Whole Flow of VNE-MOPSO Algorithms

For each virtual network request, the VNE-MOPSO algorithm firstly generates the initial position of particles randomly, then judges whether it is feasible for each new location and updates the external archive, so as to obtain the difference of entropy and evolution state. Then it updates the particle's speed and position according to the adaptive parameter strategy until the end of the iteration. The final mapping scheme is selected randomly from the external archive.

In Eq. (9), we transform the velocity vector $V_{i+1}$ into binary value by probability mapping. The specific method is to use sigmoid function to map $V_{i+1}$ to [0, 1] intervals as probability. If the probability is greater than or equal to the decimal of a randomly generated [0, 1] intervals, the next step speed is 1, otherwise the value is 0, as shown in Eq. (12). We use Eq. (10) to update the position of the particles, and randomly select a new physical node that satisfies host resource constraints for a virtual node whose velocity component is 1.

$$V_{i+1}^* = \begin{cases} 1, & \text{if rand}() \leq \text{Sigmoid}(V_{i+1}); \\ 0, & \text{otherwise.} \end{cases} \tag{12}$$

The pseudo-code of VNE-MOPSO algorithm is shown in algorithm 2.

**Algorithms 2.** Virtual Network Embedding Based on Pareto Entropy (VNE-MOPSO)

---

**Input:** Virtual Network $G_v$, Physical Network $G_s$;
**Output:** Mapping solution.

---

1: get the node queue and link queue of real-time idle resources in $G_s$;
2: for each particle instance in the population, initialize position vector;
3: initialize global external archive $gArchive = \varnothing$, initialize Individual archive $pArchive = \varnothing$;
4: for ( int i = 0; i < MaxItCount ; i++ ){
5:    if (current position is feasible){
6:        use the shortest path method to generate the mapping scheme and calculate the values $f_1, f_2$ of objective functions;
7:        for each particles, call algorithm 1 to update $gArchive$, preserve the evolutionary state and the difference of entropy at present;
8:        for each particles, call algorithm 1to update $pArchive$;
9:        randomly select a solution from $gArchive$ ,and take it as a group optimal solution $gBest$
10:       select the nearest solution to the group optimal solution from $pArchive$ , and take it as an individual optimal solution $gBest$;
11:       according to Eq. (11), calculate the values of $\omega, c_1, c_2$ by evolutionary state and the difference of entropy
12:       according to Eq. (9-12), update the velocity vector and position vector of particles; }
13:    else if (current position is not feasible){
            randomly adjust the particle position; }
14:    if ( $gArchive$ remains unchanged consecutively for 8 rounds){
            algorithm terminates; }
15: if ( $gArchive \neq \varnothing$ ){
        randomly choose a solution as the mapping scheme from $gArchive$. }

## 4  Simulation Results

The performance of VNE-MOPSO is compared with that of VNE-UEPSO in reference [4]. All the algorithms in this paper are implemented in Java language under Windows system. The simulation module of VNE algorithm is compiled on the platform of CloudSim3.0.3 [14]. In order to ensure the diversity of topologies, the probability of connectivity, the boundary of resource demand and the number range of nodes are used as input parameters to generate random topologies.

In the experiment, the termination condition of discrete particle swarm optimization (DPSO) is that the global optimal position is not changed consecutively for 8 rounds or the total number of iterations exceeds 30 rounds. Each experiment tests 2000 virtual network requests. The algorithm maps 70 virtual network requests in the search waiting queue for the first time to ensure that the physical network resource occupancy reaches full as soon as possible. Later it searches for the first 20 requests in the queue at a time. The relative weight ratio of computing resources and bandwidth resources in Eq. (1) is set to 1:1. The initial values of motion parameters $\omega, c_1, c_2$ are 0.85, 0.7, 2.3. The energy consumption parameters $P_m, P_b, P_n$ are 300, 150, 150. The maximum capacity of external archive $K$ is 5. The physical network and virtual network parameters in the experiment are shown in Table 1.

**Table 1.**  Basic experimental parameter setting

| Parameters | Substrate network | Virtual network |
|---|---|---|
| Number of nodes | 80 | 2–10 |
| Connectivity | 0.2 | 0.4 |
| CPU capacity | 10000 | 250–2500 |
| Bandwidth capacity | 10000 | 200–1000;600–3000 |
| Survival time | | 50–500 |

In experiment 1, we compared the change trend of physical network revenue-cost ratio and energy consumption when receiving 2000 virtual network requests under different virtual link bandwidth parameters. As can be seen from Fig. 1 and Fig. 2, the VNE-MOPSO algorithm can reduce the mapping cost and energy consumption of physical networks. When the substrate network is idle (VNR count is around 0–800), there is little difference between the two algorithms. The main reason is that the physical network resources are sufficient in the early stage, and the search ability of particle swarm optimization is strong. The optimization effect of the VNE-MOPSO algorithm proposed in this paper is not obvious. However, with the increasing of the number of virtual network requests accepted, the optimization effect of the VNE-MOPSO algorithm continues to increase, and finally saves about 3.63%, 6.88% of the mapping cost and 9.81%, 4.64% of the energy consumption, respectively.

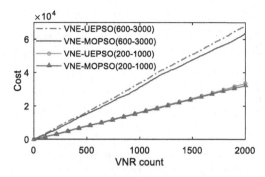

**Fig. 1.** Cost of substrate network under different virtual request parameters

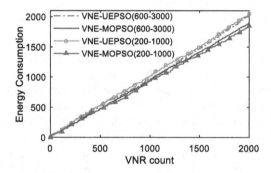

**Fig. 2.** Energy consumption of substrate network under different virtual request parameters

In addition, the larger the bandwidth of virtual link, the higher the corresponding mapping cost, the better the optimization effect on mapping cost the VNE-MOPSO algorithm is. Energy consumption is determined by the number of physical nodes and links that have been opened and the load of physical nodes. Energy consumption has nothing to do with the bandwidth of virtual links. Thus, the energy consumption of virtual links with bandwidth of 200–1000 is higher than that of 600–3000.

In experiment 2, we compared the trends of long-term average revenue over time when virtual link bandwidth is in 600–3000. As can be seen from Fig. 3, the VNE-MOPSO algorithm proposed in this paper has advantages over VNE-UEPSO in terms of long-term average revenue in physical networks. This is because the VNE-MOPSO algorithm takes mapping cost and energy consumption as optimization objectives, takes into account the diversity and convergence of Pareto front-end in the evolutionary process, saves physical network resources, and speeds up mapping speed, thus it can improve the long-term average revenue. In Fig. 3, the long-term average revenue is relatively high in the early stage, and then shows a downward trend. This is due to the abundant physical network resources, high mapping efficiency and the large number of virtual network requests accepted in the early stage, so the revenue is also high. However, with the increasing number of accepted virtual network requests, the process of virtual network mapping gradually reaches a stable state in the process of occupying and releasing physical resources, and the average revenue tends to be stable.

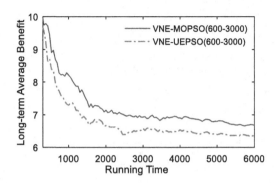

**Fig. 3.** The trend of long-term average benefit over time

## 5   Conclusions

Aiming at the VNE problem, considering the cost of mapping and energy consumption, and combining with the theory of multi-objective optimization, a multi-objective particle swarm optimization VNE algorithm based on Pareto entropy is proposed. According to the update status of the external Pareto solution set, the difference of entropy is calculated and the evolution status of the population is evaluated. Combining with the dynamic adaptive particle velocity update strategy, the mapping cost and energy consumption are reduced, and the long-term average benefit is also obtained. Experiments show that the algorithm proposed in this paper has certain advantages over other similar algorithms in terms of revenue, energy consumption and solution efficiency.

Virtual network mapping is a key problem in cloud resource allocation, and many factors need to be considered in the designing of algorithm. This paper considers multi-objective optimization from the perspective of mapping cost and energy consumption of substrate physical network. The next step is to try to optimize more objectives simultaneously and introduce virtual network quality of service as an additional parameter.

## References

1. The 42nd China statistical report on Internet development. http://www.cnnic.cn/hlwfzyj/hlwxzbg/hlwtjbg/201808/t20180820_70488.htm. Accessed 20 Mar 2019
2. Fischer, A., Botero, J., Beck, M., et al.: Virtual network embedding: a survey. IEEE Commun. Surv. Tutor. **15**(4), 1888–1906 (2013)
3. Cheng, X., Zhang, Z., Sen, S., et al.: Survey of virtual network embedding problem. J. Commun. **32**(10), 143–151 (2011)
4. Zhang, Z., Cheng, X., Sen, S., et al.: A unified enhanced particle swarm optimization-based virtual network embedding algorithm. Int. J. Commun. Syst. **26**(8), 1054–1073 (2013)
5. Wang, W., Wang, B., Wang, Z., et al.: The virtual network mapping algorithm based on hybrid swarm intelligence optimization. J. Comput. Appl. **34**(4), 930–934 (2014)

6. Wang, Q.: Research on Virtual Network Mapping Algorithm Based on Particle Swarm Optimization. Northeastern University, Shen yang (2015)
7. Wang, C., Yuan, Y., Peng, S., et al.: Fair virtual network embedding algorithm with topology pre-configuration. J. Comput. Res. Dev. **54**(1), 212–220 (2017)
8. Zheng, H., Li, J., Gong, Y., et al: Link mapping-oriented ant colony system for virtual network embedding. In: IEEE Congress on Evolutionary Computation 2017, pp. 1223–1230. IEEE, Piscataway (2017)
9. Wang, H., Yen, G.G., Zhang, X.: Multiobjective particle swarm optimization based on Pareto entropy. J. Softw. **25**(5), 1025–1050 (2014)
10. Chen, X., Li, C., Chen, L., et al.: Multiple feedback control model and algorithm for energy efficient virtual network embedding. J. Softw. **28**(7), 1790–1814 (2017)
11. Chabarek, J., Sommers, J., Barford, P., et al: Power awareness in network design and routing. In: The 27th Conference on Computer Communications IEEE 2008, pp. 457–465. IEEE, Phoenix (2008)
12. Lu, G., Guo, C., Li, Y., et al: ServerSwitch: a programmable and high performance platform for data center networks. In: The 8th USENIX Conf. on Networked Systems Design and Implementation 2011, pp. 1–14. USENIX Association, Berkeley (2011)
13. Ratnaweera, A., Halgamuge, S., Watson, H.: Self-organizing hierarchical particle swarm optimizer with time-varying acceleration coefficients. IEEE Trans. Evol. Comput. **8**(3), 240–255 (2004)
14. Calheiros, R.N., Ranjan, R., Beloglazov, A., et al.: CloudSim: a toolkit for modeling and simulation of cloud computing environments and evaluation of resource provisioning algorithms. Softw. Pract. Exp. **41**(1), 23–50 (2011)

# Measuring and Analyzing the Burst Ratio in IP Traffic

Dominik Samociuk, Marek Barczyk, and Andrzej Chydzinski[✉]

Institute of Informatics, Silesian University of Technology, Gliwice, Poland
{dominik.samociuk,marek.barczyk,andrzej.chydzinski}@polsl.pl

**Abstract.** The burst ratio is a parameter of the packet loss process, characterizing the tendency of losses to group together, in long series. Such series of losses are especially unwelcome in multimedia transmissions, which constitute a large fraction of contemporary traffic. In this paper, we first present and discuss results of measurements of the burst ratio in IP traffic, at a bottleneck link of our university campus. The measurements were conducted in various network conditions, i.e. various loads, ports/applications used and packet size distributions. Secondly, we present theoretical values of the burst ratio, computed using a queueing model, and compare them with the values obtained in the measurements.

**Keywords:** Packet loss measurement · Burst ratio · Performance evaluation

## 1 Introduction

Packet loss is an inherent property of networks based on the TCP/IP protocol stack. Their "best effort" design is on one hand flexible and resistant to large-scale failures, but on the other it does not guarantee the delivery of packets. An occasional loss of an individual packet is usually not a problem. It is retransmitted using the TCP mechanism, if completeness of data is required. Or, it can be simply ignored, if it is a part a flow, which allows for some data loss. However, the problem is much more complicated when packet losses tend to group together, i.e. occur one after another, in series. In multimedia transmissions, such series of losses may cause for instance a video image to freeze, which is the main impediment in video Quality of Experience, [1].

The best way to describe the tendency of packet losses to cluster together is by using the burst ratio parameter $B$, [2]. Formally, $B$ is the ratio of the observed average length of the series of lost packets, $\overline{G}$, to the theoretical average length of the series of losses in the case of independent packet loss, $\overline{K}$. In other words, if $\overline{K}$ denotes the average length of the series in the Bernoulli process, then

$$B = \frac{\overline{G}}{\overline{K}}. \tag{1}$$

This work was conducted within project 2017/25/B/ST6/00110, founded by National Science Centre, Poland.

Q. Li et al. (Eds.): BROADNETS 2019, LNICST 303, pp. 86–101, 2019.
https://doi.org/10.1007/978-3-030-36442-7_6

If $L$ denotes the loss ratio (loss probability), then it is easy to check that in the Bernoulli process it holds

$$\overline{K} = \frac{1}{1-L}. \tag{2}$$

This means, that $B$ is in fact a function of $\overline{G}$ and $L$, i.e.

$$B = \overline{G}(1-L). \tag{3}$$

It is very important not to confuse the burst ratio with the traffic burstiness, which has been studied earlier and has a vast literature. The burst ratio characterizes the loss process, rather than the arrival process. It has got more attention recently, due to its connection with the quality of multimedia transmissions.

To illustrate the burst ratio, we may consider a stream of 16 packets at a network node's buffer:

$$FFFFDDFFFFFDDDFFF$$

Here $F$ denotes a packet successfully transmitted/forwarded and $D$ - a packet that was lost/dropped. There are two series of lost packets of length 2 and 3, respectively. Thus $\overline{G} = \frac{2+3}{2} = 2.5$. Moreover, there are 5 lost packets out of 16, therefore $L = \frac{5}{16}$. Thus $\overline{K} = \frac{1}{1-\frac{5}{16}} = 1.45$ and $B = \frac{\overline{G}}{\overline{K}} = 1.72$. Obviously, if $B > 1$, then the losses tend to group together - the higher the $B$ value, the stronger this tendency is. When $B = 1$, the losses look purely random and independent of each other. Finally, if $B < 1$, then the losses tend to occur separately.

The most important parameter of the packet loss process is, of course, the loss ratio, $L$. It has been studied for decades now, using measurements, simulations and mathematical modeling. The burst ratio is considered to be the second important loss parameter, especially in the real-time multimedia transmissions. In such applications it makes a difference whether one single packet is lost after every 20 delivered packets, or 3 packets are lost in a row after every 60 delivered packets. This effect can even be captured in strict formulas. For instance, in [3] the impairment factor $I$ for the digitized voice has been proposed as a function of $L$ and $B$, in the following form:

$$I = I_e + \frac{100(95 - I_e)L}{100L/B + R}, \tag{4}$$

where $I_e$ and $R$ are some constants, unimportant here, with default values $I_e = 0$ and $R = 4.3$. In the latter example with $L = \frac{1}{21} = \frac{3}{63}$, we have $I = 49.9$ for $B = 1$ and $I = 76.8$ for $B = 3$. In other words, the quality may drop by far when $B$ increases and $L$ remains unaltered. The reader can easily find examples, in which the overall voice quality is worse, due to increased $B$, even though $L$ is decreased.

In this paper, measurements of the burst ratio at the bottleneck link of our university campus are described first. The link under study was the 1 Gbps link, which connects the dormitory network, with the rest of the campus network. At different times of the day, 18 samples of traffic were collected, each consisting of one milion packets. The experiment started early in the morning and ended late in the evening, which enabled us to observe variable network conditions, depending on end users activities. For each of the collected traffic sample, the burst ratio was computed and discussed, as well as the accompanying loss parameters ($L$ and $\overline{G}$). This is presented in the first part of the paper.

In the second part of the paper, theoretical values of the burst ratio are computed using a queuing model, and compared with the measured $B$ values. The main reason of the packet loss at a bottleneck link is an overflow of the buffer at the output interface. Such interface is typically organized as a simple tail-drop queue. Therefore, it seems reasonable to use a properly parameterized tail-drop queueing model for theoretical prediction of the burst ratio.

The rest of the paper is organized as follows. In Sect. 2, the literature related to the burst ratio and loss ratio is recalled. In Sect. 3, the network topology and devices, as well as the hardware and software used in the experiment, are described. Then, the measurement process is presented in detail. Section 4 is devoted to presentation and discussion of the experiment results. Firstly, observed traffic is characterized in terms of the load, loss ratio, TCP/UDP fractions, port usage and packet size distribution. Then the burst ratio values are shown and discussed, with an accent on their relations with the observed traffic characteristics. In Sect. 5, the theoretical values of the burst ratio, computed via a queueing model, are shown and compared with the measurement results. Finally, conclusions and future work directions are gathered in Sect. 6.

## 2   Related Work

To the best of the authors' knowledge, there are no published measurement results, nor its analysis, of the burst ratio in a real, operating network.

The closest paper one can find is [4], where the burst ratio measurements carried out in a networking lab, using traffic from a hardware generator, are presented. Such results, obtained for artificial traffic in a local lab, have several drawbacks, when compared to the real IP traffic measurements. Firstly, propagation times are negligible in a local lab, so multiple-flow traffic with different round trip times (RTTs) in different flows cannot be mimicked. (It is well known that RTT has a great impact on the TCP behaviour.) Secondly, due to the limitations of the generator, only very simple distribution of the packet size can be used, far from the very rich structure in real IP traffic. The same concerns diversified duration of flows, port usage and link loads induced by human users and the applications they exploit.

Analytical studies of the burst ratio were based initially on Markovian loss models (see e.g. [5–9] for more information on such models). In particular, in [2] a formula for the burst ratio in a single link, represented by a two-state Markovian loss model, was derived. In [10,11], the exact and simplified formulas for $B$ in a few concatenated links, each represented by the two-state Markovian loss

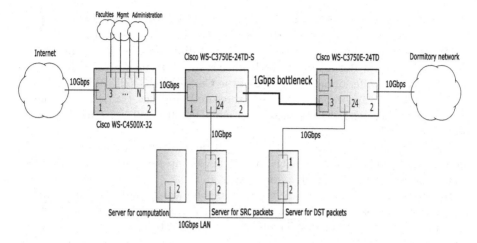

**Fig. 1.** Network topology and devices.

model, were obtained. In these studies, the finite-buffer queue, which constitutes the main reason of loss in the wired networking, was not modeled. Instead, it was assumed that the loss process follows some Markov chain, which was then parameterized and solved. Such approach gives the fundamental insight into the behaviour of the burst ratio, but its depth is limited by known limitations of Markov modeling – e.g. the series of losses cannot have a large variance. Recently, the burst ratio has been derived for two classic queueing models with single and batch arrivals, see [12,13], respectively. Moreover, it has been simulated for an active queue management scheme, in which the probability of dropping a packet is some function of the queue size, [14]. The best results were obtained for a convex, exponential dropping function, perhaps due to some general properties of convex functions (see e.g. [15]).

The main loss characteristic, the loss ratio, has been studied for many years now using measurements (see e.g. [16–20] and the references given there) and analysis (e.g. [21–23]).

Finally, there are also papers [24–26], in which the statistical structure of packet losses is studied using other characteristics, like the probability that in a block of $m$ packets exactly $n$ packets are lost, but not necessarily one after another.

## 3   Experiment Setup and Course

### 3.1   Topology, Devices and Software

The scheme of the campus network is depicted in Fig. 1.

On the left, there is an Internet connection to Cisco WS-C4500X-32. This device routes packets to different campus network segments, i.e. faculties, administration, management (on the top) and dormitories (on the right).

Then, Cisco WS-C3750E-24TD-S and Cisco WS-C3750E-24TD devices perform the distribution function in the part of the network under study.

Traffic was studied at the 1 Gbps bottleneck link depicted in the middle of Fig. 1, bound to the dormitory network on the right. The measured packet losses occurred at the buffer of output port 2 of the Cisco WS-C3750E-24TD-S device.

In order to capture, anonymize, store and analyze traffic, three high-performance servers, HP Enterprise ProLiant DL380 Gen9, were used (on the bottom of Fig. 1).

The servers were equipped with the DPDK network cards (Data Plane Development Kit), which together with the DPDK-dump application [27], enabled very efficient capturing of traffic, for further analysis.

To store the captured data in the servers, the well-known SQLite database was chosen. The data was stored in 3 separate files: source packets, destination packets and combined packets. The last one was used to compute the loss characteristics, i.e. the loss ratio and the burst ratio. Each database maintained a table with all integer fields: timestamps, IP source and destination addresses, IP IDs, IP lengths, IP protocol numbers, Layer 4 source and destination ports and sequence and acknowledgement numbers of TCP segments.

In order to take care of users' privacy, three data anonymizing mechanisms were used when collecting network traffic: Tcpmkpub [28] for anonymization of packet headers in trace files, Tcpdpriv [29] for eliminating confidential information from packets and Crypto-PAn [30–32] for anonymizing IP addresses in their traces in a prefix-preserving manner.

### 3.2   Course of the Experiment

Internet traffic bound to dormitory network was obtained on Cisco WS-C4500X-32 device on port 1. Then it was routed to port 1 of WS-C3750E-24TD-S device. In this device, it was duplicated using SPAN function to port 24, and from that port forwarded to the second server in Fig. 1.

At the same time, original traffic from port 1 of WS-C3750E-24TD-S was routed to output port 2, and through the 1Gbps link passed to input port 3 of WS-C3750E-25TD device. On that device, arriving traffic was duplicated again using SPAN to port 24, and then forwarded to the third server. At the same time, original traffic from port 3 of WS-C3750E-25TD was routed to destinations in the dormitory network through the output port 2.

Therefore, the second server collected traffic before losses could occur at the buffer of the bottleneck link, while the third server collected traffic thinned by these losses.

Pre-processed, anonymized traces from servers 2 and 3 were then copied to the first server, where source and destination packets were combined, what enabled calculations of all interesting characteristics, including the loss and burst ratio.

The buffer size of the interface of interest, i.e. port 2 of WS-C3750E-24TD-S device, was set to 100 packets.

The experiment was carried out in one day in the middle of the week (Tuesday), from 6:30 to 23:30. During this period, at different times of the day, 18 traces were collected, each containing one milion of packets.

## 4   Results

### 4.1   Traffic Characteristics

Two basic characteristics of collected traffic in time are depicted in Figs. 2 and 3. They are the link load, $\rho$, and the packet loss ratio, $L$, respectively. As can be seen, the load was small in the morning, 5% at 6:30, then it was growing until

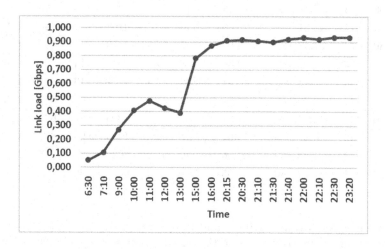

**Fig. 2.** Link load in time.

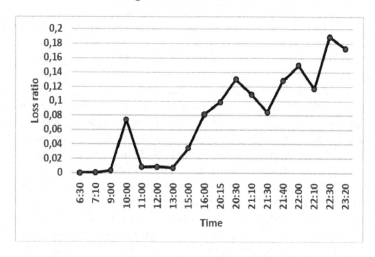

**Fig. 3.** Loss ratio in time.

**Table 1.** TOP 5 ports in TCP connections.

| Time | Port1 | % | Port2 | % | Port3 | % | Siz4 | % | Port5 | % |
|---|---|---|---|---|---|---|---|---|---|---|
| 6:30 | 443 | 52.32 | 80 | 12.91 | 1185 | 1.21 | 8000 | 0.51 | 993 | 0.10 |
| 7:10 | 443 | 58.31 | 80 | 19.39 | 8000 | 0.24 | 9212 | 0.12 | 46792 | 0.04 |
| 9:00 | 443 | 55.00 | 80 | 12.43 | 50536 | 0.47 | 8081 | 0.43 | 1935 | 0.24 |
| 10:00 | 443 | 61.74 | 80 | 17.16 | 440 | 1.01 | 8080 | 0.10 | 8000 | 0.06 |
| 11:00 | 443 | 37.23 | 80 | 19.56 | 440 | 1.40 | 282 | 0.55 | 18331 | 0.07 |
| 12:00 | 443 | 52.27 | 80 | 17.66 | 440 | 0.67 | 8081 | 0.29 | 1935 | 0.13 |
| 13:00 | 443 | 54.02 | 80 | 12.33 | 1935 | 1.44 | 440 | 0.41 | 8000 | 0.07 |
| 15:00 | 443 | 43.05 | 80 | 26.70 | 440 | 0.58 | 1935 | 0.26 | 282 | 0.11 |
| 16:00 | 443 | 38.41 | 80 | 18.18 | 440 | 0.21 | 282 | 0.15 | 1935 | 0.07 |
| 20:15 | 443 | 43.01 | 80 | 20.49 | 440 | 0.17 | 20108 | 0.14 | 1935 | 0.11 |
| 20:30 | 443 | 37.32 | 80 | 29.02 | 8080 | 0.69 | 18073 | 0.11 | 8999 | 0.10 |
| 21:10 | 443 | 38.97 | 80 | 19.06 | 53238 | 0.66 | 1935 | 0.14 | 440 | 0.14 |
| 21:30 | 443 | 45.76 | 80 | 18.61 | 53238 | 2.36 | 8777 | 0.13 | 8080 | 0.10 |
| 21:40 | 443 | 42.29 | 80 | 18.06 | 9992 | 4.12 | 1935 | 0.17 | 8999 | 0.10 |
| 22:00 | 443 | 42.79 | 80 | 21.89 | 6905 | 0.24 | 6906 | 0.22 | 6907 | 0.21 |
| 22:10 | 443 | 45.12 | 80 | 19.64 | 81 | 0.25 | 1935 | 0.25 | 8777 | 0.11 |
| 22:30 | 443 | 31.32 | 80 | 29.06 | 1935 | 0.12 | 8080 | 0.09 | 8777 | 0.09 |
| 23:20 | 443 | 31.74 | 80 | 31.65 | 81 | 0.39 | 8080 | 0.30 | 440 | 0.17 |

20:00, when it reached about 90%, and then it remained high until the end of the experiment. Similarly, the loss ratio varied from a very small 0.0025% in the morning, up to high 18.9% at 22:30 in the evening. Such characteristics were to be expected - they follow the typical human activity pattern.

In Tables 1 and  2, the five mostly used ports are presented for TCP and UDP transmissions. The largest volume of traffic was generated between the Web browsers and servers, using encrypted (port 443) or plain-text (port 80) communication.

There were also their backup or non-standard versions (port 8080) and the onion routing service for anonymous communication (port 81). The Web browsing would not be possible without the Domain Name System queries (port 53). A group of ports was devoted to direct multimedia transmissions, namely ports 8081, 8777, 8999 – for Apple's Web Service and iTunes streams, port 440 – SGP protocol used in VoIP transmissions, port 1071 – BSQUARE-VoIP, port 5100 – for Mac OS X camera and scanner sharing. Other part of traffic consisted of connections to databases/databases management systems (ports 8000 and 9212) and filesystem access (ports 50536 and 53238 – Apple's Xsan filesystem access). There were also games involved (port 1119 – Battle.net chat/game protocol) and usage of Adobe Flash or similar application with Real Time Messaging Protocol, on port 1935. A part of the throughput was used for downloading files with

**Table 2.** TOP 5 ports in UDP flows.

| Time | Port1 | % | Port2 | % | Port3 | % | Siz4 | % | Port5 | % |
|------|-------|------|-------|------|-------|------|-------|------|-------|------|
| 6:30 | 443 | 31.76 | 55743 | 0.23 | 24874 | 0.17 | 5100 | 0.14 | 53 | 0.05 |
| 7:10 | 443 | 20.82 | 24874 | 0.39 | 55743 | 0.10 | 5100 | 0.07 | 1119 | 0.07 |
| 9:00 | 443 | 28.86 | 22203 | 1.40 | 6881 | 0.18 | 27050 | 0.12 | 27053 | 0.09 |
| 10:00 | 443 | 17.99 | 22203 | 0.96 | 6881 | 0.33 | 27005 | 0.09 | 52437 | 0.05 |
| 11:00 | 443 | 24.32 | 47692 | 1.86 | 20588 | 1.16 | 22535 | 1.01 | 61266 | 0.97 |
| 12:00 | 443 | 26.72 | 16617 | 0.35 | 6881 | 0.32 | 22203 | 0.21 | 24934 | 0.19 |
| 13:00 | 443 | 28.75 | 16617 | 0.30 | 50667 | 0.21 | 6881 | 0.21 | 24934 | 0.20 |
| 15:00 | 443 | 27.64 | 11088 | 0.87 | 10001 | 0.03 | 1071 | 0.03 | 1119 | 0.03 |
| 16:00 | 443 | 42.10 | 5749 | 0.15 | 24874 | 0.04 | 26507 | 0.03 | 26523 | 0.03 |
| 20:15 | 443 | 30.01 | 8999 | 0.92 | 17501 | 0.51 | 53139 | 0.29 | 11088 | 0.25 |
| 20:30 | 443 | 29.67 | 17501 | 0.51 | 8999 | 0.40 | 30808 | 0.27 | 25231 | 0.09 |
| 21:10 | 443 | 37.51 | 8999 | 0.46 | 25231 | 0.31 | 51193 | 0.20 | 16394 | 0.12 |
| 21:30 | 443 | 29.53 | 8999 | 0.48 | 25231 | 0.33 | 51193 | 0.24 | 16394 | 0.16 |
| 21:40 | 443 | 32.39 | 8999 | 0.33 | 52909 | 0.30 | 51193 | 0.16 | 51896 | 0.08 |
| 22:00 | 443 | 31.17 | 8999 | 0.23 | 25231 | 0.17 | 6881 | 0.13 | 51193 | 0.09 |
| 22:10 | 443 | 30.08 | 8999 | 0.43 | 25231 | 0.26 | 51193 | 0.11 | 51413 | 0.10 |
| 22:30 | 443 | 35.74 | 8999 | 0.14 | 56143 | 0.14 | 51896 | 0.08 | 25231 | 0.08 |
| 23:20 | 443 | 32.28 | 12345 | 1.11 | 16393 | 0.14 | 23952 | 0.09 | 33480 | 0.09 |

BitTorrent applications (ports 6881, 6905, 6906, 6907). Finally, a fraction of traffic consisted of email transfers (port 993).

In Table 3, the distribution of the packet size in every collected trace is shown. As can be observed, large packets (over 1400 bytes) predominated at every time of the day – their share was always above 50%. The second important group consisted of packets in range 1201–1400 bytes, while the third – of small packets, up to 200 bytes. It is worth noticing, that the size distribution varied significantly during the day. For instance, the fraction of small packets ($\leq 200$ bytes) varied from 17% at 6:30 to 6% at 18:00. The fraction of large packets ($>1400$ bytes) varied from 72% at 10:00 to 50% at 21:10.

During the measurements, a surprisingly high percentage of UDP traffic was observed, from 20% up to 43%, at different times of the day. One possible explanation of this phenomenon is that a part of UDP traffic might have been in fact HTTPS traffic on port 443. This is possible when replacing multiple TCP connections with one multiplexed UDP flow by means of QUIC protocol.

## 4.2 Measured Burst Ratio

The final results of measurements are shown in Table 4. The burst ratio, $B$, and the average length of the series of losses ($\overline{G}$) are given for every collected trace.

**Table 3.** Distributions of the packet size [%]. Packet sizes are grouped in ranges: 0–200 bytes, 201–400 bytes, 401–600 bytes, etc.

| Time | 0–200 | 201–400 | 401–600 | 601–800 | 801–1000 | 1001–1200 | 1201–1400 | 1401–1600 |
|---|---|---|---|---|---|---|---|---|
| 06:30 | 16.65 | 1.48 | 0.49 | 0.37 | 0.41 | 0.30 | 27.63 | 52.67 |
| 07:10 | 16.01 | 1.26 | 0.52 | 0.32 | 0.37 | 0.27 | 21.32 | 59.94 |
| 09:00 | 12.06 | 1.02 | 1.24 | 0.38 | 0.48 | 0.37 | 27.32 | 57.13 |
| 10:00 | 7.94 | 1.14 | 0.58 | 0.28 | 0.30 | 0.25 | 17.50 | 72.01 |
| 11:00 | 8.70 | 1.25 | 1.02 | 0.36 | 0.39 | 0.30 | 23.22 | 64.75 |
| 12:00 | 8.94 | 1.59 | 1.04 | 0.60 | 0.63 | 0.47 | 25.17 | 61.55 |
| 13:00 | 16.82 | 1.62 | 0.91 | 0.50 | 0.60 | 0.53 | 26.27 | 52.75 |
| 15:00 | 6.88 | 1.00 | 0.77 | 0.28 | 0.34 | 0.23 | 27.13 | 63.36 |
| 16:00 | 6.46 | 0.97 | 0.77 | 0.33 | 0.42 | 0.27 | 40.68 | 50.11 |
| 20:15 | 10.20 | 1.56 | 0.72 | 0.66 | 0.60 | 0.43 | 28.63 | 57.20 |
| 20:30 | 9.38 | 1.51 | 0.79 | 0.47 | 0.46 | 0.58 | 28.37 | 58.43 |
| 21:10 | 10.10 | 1.63 | 1.24 | 0.55 | 0.57 | 0.56 | 34.90 | 50.45 |
| 21:30 | 12.55 | 1.67 | 0.92 | 0.57 | 0.56 | 0.52 | 27.10 | 56.11 |
| 21:40 | 11.17 | 1.81 | 1.01 | 0.55 | 0.51 | 0.41 | 30.04 | 54.51 |
| 22:00 | 10.48 | 1.57 | 0.94 | 0.46 | 0.47 | 0.46 | 30.83 | 54.78 |
| 22:10 | 11.92 | 1.67 | 1.49 | 0.59 | 0.57 | 0.46 | 28.32 | 54.98 |
| 22:30 | 10.35 | 1.44 | 0.98 | 0.46 | 0.48 | 0.58 | 33.52 | 52.19 |
| 23:20 | 9.99 | 1.24 | 1.07 | 0.52 | 0.48 | 0.40 | 30.68 | 55.62 |

They are accompanied by the corresponding link load and loss ratio. Additionally, the burst ratio evolution in time is depicted in Fig. 4.

Firstly, the fact that the burst ratio is always significantly greater than 1 can be observed. This means that the losses in real IP traffic do have tendency to group together.

This result is consistent with the previous theoretical results of [12], based on a queueing model, and the results of [4], obtained via artificially generated traffic.

It is important to note, that greater than 1 values of $B$ were observed for a wide variety of loss ratios, from 0.0025% to 18.9% and for a wide variety of loads, from 5% to 93%.

It is rather hard to single out one traffic parameter, which has the greatest influence on $B$. At the first look at Table 4, it seems that $B$ grows with $\rho$. But this is not really the truth – almost the same $B$ was obtained at 9:00 and 22:30, but the load was very different at those times: 27% versus 93%. The loss ratio is also not a good candidate - almost the same $B$ was obtained for very different values of $L$: 0.3% and 18.9%.

These results are also consistent with Theorem 1 of the next section, in which the dependence of $B$ on $\rho$ and $L$ is weak. From the same theorem it follows that the burst ratio depends strongly on the distribution of the service time (here the packet size), which in our experiment varied significantly in time (see Table 3).

**Table 4.** The measured burst ratio versus time, the link load and the loss ratio.

| Time | $\rho$ | $L$ | $\overline{G}$ | $B$ |
|---|---|---|---|---|
| 6:30 | 0.050 | $0.00251 \times 10^{-2}$ | 1.30769 | **1.20921** |
| 7:10 | 0.107 | $0.01664 \times 10^{-2}$ | 1.20750 | **1.20730** |
| 9:00 | 0.269 | $0.31523 \times 10^{-2}$ | 1.31117 | **1.30703** |
| 10:00 | 0.407 | $7.40980 \times 10^{-2}$ | 1.46823 | **1.35944** |
| 11:00 | 0.476 | $0.78170 \times 10^{-2}$ | 1.40475 | **1.39377** |
| 12:00 | 0.426 | $0.82825 \times 10^{-2}$ | 1.44072 | **1.42879** |
| 13:00 | 0.392 | $0.67354 \times 10^{-2}$ | 1.46910 | **1.45921** |
| 15:00 | 0.782 | $3.42986 \times 10^{-2}$ | 1.50438 | **1.45278** |
| 16:00 | 0.871 | $8.16498 \times 10^{-2}$ | 1.52892 | **1.40408** |
| 20:15 | 0.908 | $9.83570 \times 10^{-2}$ | 1.56778 | **1.41358** |
| 20:30 | 0.915 | $13.0220 \times 10^{-2}$ | 1.58598 | **1.37945** |
| 21:10 | 0.907 | $10.8499 \times 10^{-2}$ | 1.58065 | **1.40915** |
| 21:30 | 0.897 | $8.45893 \times 10^{-2}$ | 1.59269 | **1.45796** |
| 21:40 | 0.920 | $12.8154 \times 10^{-2}$ | 1.60651 | **1.40063** |
| 22:00 | 0.932 | $14.9776 \times 10^{-2}$ | 1.60316 | **1.36304** |
| 22:10 | 0.919 | $11.7032 \times 10^{-2}$ | 1.61177 | **1.42314** |
| 22:30 | 0.934 | $18.9270 \times 10^{-2}$ | 1.61415 | **1.30864** |
| 23:20 | 0.933 | $17.3317 \times 10^{-2}$ | 1.59776 | **1.32084** |

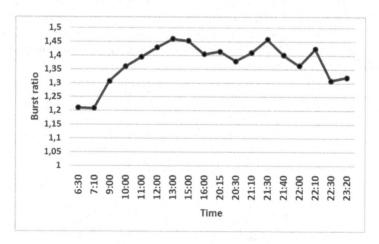

**Fig. 4.** Burst ratio in time.

From the measurements it follows also that $B$ may depend on other characteristics of IP traffic. For instance, at 22:00, a smaller $B$ than at other times
with similar $L$ and $\rho$ were observed. A the same time, an increase in BitTorrent

application usage (ports 6905, 6906, 6907) was noticed. On the other hand, during the measurement at 21:10, a high $B$ was observed. At this time, plenty of filesystem accesses (on port 53238) were performed by end users. At 9:30 and 22:10, traffic on port 443 was consuming a large fraction of the total bandwidth and $B$ assumed very high values. It is likely that usage of these and other specific applications has an impact on the traffic autocorrelation and the packet size distribution, which on the other hand influence the burst ratio.

## 5   Comparison with the Theoretical Burst Ratio

It is intuitively clear, that the simple drop-tail buffering mechanism, used commonly in networking devices, can cause clustering of packet losses. Namely, when the buffer for packets is full, a few newly arriving packets may be lost in a row, before some space in the buffer becomes available again. This effect can be described precisely using a solution of the queueing model of the buffering mechanism in a device.

Herein, the model of the queue with general distribution of the service time, given by distribution function $F(t)$, finite buffer of size $N$ and Poisson arrivals of rate $\lambda$, is used. The greatest advantage of this model is that it enables calculations for an arbitrary service time distribution (the packet size distribution), which has a deep impact on the value of $B$. Therefore, using this model the burst ratios for all packet size distributions given in Table 3 can be computed. Obviously, all the loads from Table 4 can be incorporated as well. In [12], the following two theorems on the burst ratio in the considered model were proven. (They were obtained using the potential method, used previously in analysis of the loss ratio and other loss characteristics of queueing systems, see e.g. [22,33]). The first theorem enables calculation of the exact value of $B$ as a function of the parameters of the queueing model.

**Theorem 1.** *In the queueing system with the general service time distribution the burst ratio equals:*

$$B = \frac{\sum_{k=0}^{N-1} r_k}{[1 - g(0)](1 + \rho \sum_{k=0}^{N-1} r_k)} \sum_{l=1}^{\infty} lg(l), \tag{5}$$

*where*

$$g(l) = \frac{\sum_{k=1}^{N} R_{N-k} a_{k+l} - \sum_{k=1}^{N-1} R_{N-1-k} a_{k+l}}{\sum_{k=0}^{N} R_{N-k} a_k - \sum_{k=0}^{N-1} R_{N-1-k} a_k}, \tag{6}$$

$$R_0 = 0, \qquad R_1 = \frac{1}{a_0}, \tag{7}$$

$$R_{k+1} = R_1 \left( R_k - \sum_{i=0}^{k} a_{i+1} R_{k-i} \right), \quad k = 1, 2, \ldots, \tag{8}$$

$$r_0 = 1, \qquad r_1 = \frac{1}{a_0} - 1, \tag{9}$$

$$r_{k+1} = \frac{1}{a_0}\left(r_k - \sum_{i=0}^{k-1} a_{i+1} r_{k-i} - a_k\right), \quad k = 1, 2, \ldots, \tag{10}$$

*and*

$$a_k = \int_0^\infty \frac{e^{-\lambda u}(\lambda u)^k}{k!} dF(u), \quad k = 0, 1, 2, \ldots. \tag{11}$$

Here $\rho$ denotes the load of the queueing system, i.e.:

$$\rho = \lambda \int_0^\infty t\, dF(t). \tag{12}$$

The second theorem enables calculation of the limiting value of $B$, as the buffer size grows to infinity. As the convergence is rather quick, it can be also used to obtain approximate values of $B$, even for relatively small buffer sizes.

**Theorem 2.** *In the queueing system with the general service time distribution the limiting burst ratio equals:*

$$\lim_{N \to \infty} B(N) = \frac{\min\{1, \rho^{-1}\}}{1 - h(0)} \sum_{l=1}^\infty lh(l), \tag{13}$$

*where*

$$h(l) = \frac{1}{x_0^{l+1}} f(\lambda - \lambda x_0) - \frac{1}{x_0^{l+1}} \sum_{j=0}^l x_0^j a_j, \tag{14}$$

$$f(s) = \int_0^\infty e^{-st} dF(t), \tag{15}$$

*and $x_0$ is a positive and not equal to 1 solution of the equation*

$$f(\lambda - \lambda x) = x. \tag{16}$$

In the third column of Table 5, the theoretical burst ratios are presented, computed for $N = 100$, $\rho$ values as in Table 4 and the packet size distributions as in Table 3. They were obtained using Theorem 1, but they could have been obtained as well from Theorem 2 (for $N = 100$, Theorem 2 gives results very close to Theorem 1). For comparison, in the second column of Table 5, the measured values of the burst ratio are recalled. Finally, in the last column of Table 5, the relative error is calculated as a percentage of the measured burst ratio.

As can be seen, after 15:00, when the load is high, the theoretical burst ratio agrees very well with its measured counterpart – the relative error is around a few percent. The results for earlier traces, when the load is low or moderate, are worse – the relative error is in the range of 11–22%. This error is most likely caused by the much more complicated statistical structure of real IP traffic, compared with the Poisson traffic in the model. Apparently, this difference is not very important, when the load is high, but gets the more important, the lower the load is.

**Table 5.** Measured burst ratio, $B_m$, theoretical burst ratio, $B_t$, and the relative error.

| Time | $B_m$ | $B_t$ | Error, $\frac{|B_t - B_m|}{B_m}$ |
|------|-------|-------|-------|
| 06:30 | 1.20921 | 1.01191 | 16.3% |
| 07:10 | 1.20730 | 1.02895 | 14.7% |
| 09:00 | 1.30703 | 1.08371 | 17.0% |
| 10:00 | 1.35944 | 1.13116 | 16.7% |
| 11:00 | 1.39377 | 1.16085 | 16.7% |
| 12:00 | 1.42879 | 1.14167 | 20.1% |
| 13:00 | 1.45921 | 1.13955 | 21.9% |
| 15:00 | 1.45278 | 1.28959 | 11.2% |
| 16:00 | 1.40408 | 1.33042 | 5.2% |
| 20:15 | 1.41358 | 1.36465 | 3.4% |
| 20:30 | 1.37945 | 1.36449 | 1.0% |
| 21:10 | 1.40915 | 1.36529 | 3.1% |
| 21:30 | 1.45796 | 1.36938 | 6.0% |
| 21:40 | 1.40063 | 1.37616 | 1.7% |
| 22:00 | 1.36304 | 1.37827 | 1.1% |
| 22:10 | 1.42314 | 1.37968 | 3.0% |
| 22:30 | 1.30864 | 1.37855 | 5.3% |
| 23:20 | 1.32084 | 1.37601 | 4.1% |

It is also worth noticing, that the theoretical value of $B$ is typically smaller than its measured counterpart. Therefore, it can be used as a lower bound for the real value.

As discussed before, a higher usage of some specific applications may influence the traffic autocorrelation and, eventually, $B$. Therefore, to further decrease the error of the queueing model, it might be necessary to incorporate the traffic autocorrelation into the model.

This supposition can be further strengthened by studying the autocorrelation function. In Fig. 5, the autocorrelation function of packet interarrival times is presented for the trace collected at 15:00. (Similar results were obtained for other times). As can be seen in the figure, the autocorrelation function decreases very slowly, and is still non-negligible for the lag of $10^4$. Its is well known, that such a long-range autocorrelation of the arrival process may significantly influence the queueing performance. This constitutes, of course, a strong motivation for deriving the burst ratio in models with more complicated traffic, incorporating the autocorrelation of interarrival times.

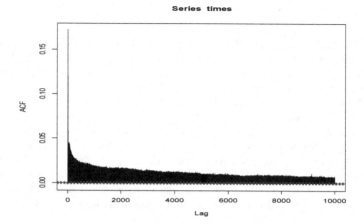

**Fig. 5.** Autocorrelation function of interarrival times for the trace collected at 15:00.

## 6   Conclusions and Future Work

In this paper, the results of measurements of the burst ratio, a parameter expressing the tendency of packet losses to occur in series, rather than as separate objects, were presented. Among other things, the study was motivated by the fact that clustering of packets losses is especially bad for multimedia transmissions. The measurements were conducted in real IP traffic, in an operational network.

Firstly, the obtained burst ratio was significantly higher than 1 in all measurements. This means that losses in real IP traffic do have tendency to group together. This observation is consistent with the previous results obtained in a queueing model and in a networking lab, on artificially generated traffic.

Secondly, the measurements showed that the dependence of the burst ratio on the load and the loss ratio is small. On the other hand, they demonstrated the possible dependence of the burst ratio on other traffic characteristics, including the application usage and the resulting traffic autocorrelation.

Thirdly, when comparing the measured burst ratio with its theoretical counterpart computed for the queueing model with Poisson arrivals, a relatively high accuracy of the model was observed for high load values. For a low or moderate load, the relative error was higher, with the maximum of 22%. As observed, the theoretical value of $B$ for the Poisson traffic model is typically smaller than its real counterpart, thus it can be used as a lower bound of the real value.

To decrease further the discrepancy between the queueing model and measurements, the autocorrelation of the interarrival times can be incorporated into the model, e.g. using MMPP [34] or BMAP [35,36] processes.

# References

1. Mongay Batalla, J., et al.: Adaptive video streaming: rate and buffer on the track of minimum re-buffering. IEEE J. Sel. Areas Commun. **34**(8), 2154–2167 (2016)
2. McGowan, J.W.: Burst ratio: a measure of bursty loss on packet-based networks, 16 2005. US Patent (2005)
3. ITU-T Recommendation G.113: Transmission impairments due to speech processing. Technical report (2007)
4. Samociuk, D., Chydzinski, A., Barczyk, M.: Experimental measurements of the packet burst ratio parameter. In: Kozielski, S., Mrozek, D., Kasprowski, P., Małysiak-Mrozek, B., Kostrzewa, D. (eds.) BDAS 2018. CCIS, vol. 928, pp. 455–466. Springer, Cham (2018). https://doi.org/10.1007/978-3-319-99987-6_35
5. Sanneck H.A., Carle, G.: Framework model for packet loss metrics based on loss runlengths. In: Proceedings of Multimedia Computing and Networking, pp. 1–11 (2000)
6. Jiang, W., Schulzrinne, H.: Modeling of packet loss and delay and their effect on real-time multimedia service quality. In: Proceedings of NOSSDAV, pp. 1–10 (2000)
7. Veeraraghavan, M., Cocker, N., Moors, T.: Support of voice services in IEEE 802.11 wireless LANs. In: Proceedings of IEEE INFOCOM, vol. 1, pp. 488–497 (2001)
8. Hasslinger, G., Hohlfeld, O.: The Gilbert-Elliott model for packet loss in real time services on the Internet. In: Proceedings of Measuring, Modelling and Evaluation of Computer and Communication Systems Conference, pp. 1–15 (2008)
9. Clark, A.: Modeling the effects of burst packet loss and recency on subjective voice quality. In: Proceedings of Internet Telephony Workshop, pp. 123–127 (2001)
10. Rachwalski, J., Papir, Z.: Burst ratio in concatenated Markov-based channels. J. Telecommun. Inf. Technol. **1**, 3–9 (2014)
11. Rachwalski, J., Papir, Z.: Analysis of burst ratio in concatenated channels. J. Telecommun. Inf. Technol. **4**, 65–73 (2015)
12. Chydzinski, A., Samociuk, D.: Burst ratio in a single-server queue. Telecommun. Syst. **70**, 263–276 (2019)
13. Chydzinski, A., Samociuk, D., Adamczyk, B.: Burst ratio in the finite-buffer queue with batch Poisson arrivals. Appl. Math. Comput. **330**, 225–238 (2018)
14. Samociuk, D., Chydzinski, A.: On the impact of the dropping function on the packet queueing performance. In: Proceedings of International Convention on Information and Communication Technology, Electronics and Microelectronics, pp. 473–478 (2018)
15. Smolka, B., et al.: New filtering technique for the impulsive noise reduction in color images. Math. Probl. Eng. **2004**(1), 79–91 (2004)
16. Benko, P., Veres, A.: A passive method for estimating end-to-end TCP packet loss. In: Proceedings of IEEE GLOBECOM 2002, pp. 2609–2613 (2002)
17. Bolot, J.: End-to-end packet delay and loss behavior in the Internet. In: Proceedings of ACM SIGCOMM 1993, pp. 289–298 (1993)
18. Coates, M., Nowak, R.: Network loss inference using unicast end-to-end measurement. In: Proceedings of ITC Conference on IP Traffic, Measurement and Modeling, pp. 282–289 (2000)
19. Duffield, N., Presti, F.L., Paxson, V., Towsley, D.: Inferring link loss using striped unicast probes. In: Proceedings of IEEE INFOCOM 2001, pp. 915–923 (2001)
20. Sommers, J., Barford, P., Duffield, N., Ron, A.: Improving accuracy in end-to-end packet loss measurement. ACM SIGCOMM Comput. Commun. Rev. **35**(4), 157–168 (2005)

21. Takagi, H.: Queueing Analysis - Finite Systems. North-Holland, Amsterdam (1993)
22. Chydzinski, A., Wojcicki, R., Hryn, G.: On the number of losses in an MMPP queue. In: Koucheryavy, Y., Harju, J., Sayenko, A. (eds.) NEW2AN 2007. LNCS, vol. 4712, pp. 38–48. Springer, Heidelberg (2007). https://doi.org/10.1007/978-3-540-74833-5_4
23. Chydzinski, A., Mrozowski, P.: Queues with dropping functions and general arrival processes. PLoS One **11**(3), e0150702 (2016)
24. Yu, X., Modestino, J.W., Tian, X.: The accuracy of Gilbert models in predicting packet-loss statistics for a single-multiplexer network model. In: Proceedings of IEEE INFOCOM 2005, pp. 2602–2612 (2005)
25. Cidon, I., Khamisy, A., Sidi, M.: Analysis of packet loss processes in high-speed networks. IEEE Trans. Inf. Theory **39**(1), 98–108 (1993)
26. Bratiychuk, M., Chydzinski, A.: On the loss process in a batch arrival queue. Appl. Math. Model. **33**(9), 3565–3577 (2009)
27. DPDK-Dump application for capturing traffic using DPDK. https://github.com/marty90/DPDK-Dump
28. TCPmkpub. http://www.icir.org/enterprise-tracing/tcpmkpub.html
29. TCPdpriv. http://ita.ee.lbl.gov/html/contrib/tcpdpriv.html
30. Jinliang, F., Jun, X., Mostafa, H.A.: Prefix-preserving IP address anonymization. Comput. Netw. **46**(2), 253–272 (2004)
31. Jinliang, F., Jun, X., Mostafa, H.A.: On the design and performance of prefix-preserving IP traffic trace anonymization. In: Proceedings of ACM SIGCOMM Internet Measurement Workshop, San Francisco (2001)
32. Jinliang, F., Jun, X., Mostafa, H.A.: Prefix-preserving IP address anonymization: measurement-based security evaluation and a new cryptography-based scheme. In: Proceedings of IEEE International Conference on Network Protocols, Paris (2002)
33. Chydzinski, A.: Duration of the buffer overflow period in a batch arrival queue. Perform. Eval. **63**(4–5), 493–508 (2006)
34. Fischer, W., Meier-Hellstern, K.: The Markov-modulated Poisson process (MMPP) cookbook. Perform. Eval. **18**(2), 149–171 (1992)
35. Lucantoni, D.M.: New results on the single server queue with a batch Markovian arrival process. Commun. Stat. Stoch. Models **7**(1), 1–46 (1991)
36. Chydzinski, A.: Queue size in a BMAP queue with finite buffer. In: Koucheryavy, Y., Harju, J., Iversen, V.B. (eds.) NEW2AN 2006. LNCS, vol. 4003, pp. 200–210. Springer, Heidelberg (2006). https://doi.org/10.1007/11759355_20

# WiCLR: A Sign Language Recognition System Framework Based on Wireless Sensing

Wang Lin[1,3(✉)], Liu Yu[1,2,3], and Jing Nan[1,3]

[1] School of Information Science and Engineering, Yanshan University, Qinhuangdao 066004, China
wangllinn@gmail.com
[2] Army Military Transportation University, Tianjin 300171, China
[3] Key Lab of Software Engineering in Hebei Province, Qinhuangdao 066004, Hebei, China

**Abstract.** The non-intrusion and device-free sign language recognition (SLR) is of great significance to improve the quality of life, broaden living space and enhance social service for the deaf and mute. In this paper, we propose a SLR system framework, called WiCLR, for identifying isolated words in Chinese sign language exploring the channel state information (CSI). WiCLR is made up entirely of commercial wireless devices, which does not incur significant deployment and maintenance overhead. In the framework we devise a signal denoising method to remove the environment noise and the internal state transitions in commercial devices. Moreover, we propose the multi-stream anomaly detection algorithm in action segmentation and fusion. Finally, the extreme learning machine (ELM) is utilized to meet the accuracy and real-time requirements. The experiment results show that the recognition accuracy of the approach reaches 94.3% and 91.7% respectively in an empty conference room and a laboratory.

**Keywords:** CSI · Isolated sign language · Activity recognition · Wireless sensing

## 1 Introduction

Sign language recognition is a vital application of human-computer interaction (HCI) with social value and technical challenge for deaf and dumb people to communicate with others. The automatic recognition of sign language plays an irreplaceable role in improving the quality of life, expanding living space, and perfecting the service function of public facilities and public institutions. The main data bulletin of the second national disabled sampling survey [1] shows that the number of deaf and dumb people for about 2.66% of the total population in China. However, they have great difficulties in communicating due to the low popularity of standard Chinese sign language and the lack of professional sign language translation. Sign language recognition is used to reduce the communication barriers in deaf and dumb people by detecting, analyzing and explaining sign language actions with computer. In sign language recognition, isolated words are the basis and prerequisite for constructing continuous sentences.

© ICST Institute for Computer Sciences, Social Informatics and Telecommunications Engineering 2019
Published by Springer Nature Switzerland AG 2019. All Rights Reserved
Q. Li et al. (Eds.): BROADNETS 2019, LNICST 303, pp. 102–117, 2019.
https://doi.org/10.1007/978-3-030-36442-7_7

Sign language recognition methods mainly include sign language recognition based on sensor [2–6] and computer vision [7–9]. Sensor-based systems need wearable device, which this is an intrusion perception that requires the user to carry a specific device (e.g., data glove [2, 3], EMG sensor [4, 5], PPG sensor [6], etc.). Vision-based systems on computer require sufficient lighting and LOS conditions in active environment, which may cause spatial-temporal constraints and privacy issues. With the guarantee of precision, an ideal method of symbolic language recognition is device-independent and privacy-protected.

In recent years, wireless networks have been widely deployed. Wireless signals are not only used for communication, but also can be used to realize environmental sensing by refraction and reflection of the individual in the environment. Especially in the indoor environment, wireless sensing forms a wide spectrum of applications, such as indoor localization [10–14], intrusion detection [15–18], activity recognition [19–24], and so on. At present, received signal strength indicator (RSSI) and channel state information (CSI) are the most common data of wireless sensing in activity recognition. The instability of RSSI greatly restricts the recognition accuracy, but CSI is favored by researchers for its high stability and sensitivity to action. Thus CSI-based approach is a completely new idea for fine-grained micro-movement recognition. E-eyes [19] employs CSI histogram as the fingerprint to identify nine behaviors related to location in daily life. WiHear [20] is exploited to recognize lip language by obtaining the change of CSI with specially-made directional antenna. WiFall [21] uses a feature sequence of CSI to detect falls. Smokey [22] detects the smoking by using the periodicity of CSI influenced by smoking movement. WiFinger [23] utilizes directional antennas to identify ten finger gestures under the 5 GHz. CARM [24] proposes CSI-activity model and CSI-speed model, then uses HMM and feature fusion to identify nine large-scale movements. All these work focus on profiling the influence of the movement of the body on the radio signal propagation, and use the feature template to identify the specific activity. However, the lack of two arms to participate in the identification of the movement, especially the two arms may interfere with each other, greatly increasing the difficulty of identification.

In this paper, we present a sign language isolated word recognition system framework exploiting the channel state information, called WiCLR. Firstly, a signal denoising method using discrete wavelet transform and the principal component analysis is proposed to remove the noise caused by the environment and the internal state transitions in commercial WiFi devices. Moreover, the interpolation method is combined to overcome the sampling defect caused by hardware imperfect and preserve the details of all the subcarriers. Secondly, the multi-stream anomaly detection algorithm is designed to analyze all the CSI streams synthetically in sign language action segment and fusion. Furthermore, the time domain features are extracted to classify different sign language actions. Thirdly, the extreme learning machine (ELM) is utilized as the fine granularity activity classifier via wireless signals for the first time to meet the accuracy and real-time requirements. Finally, CSI data of different sign language actions are collected in both the empty conference room and the laboratory. The experiment results show that the recognition accuracy of WiCLR reaches 94.3% and 91.7% respectively in the two scenarios.

The main contributions are summarized as follows.

(1) A non-invasive and device-free sign language recognition system framework based on physical layer CSI is proposed on COTS wireless devices.
(2) Useful signal denoising method and multi-stream anomaly detection based action segmentation and fusion algorithm are proposed. We analyze all the CSI streams and subcarriers to improve the recognition accuracy and robustness of the system.
(3) ELM is firstly used for wireless sensing-based activity classification and recognition, which greatly improves the recognition efficiency.

## 2  Preliminaries

### 2.1  CSI

We can get a set of CSI data by WiFi NICs. The received signal can be expressed as:

$$y = Hx + n \tag{1}$$

where $y$ is the received signal vector, H is the channel gain matrix, x is the transmitted signal vector and n is the noise vector.

Each CSI data represents one of the amplitude and phase of an OFDM subcarrier just as Eq. 2:

$$H(k) = \|H(k)\|e^{j\angle H(k)} \tag{2}$$

where $\|H(k)\|$ and $\angle H(k)$ are the amplitude and phase of the kth subcarrier respectively.

In Fig. 1, the phase in static environment is evenly distributed in $[-\pi, \pi]$. The irregularity is mainly caused by phase shift. Moreover, the deviation of hardware equipment also brings varieties of errors [24], which requires precise modulation and complex error analysis to eliminate the effect on the phase partly. However, the amplitude information remains relatively stable and has no obvious temporal variability. Therefore, we extract the amplitude information of CSI for further observation.

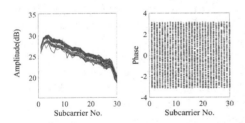

**Fig. 1.** CSI amplitude and phase in a static environment

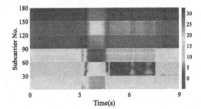

**Fig. 3.** CSI streams with the same action

The amplitude with the three different sign language activities show in Fig. 2b–d can be identified comparing with that in Fig. 2a in static environment. The amplitude of CSI in dynamic environment fluctuates violently and has quite difference in different actions, since the change of actions causes the continuous change of signal propagation path. The sensitivity of CSI amplitude to dynamic environment is the theoretical foundation for identifying sign languages in this paper. Figure 3 shows the fluctuations of all the subcarriers is correlated. The amplitudes change smoothly across different subcarriers in the same antenna pair and the amplitudes fluctuate violently between different antenna pair. However, they have strong correlation as well, for example, the peak of the CSI stream corresponds to the valleys received by different transceiver antennas.

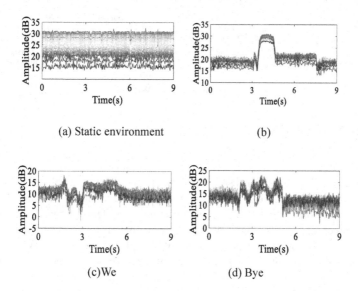

(a) Static environment                    (b)

(c)We                                (d) Bye

**Fig. 2.** CSI amplitude of static environment and different sign language actions

## 2.2   MIMO

MIMO is a method for multiplying the capacity of a radio link using multiple transmit and receive antennas to exploit multipath data transmission. Assuming the transmitter has M antennas and the receiver has N antennas, all the data streams can be expressed as below:

$$H = \begin{bmatrix} H_{11} & H_{12} & \cdots & H_{1M} \\ H_{21} & H_{22} & \cdots & H_{2M} \\ \vdots & \vdots & \ddots & \vdots \\ H_{N1} & H_{N2} & \cdots & H_{NM} \end{bmatrix} \tag{3}$$

In the indoor environment, the signal transmission is affected by the multipath effect, thus,the transmission paths of different data streams are different which causes different time delay, amplitude attenuation and phase shift. As a result, the data streams received by different receiving antennas is different.

# 3   System Design

As shown in Fig. 4, WiCLR consists of four modules: data collection, data preprocessing, action segmentation and extraction and sign language recognition module.

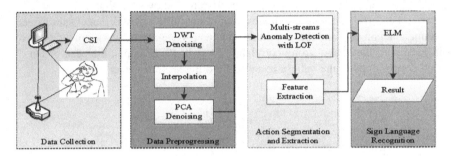

**Fig. 4.** Overview of WiCLR

## 3.1   Data Preprocessing

**DWT Denoising**
The discrete wavelet transform (DWT) has fine time-frequency localization and multiscale transformation which can process and separate the signal in time and frequency domain simultaneously to get approximate signal and detail signal. Therefore, the denoising effect of discrete wavelet transform is better for non-stationary signals. Comparing with the Butterworth low pass filter and median filter, DWT denoising eliminates the noise effect and preserves the details of the CSI effectively in Fig. 5.

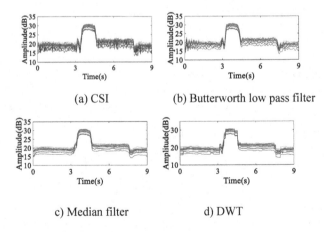

**Fig. 5.** Different denoising methods

**Interpolation**

In the transmission of wireless signal, the receiver is affected by data packet loss, transmission delay and other processing delay inevitably. Therefor, even though the transmitter sends packets with a fixed rate, we cannot ensure that the receiver obtains the data packets with the same rate. In other words, the CSI sequence is unevenly sampled. Hence, interpolation is used to get the accurate CSI sequences in time domain. We take advantage of spline interpolation to construct CSI sequences with uniform time intervals by adding approximate values in missing sampling intervals.

**PCA Denoising**

As shown in Fig. 6, after denoising, the action of sign language has the greatest impact on CSI, the second and third principal component may be the noise caused by the absence of external environment or the internal state of the equipment. Therefore, we choose the first principal component for further analysis. The Fig. 7 shows the CSI sequence after PCA denoising. By preserving the first principal component of the 30 subcarriers, PCA denoising can not only eliminates the uncorrelated noise components that can not be filtered by the wavelet denoising, but also retains the features of the CSI to the maximum to some extent. At the same time, PCA denoising reduces the dimension of CSI and the computational complexity. PCA denoising synthetically calculates the 30 subcarriers without artificially selecting subcarriers, which improves the accuracy of recognition.

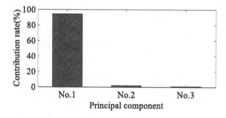
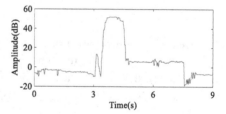

**Fig. 6.** Contribution rate                    **Fig. 7.** PCA denoising

## 3.2 Action Segmentation and Extraction

In MIMO system, there are significant differences in the characteristics of different motions at six CSI data streams. To solve this problem, we propose an action segmentation method based on multi-streams anomaly detection with LOF which analysis of all CSI data streams without the manual selection. The method is more accurate and effective to segment CSI sequences and extract actions. In Fig. 8, the fluctuations of different CSI streams are not the same. The action segmentation as the following signal processing:

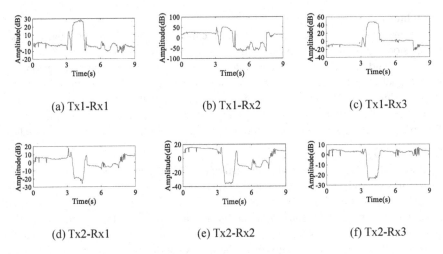

**Fig. 8.** Different CSI streams

The data stream after wavelet denoising, interpolation and PCA denoising is represented as $S_{t,r}$

$$S_{t,r} = \left[ S_{t,r}(1) | S_{t,r}(2) | \cdots | S_{t,r}(N) \right]^T \tag{4}$$

(1) Detecting outliers based on LOF for $S_{t,r}(i)$, and get the start point and the end point $t_{begin}(i)$, $t_{end}(i)$ respectively.

(2) Selecting the earliest starting point $\min(t_{begin}(i))$ and the last termination point $\max(t_{begin}(i))$ as the $t_{begin}$ and $t_{end}$ of the action, and get the CSI sequence $P_{t,r}$:

$$P_{t,r} = S_{t,r}^{t_{end}-t_{begin}} = \left[ P_{t,r}(1) | P_{t,r}(2) | \cdots | P_{t,r}(N) \right]^T \tag{5}$$

(3) Calculating the mean absolute deviation (MAD) is as followed:

$$\text{MAD}(i) = \frac{\sum_{t=1}^{k} \left| P_{t,r}^t(i) - m \right|}{k}, i = 1, 2, \cdots, N \tag{6}$$

where $m$ is the average value of sequence $P_{t,r}(i)$. The weight value $w_i$ of the MAD ratio as each piecewise sequence:

$$w_i = \frac{\text{MAD}(i)}{\sum_{i=1}^{N} \text{MAD}(i)}, i = 1, 2, \cdots, N \tag{7}$$

(4)  The piecewise sequence H after the fusion is obtained.

$$H = \sum_{i=1}^{N} w_i P_{t,r}(i), i = 1, 2, \cdots, N \qquad (8)$$

Since different CSI data streams have different sensitivity to the same action, we get different start point and end point for different data streams with LOF. It is more accurate and effective to segment the action sequence with the start and end of the fluctuation at the earliest and the last. The average absolute deviation depicts the discrete degree of data, and determines the weights of different data streams with MAD which can not only retain the information of each data stream, but also select the most sensitive data flow characteristics. The Fig. 9 shows the CSI with action segmentation method based on multi-streams anomaly detection with LOF.

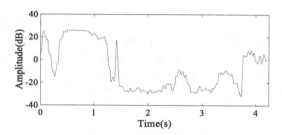

**Fig. 9.** Multi-streams anomaly detection with LOF

The action segmentation method can segment sign language actions in time domain. However, it can not recognize the category of sign language. At the same time, the detected CSI streams can not be identified by the classifier directly. In order to distinguish the different sign language movements, it is necessary to extract the sensitive action features of the CSI streams. We decide to choose the maximum value, minimum value, mean value, standard deviation, absolute middle difference, four division distances, action duration and signal change rate to reflect the features of CSI completely and effectively.

## 3.3  Sign Language Recognition

The extracted CSI features are sent to the ELM classifier for recognition. As shown in Fig. 10, the feature sequences of the signed sign language are used as the input of the classifier, and the model is trained by ELM. The number of ELM hidden nodes is adjusted at the test stage until the recognition rate is optimal. Finally, the optimized classification model is used to obtain the recognition results of the test sign language actions.

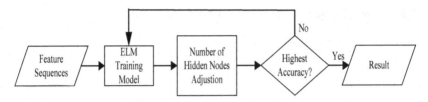

**Fig. 10.** ELM classification

(1) Data normalization: In this step, we normalize the extracted 8 time domain features between [−1, 1], then put the processed feature sequence and its corresponding label into the ELM classifier as the input.

(2) Training model initialization: we select an appropriate activation function and set up a small positive integer artificially as the number of nodes in the hidden layer to obtain an initial training model. The activation function of ELM classifier can be any infinitely differentiable function. In the laboratory environment, we set the number of hidden layer nodes as 500 and compare four different activation functions as shown in Table 1: Sigmoid, Sine, Hardlim and Trigonometric function, as a result we choose Sigmoid function for its highest recognition accuracy and less volatile (due to the weight $a_i$ and threshold $b_i$ are determined randomly in the classification, the accuracy is volatile).

**Table 1.** Performance of different activation functions

| Function | Testing accuracy | Recognition accuracy |
|---|---|---|
| Sigmoid | 82.0% | 78.2% |
| Sine | 100% | 62.5% |
| Hardlim | 69.4% | 64.6% |
| Trigonometric | 55.2% | 45.8% |

(3) ELM classifier optimization: The recognition accuracy of the initial training model is not enough because the number of hidden layer nodes is less and the network structure is simple. In this paper, we optimize the ELM classifier to achieve the best recognition accuracy by increasing the number of hidden layer nodes. In initial, we set the number of hidden layer nodes $L_0 = 10$, then add $\delta$ nodes randomly, the total number of nodes $L_k = L_{k-1} + \delta$. Calculating the recognition accuracy test_Acc$_k$, if $0 < \text{test\_Acc}_k - \text{test\_Acc}_{k-1} < 1\%$, then increase the step length; if $\text{test\_Acc}_k - \text{test\_Acc}_{k-1} < 0$, then reduce the step length, we will recalculate test_Acc$_k$ until the recognition accuracy reaches the best and select $L_{k-1}$ as the number of hidden layer nodes. As Fig. 11 shown, this paper sets the number of hidden layer nodes 1500 to get the best accuracy since the continued increase will cause overfitting and the performance decreases.

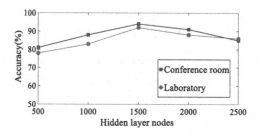

**Fig. 11.** Effects in different counts of hidden layer nodes

(4) Recognition results: Put the test data set into the optimized ELM classifier and the corresponding expected output as T, the recognition result is shown in following formula:

$$Label = \arg \max_{1 \leq i \leq m} (Ti) \tag{9}$$

## 4   Experimental Results

Although sign language has different forms in different countries and regions, these sign gestures are mainly performed with arms, palms and fingers. In our evaluation, we select the standard Chinese sign language isolated words as an object of recognition.

### 4.1   Experimental Setup

The CISCO WRVS4400N wireless router with 2 antennas is used as the transmitter, and the Intel 5300 NICs with 3 antennas is used as the receiver, and the data processing is analyzed with open source CSI Tools and MATLAB. As shown in Fig. 12, the experimental environment is conference room (3.6 m × 6.6 m) and laboratory (7.4 m × 8.2 m). The identified object sits in the middle of the transceiver terminal, and the receiver is 50 cm away from the transmitter. The height of gesture is as the same as the antennas.

We have 5 volunteers for a week of experimental data collection. They are 3 boys and 2 girls which in 21–28 years old, the weight in 48–82 kg, and the height of 160–180 cm.

### 4.2   Performance Metrics

To test the performance of the WiCLR method, we focus on the following indicators:

**Confusion Matrix:** Each column represents sign language actions classified by the system, and each line represents sign language actions made by the test subjects. If $D_{i,j}$

is used to represent cells in column J of column I, then the sign language on behalf of line I $P_{i,j}$ is identified as the probability of sign language in J.

**Precision:** $P_i = \frac{N_{TP}^i}{N_{TP}^i + N_{FP}^i}$, $N_{TP}^i$ is to correctly identify the number of I, $N_{TP}^i$ is for all the other actions that are mistakenly identified as sign language I, the precision rate indicates the accuracy rate of sign language I.

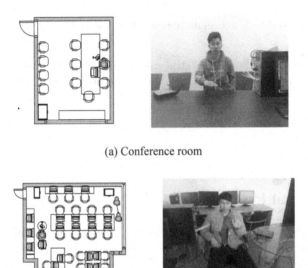

(a) Conference room

(b) Lab

**Fig. 12.** Experimental environments

**Recall:** $R_i = \frac{N_{TP}^i}{N_{TP}^i + N_{FN}^i}$, $N_{FN}^i$ is the number of sign language I is mistaken for other actions. Recall rate indicates the recall rate of sign language I.F1.

**F1-Score:** $F_1^i = 2 \times \frac{P_i \times R_i}{P_i + R_i}$, which is the correlation between accuracy and recall is expressed. From 0 to 1, the recognition performance is getting better and better.

**Accuracy:** $A = \frac{\#Correct}{\#Sum}$, #Correct is the number of the correct signal language, #Sum is the summary of the samples.

### 4.3    Experimental Analysis

**Accuracy of Anomaly Detection**
In order to measure the accuracy of the segmentation, the following conditions are considered to be incorrectly segmented: Firstly, the static environment is detected as an exception. Secondly, the action is not detected as an exception. Thirdly, a movement is split into multiple actions; and fourthly, the error of the action endpoint is above 0.5 s. As Fig. 13 shown, the accuracy rate of WiCLR is basically over 90%, while the anomaly detection of single stream is very unstable, and the accuracy rate of different sign language movements is different. The joint multi-streams can improve the accuracy of action segmentation and the performance of sign language recognition.

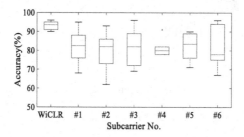

**Fig. 13.** Segmentation Accuracy

**Recognition Accuracy of Sign Language Isolated Words**
We test the accuracy of our system in the two typical indoor environments of laboratory and conference room respectively. The experimental results are shown in Fig. 14. In the conference room, the average recognition accuracy of 6 sign language actions is 94.3%, and in the laboratory is 91.7%. In this part, the recognition accuracy of large-scale sign language actions we, sorry and goodbye are better because of the large amplitudes and long durations of signal changing. However the accuracy of micro-scale actions I, know and thank has decreased. The accuracy rate, recall rate and F1 score of each sign language are above 80%, where the action I has the lowest accuracy. Because the action is not only small in action, and the duration is close to thank and know, and the action style or the subaction of we is easy to confused.

In the laboratory, radio signal attenuation is obvious because of many obstacles and rich multipath which cause signals are reflected, refracted and diffracted to the receiver. Thus, the recognition performance is not as good as that in the conference room. However, the recognition accuracy of sign language in the rich or sparse multipath indoor environment is more than 90%. It is proved that our system can be used to extract the feature information of CSI by COTS wireless devices to research the fine grained sign language recognition, and it has certain robustness.

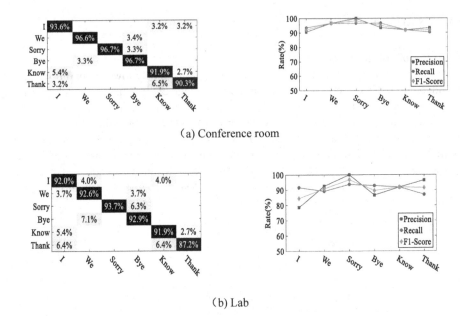

(a) Conference room

(b) Lab

**Fig. 14.** WiCLR recognition performance

## Effects of Different System Classifications

Our classification method is compared with the methods of WiFall [21] and WiFinger [23]. The results of recognition performance of the three methods just as shown in Fig. 15, using ELM to classify sign language will not only have higher recognition accuracy, simple method and less recognition time, but also have a more significant advantage in recognition performance when the sign language is increased.

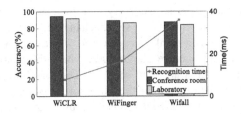

**Fig. 15.** Different system classifications

## Effects of Different Features

In this experiment, we extract one feature: action duration, two features: maximum and minimum, four features: maximum, minimum, standard deviation, duration and all characteristics to compare. As shown in Fig. 16, the accuracy of extracting all of the 8 features is the highest. This is because each feature can only describe a part of the CSI sequence. Only by taking into account all the features can we fully describe the effects of different sign language actions on the wireless signal.

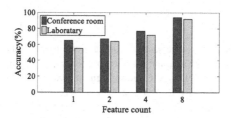

**Fig. 16.** Different features

**Effects of Different Volunteers**

For each volunteer, the accuracy rate of sign language recognition in the conference room and the laboratory is calculated respectively. From Fig. 17, the accuracy of the sign language recognition of different volunteers is different. The different habits and proficiency of the sign language action are the main factors that affect the accuracy, and non-standard sign language affects the accuracy of sign language recognition as well.

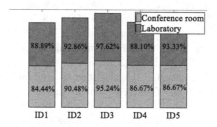

**Fig. 17.** Different volunteers

## 5 Conclusion

In this paper, we propose a low-cost, free-device and high precision system to recognize Chinese isolated sign language with the amplitude information of CSI. Firstly, we choose combined denoising method based on DWT, interpolation algorithm and PCA to preprocess the CSI amplitude information which remove the high frequency noise effectively and preserve the effect of the sign language action on the amplitude of all CSI streams to the maximum extent. Secondly, action segmentation method based on multi-streams anomaly detection with LOF is proposed to determine the start and end points of the sign language action. The CSI streams are fused with the weight value and 8 time domain features are extracted. Finally, ELM is used as classifier which is more accurate and faster than other classification methods In the future, our research will combine the phase of CSI and the existing amplitude features to improve the recognition accuracy and the scope of the isolation of isolated words. The hidden Markov model and neural network will also be researched to recognize the continued sentences on the basis of recognition the isolated words of the sign language.

# References

1. National Bureau of Statistics: The second leading group for the national sampling survey of disabled people. The second national disabled people sampling survey main data bulletin
2. Gao, W., Fang, G., Zhao, D.: A Chinese sign language recognition system based on SOFM/SRN/HMM. Pattern Recogn. **37**(12), 2389–2402 (2004)
3. Li, K., Zhou, Z., Lee, C.H.: Sign transition modeling and a scalable solution to continuous sign language recognition for real-world applications. ACM Trans. Access. Comput. **8**(2), 1–23 (2016)
4. Li, Y.: Research of Chinese sign language recognition and exploration of rehabilitation application based on surface Electromyogram. University of Science and Technology of China (2013)
5. Yang, X.: Research of Chinese sign language recognition technology based on fusion of surface electromyography and inertial sensor. University of Science and Technology of China (2016)
6. Zhao, T., Liu, J., Wang, Y., Liu, H., Chen, Y.: PPG-based finger-level gesture recognition leveraging wearables. In: IEEE International Conference on Computer Communications. IEEE (2018)
7. Sun, C., Zhang, T., Bao, B.K., et al.: Discriminative exemplar coding for sign language recognition with kinect. IEEE Trans. Cybern. **43**(5), 1418–1428 (2013)
8. Chai, X., Li, G., Lin, Y., Xu, Z., Tang, Y., Chen, X.: Sign language recognition and translation with kinect. In: The 10th IEEE International Conference on Automatic Face and Gesture Recognition, pp. 22–26. IEEE (2013)
9. Marin, G., Dominio, F., Zanuttigh, P.: Hand gesture recognition with leap motion and kinect devices. In: IEEE International Conference on Image Processing, pp. 1565–1569. IEEE (2014)
10. Wu, K., Xiao, J., Yi, Y., et al.: FILA: fine-grained indoor localization. Proc. IEEE INFOCOM **131**(5), 2210–2218 (2012)
11. Xiao, J., Wu, K., Yi, Y., et al.: FIFS: fine-grained indoor fingerprinting system. In: International Conference on Computer Communications and Networks, pp. 1–7. IEEE (2012)
12. Xiao, J., Wu, K., Yi, Y., et al.: Pilot: passive device-free indoor localization using channel state information, pp. 236–245 (2013)
13. Chapre, Y., Ignjatovic, A., Seneviratne, A., et al.: CSI-MIMO: an efficient Wi-Fi fingerprinting using channel state information with MIMO. Pervasive Mob. Comput. **23**, 89–103 (2015)
14. Wu, C.: Wireless indoor positioning via crowd sensing. Tsinghua University (2015)
15. Kosba, A.E., Saeed, A., Youssef, M.: RASID: a robust WLAN device-free passive motion detection system. In: 2012 IEEE International Conference on Pervasive Computing and Communications (PerCom), pp. 180–189. IEEE (2012)
16. Xiao, J., Wu, K., Yi, Y., et al.: FIMD: fine-grained device-free motion detection. In: 2012 IEEE 18th International Conference on Parallel and Distributed Systems (ICPADS), pp. 229–235. IEEE (2012)
17. Qian, K., Wu, C., Yang, Z., et al.: PADS: passive detection of moving targets with dynamic speed using PHY layer information. In: IEEE International Conference on Parallel and Distributed Systems, pp. 1–8. IEEE (2015)
18. Liu, W., Liu, Z., Wang, L., et al.: Human movement detection and gait periodicity analysis via channel state information. Int. J. Comput. Syst. Sci. Eng. (2017)

19. Wang, Y., Liu, J., Chen, Y., et al.: E-eyes: device-free location-oriented activity identification using fine-grained WiFi signatures. In: International Conference on Mobile Computing and Networking, pp. 617–628. ACM (2014)
20. Wang, G., Zou, Y., Zhou, Z., et al.: We can hear you with Wi-Fi! In: ACM International Conference on Mobile Computing and Networking, pp. 593–604. ACM (2014)
21. Han, C., Wu, K., Wang, Y.: WiFall: device-free fall detection by wireless networks. In: 2014 Proceedings IEEE INFOCOM, pp. 271–279. IEEE (2014)
22. Zheng, X., Wang, J., Shangguan, L., et al.: Smokey: ubiquitous smoking detection with commercial WiFi infrastructures. In: IEEE INFOCOM 2016 - IEEE Conference on Computer Communications, pp. 1–9. IEEE (2016)
23. Li, H., Yang, W., Wang, J., et al.: WiFinger: talk to your smart devices with finger-grained gesture. In: ACM International Joint Conference on Pervasive and Ubiquitous Computing, pp. 250–261. ACM (2016)
24. Wang, W., Liu, A.X., Shahzad, M., et al.: Device-free human activity recognition using commercial WiFi devices. IEEE J. Sel. Areas Commun. 35, 1118–1131 (2017)
25. Xie, Y., Li, Z., Li, M.: Precise power delay profiling with commodity WiFi. In: International Conference on Mobile Computing and Networking, pp. 53–64. ACM (2015)
26. Wang, H., Zhang, D., Ma, J., et al.: Human respiration detection with commodity WiFi devices: do user location and body orientation matter? In: ACM International Joint Conference on Pervasive and Ubiquitous Computing, pp. 25–36. ACM (2016)

# High-Resolution Image Reconstruction Array of Based on Low-Resolution Infrared Sensor

Yubing Li[1(✉)], Hamid Hussain[1], Chen Yang[1], Shuting Hu[2],
and Jizhong Zhao[1]

[1] Xi'an Jiaotong University, Xi'an 710049, Shaanxi, People's Republic of China
yubingli0513@sina.com
[2] Qinghai University, Xi'ning 810016, Qinghai, People's Republic of China

**Abstract.** As the time is progressing the number of wireless devices around us is increasing, making Wi-Fi availability more and more vibrant in our surroundings. Wi-Fi sensing is becoming more and more popular as it does not raise privacy concerns in compare to a camera based approach and also our subject (human) doesn't have to be in any special environment or wear any special devices (sensors).

Our goal is to use Wi-Fi signal data obtained using commodity Wi-Fi for human activity recognition. Our method for addressing this problem involves capturing Wi-Fi signals data and using different digital signal processing techniques. First we do noise reduction of our sample data by using Hampel filter then we convert our data from frequency domain into time domain for temporal analysis. After this we use the scalogram representation and apply the above mentioned steps to all our data in terms of sub carriers. Finally we use those sub carriers in combined for one activity sample as all the sub carriers combined form up an activity so we shall use the combined signal in the form of power spectrum image as input for the neural network.

We choose Alexnet for classification of our data. Before feeding our data into pre-trained CNN for training we first divided the data into two portions first for training which is 85% secondly for validation which is 15%. It took almost 18 h on single CPU and finally achieved an accuracy of above 90%.

**Keywords:** Wi-Fi sensing · Human activity recognition · CSI · CNN

## 1 Introduction

Human activity recognition in digital signal processing has become a subject of great importance in research community especially in recent few years with the advancement of technology. These days more and more new methods are discovered and studied to achieve the goal of Human activity recognition as it has diverse range of applications in law in order, security, health, and many other fields. There have been many methods used for HAR (human activity recognition), human counting, localization, gesture recognition, human identification, respiration monitoring, heart rate monitoring which

This work is supported by NSFC Grants No. 61802299, 61772413, 61672424.

Q. Li et al. (Eds.): BROADNETS 2019, LNICST 303, pp. 118–132, 2019.
https://doi.org/10.1007/978-3-030-36442-7_8

required some wearable sensors for the subject or cameras, which require certain conditions to work but in recent times radio frequency signals have been used for sensing purposes and performs the above mentioned tasks with the help of these radio signals. In this paper we will study the classification of human activity recognition. The reason we are motivated to use Wi-Fi/RF signals for sensing purposes is that it does not require any wearable sensors or special systems to deploy with extra cost. Cameras are not that good solution anymore as they require line of sight (LOS) with good light conditions to give a good result also it raises many privacy. There are other sensing techniques like ultrasound, radar and vision techniques but normally they are quite expensive need specialized equipment and environment sometimes while mentioned before with Wi-Fi/RF signals we don't need our subject (Human) to wear any sensors or be in a special environment rather our goal of human activity recognition can be achieved with off the shelf commodity Wi-Fi router that are easily available in markets so that the human does not need to wear sensors like gyroscope, accelerometer etc.

In this work, we have exploited Wi-Fi-CSI data for Human activity recognition we did theoretical and experimental work with CSI data in our lab to demonstrate that commodity Wi-Fi can be used for human activity recognition with very high precision. We shall use 5 sample activity classes for our research case. We will take data sample of an activity using the Wi-Fi (MIMO) along with its Orthogonal Frequency-Division Multiplexing (OFDM) making different sub carriers and later make use of different digital processing techniques to extract a refined signal that we shall use as input for our neural network to train our network for the classification of these activities. Our results from the use of neural networks indicate very high accuracy of above 90% in terms of classifying human activity. This would make it possible to use this research in remote healthcare, surveillance and more with no special environment or deployment of specialized equipment.

## 2   Related Work

In recent times more and more work has been carried out on activity recognition [1–3] using radio frequency signals such as Wi-Fi signals. Wi-Fi-CSI has been hot source for Wi-Fi sensing these days as it has many benefits over the traditional systems. These are few papers that presents some work on indoor localization using RF signals [4–6] localization means tracking the positon of a person or device, gesture recognition [7] this is a survey paper which describe some of the work related to gesture recognition which is important for our research as gestures are specific pattern movements of body parts that the RF signals have to capture and have a meaning to it, behavior recognition [4, 8–12] these papers are closely related to our work as our prime goal is human activity recognition with WiFi-CSI data and these paper are mostly about human behavior recognition with Wi-Fi CSI data. Our focus is on wearable-free sensing so in [4] writer exploits wireless signals for localization further more in [8] the writer uses CSI data for human identification as well as motion recognition of humans. These papers [9–12] discuss different methodologies for human activity recognition using CSI, these approaches include 'big data analytics', 'pattern based model based', 'deep learning models', 'signal processing'.

It is also important to mention that there are number of apps that rely on phase shifts so it should be accurate for this purpose we use AoA/ToF which are basically used for localization and tracking purposes. SopFi [13] gives us technique in which sampling time offsets/sampling frequency offsets are subtracted via linear regression. Wisee [14] is first uses raw data and applies Doppler phase shift to get motion data later it does recognition of gesture and hand with KNN similarly in WiAG [15] first use channel frequency ration to orientation later it uses KNN for gesture recognition.

As research in Wi-Fi sensing progressed researchers started working on human event detection in which different events experienced by humans were studied with the help of wife signals like falling down which will be referred as fall detection. Some of the work in literature related to fall detection is [16–20] this work is important for medical care in hospitals specially and in case of patients monitoring remotely with no extra wearable sensors. Some of the work related to human motion detection is [21–24]. Similarly human walking [25], Posture change [26], Sleeping [27], Key stroke [28], Intrusion detection [23, 29]. There have been work and papers on driving fatigue detecting, lane change, school violence, smoking, attack tamper, abnormal activity, all these using Wi-Fi sensing.

## 3   System Overview

Our working model involves capturing the CSI data from Wi-Fi signals in the indoor environment where the subject (Human) performs 5 different types of activities that are caused due to different body parts movements has effect on the channel state information in terms of the signals which are propagating form transmitter to receiver will have certain effects like amplitude attenuation, phase shift for OFDM-WiFi signals, scattering, fading, multipath effect and more so we are going to use this recorded CSI data to analyze the signal in a way it gives us information about the human activities performed. Now different signal processing techniques are used to refine a signal to ones needs that it gives some useful information for this first we shall do some pre-processing on the signal like removing the noise from the signal. In our work we shall use Hampel filter for this purpose next after noise reduction we shall do down sampling of our signal but in order to do so first we should convert our signal in to time domain so after we did these process on our given signal data we shall now apply continuous wavelet transform and get scalograms in the form of images for each sub carrier but as our data is concerned with motion and it the body movement causes each sub carrier to have some changes so all the sub carriers are important for classification hence for this reason we shall combine all the subcarriers of one sample together and generated a new image that reflect the whole signal for one complete activity. Now that we have images of each complete one activity we shall use machine learning for classification of our dataset.

We shall use deep learning and convolution neural network are a good resource to work with in this domain in our case we shall prefer to use a pre-trained neural network which means that it is a neural network which has already been trained on some or many different datasets already and has learned many things and over the time became very efficient. Some of the famous pre- trained CNN are Alexnet and google net. We shall feed our image data into these networks for classification. Using this approach has advantages over tradition camera/sensor/radar based approach as it is inexpensive because it does not require some expensive equipment rather simple Wi-Fi router form the market can capture CSI data with Wi-Fi AP also it does not need to deploy special systems as Wi-Fi is almost everywhere plus it does not require line of sight. In our work we shall use Wi-Fi CSI fingerprints for human activity. Also unlike radar systems which require high bandwidth, Wi-Fi signals use narrow bandwidth it leverages the concepts of ToF, AoA, azimuthal angle, elevation angle, look angle, phase information, channel amplitude, for signal capturing and analysis. We are going to give the working model approach in Fig. 1.

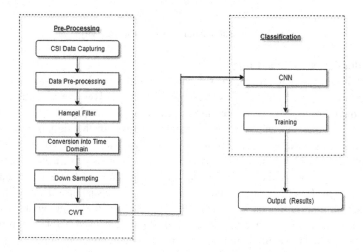

**Fig. 1.** Our working model approach

## 4   Methodology

### 4.1   Subcarrier Pre-processing

We shall now pre-process our signal data sample as the sub carriers have different frequencies and every sub carrier exhibit different response towards the changes in the environment in our case the changes are the body movements by human making up an activity so it is important to take all the subcarriers into account for the classification task but the problem is that it is difficult to do different analysis on the multivalued signals together so what we are going to do is take few signals for representing our analysis and perform different analysis on them and based on the better results we shall apply it to all the subcarriers so the response to every subcarrier can be taken into account. For this purpose we shall first use the subcarrier of bend knee activity as shown in Fig. 2, it is the raw sub carrier now we shall start pre-processing it with smoothing and noise reduction.

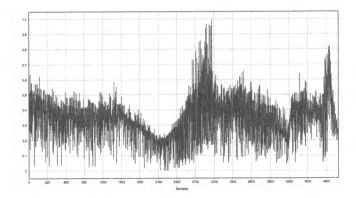

**Fig. 2.** One sub-carrier of bend knee activity

## Noise Reduction and Smoothing

In this part, we are going to process the sub carrier signals to make the signals smooth and reduce the noise in it. Smoothing is the process that involves the reduction of extra unwanted data and makes the whole signal into an approximated signal and uses that approximation which shows only the prominent features in the signal. By using this approach we can get the prominent trends in our signal data. The whole signal is converted or reduced keeping approximation in view. The advantage of using smoothing is that it reduces the unwanted data (noise) and the sudden unrelated changes in the signal. Smoothing aligns the signal into an approximation. Now we shall use 'Hampel filter' for noise reduction and outlier removal. Hampel filter is very useful in signal processing specially in context to noise and outlier removal. It has been used before when we view literature and proved to be efficient in conditions related to movement detections like for breath detection. Hampel filter is categorized in the general class of decision filters. It works on the basis of a running window which runs over the data and at any instance if the middle value of the window is greater than the median value it would be treated as outlier and replaced with the median value this is the main mechanism behind this filter and it makes it so efficient. As for noise it could be of different patterns like it could be Gaussian noise or impulsive noise, Hample filter is good for both cases (Fig. 3).

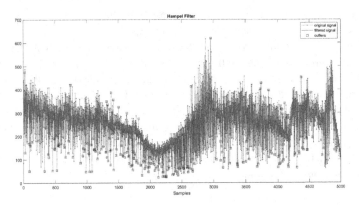

**Fig. 3.** Hampel filter applied on signal

## Conversion into Time Domain

After using Hampel filter we want to down sample our signal but for that we first have to convert the signal into time domain for that we shall divide the total number of samples by 3.5 which is average time of the activity and the resultant will become the frequency of the signal with 0 s as starting time. Our signal after changing into time domain looks as follows. Time domain analysis describes the changes in the signal over the course of time. For conversion of frequency domain into time domain inverse flourier transform function is used. Our default CSI data recorded is in frequency domain, it is because mostly the transmitter use IFFT during modulation while on the receiver ends the signal is demodulated using FFT but sometimes we need to analyze our signal in time domain so we again convert it into time domain with IFFT and a sampling frequency (Fig. 4).

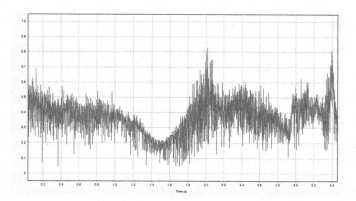

**Fig. 4.** Signal converted into time domain

## Down Sampling

After converting the signal into time domain we shall now do down sampling of our signal. Sampling is regarded as collection of data either in signal or image form, sampling could make the recorded data as smaller (down sampling) or bigger (up sampling). Normally down sampling is done by decreasing the sampling rate this process is also called decimation. Down sampling decimates the sample frequency but the pattern remains almost same the new sample is an approximation of the original signal it has some benefits like it reduces the computation cost increasing the processing efficiency (Fig. 5).

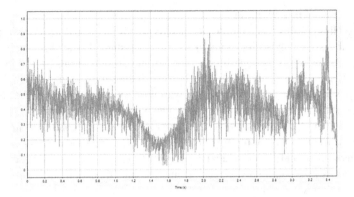

**Fig. 5.** Signal down sampled

## 4.2    Continuous Wavelet Transform

Wavelet filters are of so many different types and have so many uses. Some wavelet filters are used for noise reduction or smoothing of signal. We are going to use continuous wavelet transform what it does is that it scales the original signal into wavelets which make up the whole signal these wavelets could be small scale or big scale. In case of small scaled wavelets the oscillation is more where as in case of long scale, the oscillation would be less. One of the advantage of CWT is that it is localized in time in the wavelets are localized in time. It is quite useful for mapping the evolving properties of data related to motion. It mostly works on the basis of some mother wavelet chosen. Mother wavelets are of different types both for continuous or discrete signal analysis since are data is continuous so we can use the following mother wavelets which are for CWT: Morlet, Meyer, derivative of Gaussian, and Paul wavelets. We are going to choose Morlet wavelet as our mother wavelet as it does not require the scaling parameter and does that by itself. The final convolved result Morlet wavelets have the same time features as on in the original signal. Morelet wavelet is also more efficient in term of computation as compared to other. Below is the figure which shows both the signal after CWT and its scalogram (Fig. 6).

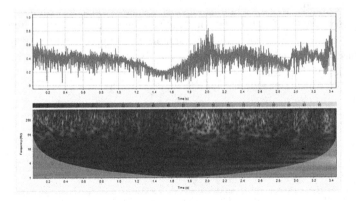

**Fig. 6.** Continuous wavelet transform

## 4.3  Using Combined Subcarriers

After applying all the mentioned signal processing techniques we finally have different form of signals including pictorial forms but the problem with using each sub carrier signal separately is that the data sample as a whole represents an activity which is the reflection of different body parts movements each movement attenuating the sub carrier differently so in order to describe the whole activity all the sub carriers should be considered in combined form as one so for this we shall combine all the sub carriers and treat them as one. After the pre-processing of the subcarriers the signals in the combined form are shown below. We are representing all the activity signals for better idea (Fig. 7).

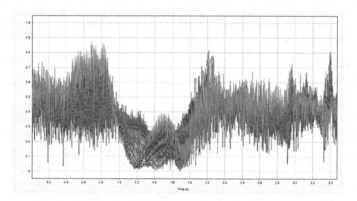

**Fig. 7.** Combined processed subcarriers - bend knee

As discussed above that we will consider effect of all the sub carriers together for this purpose we shall generate power spectrum of every data sample that represents one complete activity and is constituted by all the processed sub carriers combined together. Below we share the power spectrum of processed bend knee signal (Fig. 8).

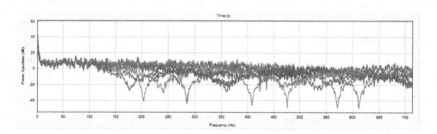

**Fig. 8.** Power spectrum of one data sample – bend knee

## 4.4    Classification Using CNN

We are going to use a deep learning neural network for classification. Convolution neural network is a type of deep learning neural network, there are many different convolution neural networks available but we are going to use transfer learning which means we are going to use an already trained network for our case i.e. human activity recognition, this classification task if performed by an already trained network like alexnet becomes more efficient and robust. Alexnet is a pre trained network that has been trained on ImageNet and has 25 layers with 24 connections further it is 8 layer deep neural network. It was trained to classify 1000 categories of data and is considered very efficient for image classification tasks. We shall input the power spectrum image of each signal making 2500 different sample images. The input size for the alexnet is $227 \times 227 \times 3$ where first two values are the size making up a 2 dimensional matrix and third value represents the color channel making it an image. There are three types of layers in alexnet CNN.

1. Convoluiton Layer
2. Max Pooling Layer
3. ReLu Layer (Rectified linear unit).

The first one that is convolution layer is focuses on searching for patterns using different kernels these kernels traverse the image and look for patterns like edges, corners etc. Second layer max pooling its main function is to do down sampling gradually it decrease the dimensionality on each iteration. Lastly Relu layer it simply converts the negative values into zero. The last three layers of Alexnet are fully connected layer, softmax, classification layer respectively. Classification layer is used to classify the dataset and give output whereas fully connected layer works a referee that it counts the score of every category and softmax changes that score into more of statistical interpretation that is probabilities.

After this we shall first load the pre trained network and change the number of class parameter which is by default 1000 in fully connected layer into 5 as in our case the number of classes we have are 5. Next we shall set the parameters to begin the training like learning rate, mini batch size, epoch etc. After that we shall start training and finally get our results and if they are not so good we can do hyper parameter tuning.

## 4.5    Algorithm

In our algorithm after the capturing of data the first step is the pre-processing of the raw CSI data that has noise for this we used different filters there sequence of application is as follows:

Frist we shall **combine** all 30 subcarriers **for** $i = 1: 30$ *in one data smaple.*
**Noise Reduction,** now we shall apply **Hampel Filter** on all subcarriers.

We shall now convert signal into **Time Domain.**
We shall now **down sample** our signal.
Now we shall transform the signal into **continuous wavelet transform.**

After performing all the above mentioned steps on all the subcarriers. We can not consider scalograms as they only represent the effect of one single subcarrier so we perform the following step;

Now we take the **Power spectrum** of all the subcarriers combined.
We generated 2500 power spectrum pictures

Now we randomly select 15% of the above mentioned data as validation data and 85% of the data as training data and finally feed it to our pre trained convolution neural network. In our case it is Alexnet.

**Select** training and testing dataset with labels
**Load** pre-trained deep learning neural network, ALEXNET
**Define** learning rate, batch size and fully connected layers
**Run** training process over power spectrum with ground truth

After the training is complete on 85% of power spectrum data then we test the classification accuracy.

**Classify** test data with learned attributes
**Output** Accuracy and Confusion Matrix
**end.**

# 5   Evaluation

## 5.1   Experimental Setup

The device we are going to use for CSI data acquisition is intel NIC 5300, the transmitter we are going to use would have directional antenna with 2.4 GHz transmitting frequency and the receiver would be a 2D antenna array receiver. Now we shall transmit the OFDM signals, where it has numerous orthogonal subcarriers with different frequencies which will interact with the subject (human) at different body parts each producing some change in the signals finally the signals would be received at the 2D antenna array. Now we shall use data collected at antennas array with the phase difference data to construct our data sample which will later be used for activity recognition.

We shall set up a directional antenna with 2.4 GHz of frequency for NIC 5300 on the transmitter side and receiver would be a 4 × 4 2D array. The phase difference between origin antenna and first antenna was π/2 similarly the phase difference between first and second antenna is also π/2 and same is between any two consecutive antennas. To get an idea of the phase difference graphically we share the below figure.

## 5.2  Accuracy

As discussed in previous section after feeding the input data in our neural network which was trained on 85% data and was validated or tested on 15% which was randomly partitioned, our model after training generated a high accuracy of 94.2%. The training also generated a confusion matrix which is given below.

**Table 1.** Confusion matrix (A)

| %         | Bend knee | Left leg | Right leg | Left arm | Right arm |
|-----------|-----------|----------|-----------|----------|-----------|
| Bend knee | 94%       | 1%       | 3%        | 1%       | 1%        |
| Left leg  | 4%        | 96%      | 0%        | 0%       | 0%        |
| Right leg | 1%        | 2%       | 91%       | 1%       | 5%        |
| Left arm  | 0%        | 6%       | 0%        | 94%      | 0%        |
| Right arm | 1%        | 1%       | 2%        | 0%       | 96%       |

As our transmitter is directional antenna and the object is just 2 m away thus reducing the noise also the movements of body cause the subcarriers to have variation in their values but if the direction of motion is same that is the direction of subcarriers is same they could be treated as one due to the fact that the difference of frequencies between these sub carriers is so small that the effect of subcarriers in one direction is almost negligible making the classification task more better as combined unidirectional subcarriers will be more prominent.

## 5.3  Training Process

As discussed in the previous section after all the pre-processing we generate power spectrums of our dataset and make a pictorial dataset consisting of (2500) samples with 5. We will use datagram function to load our data as mentioned 15% data is for validation purpose and rest is for training next we use a pre-defined function to resize our images as the input size for Alexnet is 227*227 after this we will load our pre-trained network which is Alexnet. By default the number of categories in Alexnet is 1000 we shall change it to 5 as we have 5 classes, this is done in the 23rd layer in number of classes. Next we shall set parameters for our training. First is initial learning rate which should be normally small as it is the start of training and it needs to look into data in details in beginning. Next we will set the mini batch size to 30, this refers to the number of images in each iteration. Now we shall set a maximum epoch since we are using transfer learning so the epoch size should not be large so we select epoch size as 15. Epoch size means the number of times the network will traverse the whole dataset. We did this training on a single CPU and it took 19 long hours to complete the training. It had 2500 iteration with 15 epochs. Below is the figure of training graph show the loss function and accuracy function graph which seems quite good (Fig. 9).

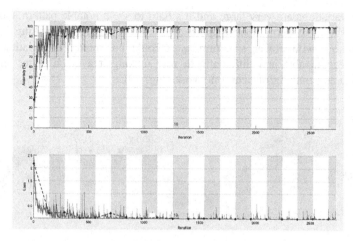

**Fig. 9.** Training graph

At the end of the training we achieved accuracy of 94%. We also generated confusion matrix as shown in Table 1.

## 5.4 Using Combined Subcarriers

The sub carrier alone do not reflect the complete activity for which they are recorded so for this reason we shall use the combined form of subcarriers that is one signal sample as a whole note that all the subcarriers are pre-processed. We used the similar method as discussed above for training our data but this time on a different image form the results it generated were not as good as the above discussed method. After training it generated an accuracy of 68%. The confusion Matrix of the above said method is shown below in Table 2.

**Table 2.** Confusion matrix (B)

| % | Bend knee | Left leg | Right leg | Left arm | Right arm |
|---|---|---|---|---|---|
| Bend knee | 68% | 6% | 8% | 10% | 8% |
| Left leg | 7% | 71% | 9% | 7% | 6% |
| Right leg | 7% | 6% | 72% | 8% | 7% |
| Left arm | 4% | 13% | 8% | 66% | 9% |
| Right arm | 9% | 8% | 9% | 7% | 67% |

This difference in result is due to the fact that there is a lot of variations in the second dataset and it is highly more likely to get the neural network confused due to similar variations in different class signal hence the approach we opted first is more suitable for this task. For a better view and idea below we are sharing a comparative picture of both the signals in discussed forms. It could be noted that in the above picture the variations are very large and it's hard to see a notable pattern and even if there exist similar patterns appear up in other class data signals making it hard to differentiate but with power spectrum as it could be noted the variations are in somewhat recognizable and differentiable way making it a much better choice for classification and training of neural network (Fig. 10).

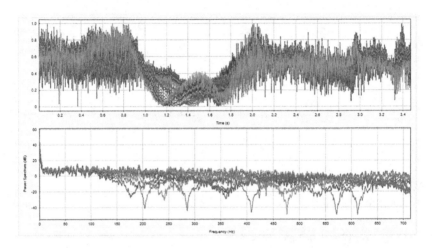

**Fig. 10.** Combined sub carriers with power spectrum (bend knee)

## 6 Conclusion

Our research suggests that human Activity Recognition with Wi-Fi is not only very efficient and cost effective method but it is the best solution considering the abundance of Wi-Fi signals around our environments. Our results successfully demonstrate that WiFi-CSI data along with phase information can successfully recognize human activities with an accuracy of more than 94% these could be very useful not only in activity recognition but in other sensing tasks, with a number of applications in healthcare, security, surveillance and much more. Our model can play an important role in many applications of Wi-Fi sensing. With our given dataset we achieved an accuracy of 94% but we want to try it on more and diverse dataset in different environments. For now our model proves to be very efficient compare to many other models we discussed in this thesis with respect to accuracy.

## References

1. Sigg, S., Blanke, U., Tröster, G.: The telepathic phone: frictionless activity recognition from WiFi-RSSI. In: IEEE PerCom (2014)
2. Wang, W., Liu, A.X., Shahzad, M., Ling, K., Lu, S.: Understanding and modeling of wifi signal based human activity recognition. In: ACM MobiCom (2015)

3. Wang, Y., Liu, J., Chen, Y., Gruteser, M., Yang, J., Liu, H.: E-eyes: device-free location-oriented activity identification using fine-grained wifi signatures. In: ACM MobiCom (2014)

4. Wengrowski, E.: A survey on device-free passive localization and gesture recognition via body wave reflections. Technical report (2014). https://pdfs.semanticscholar.org/24c6/5db8fd18a29037147ccabca09e2196ea87e5.pdf

5. Xiao, J., Zhou, Z., Yi, Y., Ni, L.M.: A survey on wireless indoor localization from the device perspective. ACM Comput. Surv. **49**(2), 31 p. (2016). https://doi.org/10.1145/2933232. Article 25

6. Yang, Z., Zhou, Z., Liu, Y.: From RSSI to CSI: indoor localization via channel response. ACM Comput. Surv. **46**(2), 32 p. (2013). https://doi.org/10.1145/2543581.2543592. Article 25

7. Wengrowski, E.: A survey on device-free passive localization and gesture recognition via bodywave reflections. Technical report (2014). https://pdfs.semanticscholar.org/24c6/5db8fd18a29037147ccabca09e2196ea87e5.pdf

8. Wang, Z., Guo, B., Yu, Z., Zhou, X.: Wi-Fi CSI based behavior recognition: from signals, actions to activities (2017). arXiv:1712.00146

9. Wu, D., Zhang, D., Xu, C., Wang, H., Li, X.: Device-free WiFi human sensing: from pattern-based to model-based approaches. IEEE Commun. Mag. **55**(10), 91–97 (2017). https://doi.org/10.1109/MCOM.2017.1700143

10. Yousefi, S., Narui, H., Dayal, S., Ermon, S., Valaee, S.: A survey on behavior recognition using wifi channel state information. IEEE Commun. Mag. **55**(10), 98–104 (2017). https://doi.org/10.1109/MCOM.2017.1700082

11. Zou, Y., Liu, W., Wu, K., Ni, L.M.: Wi-Fi radar: recognizing human behavior with commodity Wi-Fi. IEEE Commun. Mag. **55**(10), 105–111 (2017). https://doi.org/10.1109/MCOM.2017.1700170

12. Guo, X., Liu, B., Shi, C., Liu, H., Chen, Y., Chuah, M.C.: WiFi-enabled smart human dynamics monitoring. In: Proceedings of the 15th ACM Conference on Embedded Network Sensor Systems (SenSys 2017), Article 16, 13 p. (2017). https://doi.org/10.1145/3131672.3131692

13. Bagci, I.E., Roedig, U., Martinovic, I., Schulz, M., Hollick, M.: Using channel state information for tamper detection in the internet of things. In: Proceedings of the 31st Annual Computer Security Applications Conference (ACSAC 2015), pp. 131–140. ACM (2015). https://doi.org/10.1145/2818000.2818028

14. Gong, L., Yang, W., Man, D., Dong, G., Yu, M., Lv, J.: WiFi-based real-time calibration-free passive human motion detection. Sensors **15**(12), 32213–32229 (2015)

15. Lv, J., Man, D., Yang, W., Du, X., Yu, M.: Robust WLAN-based indoor intrusion detection using PHY layer information. IEEE Access **6**(99), 30117–30127 (2018). https://doi.org/10.1109/ACCESS.2017.2785444. ACM Computing Survey, vol. 1, no. 1, Article 1. Publication date: January 2019. WiFi Sensing with Channel State Information: A Survey 1:31

16. Li, H., Yang, W., Wang, J., Xu, Y., Huang, L.: WiFinger: talk to your smart devices with finger-grained gesture. In: Proceedings of the 2016 ACM International Joint Conference on Pervasive and Ubiquitous Computing (UbiComp 2016), pp. 250–261. https://doi.org/10.1145/2971648.2971738

17. Tan, S., Yang, J.: WiFinger: leveraging commodity WiFi for fine-grained finger gesture recognition. In: Proceedings of the 17th ACM International Symposium on Mobile Ad Hoc Networking and Computing (MobiHoc 2016), pp. 201–210 (2016). https://doi.org/10.1145/2942358.2942393

18. Yun, S., Chen, Y.-C., Qiu, L.: Turning a mobile device into a mouse in the air. In: Proceedings of the 13th Annual International Conference on Mobile Systems, Applications, and Services (MobiSys 2015), pp. 15–29. ACM (2015). https://doi.org/10.1145/2742647.2742662

19. Zhu, D., Pang, N., Li, G., Liu, S.: NotiFi: a ubiquitous wifi-based abnormal activity detection system. In: 2017 International Joint Conference on Neural Networks (IJCNN), pp. 1766–1773 (2017). https://doi.org/10.1109/IJCNN.2017.7966064

20. Zhou, Q., Wu, C., Xing, J., Li, J., Yang, Z., Yang, Q.: Wi-Dog: monitoring school violence with commodity WiFi devices. In: Ma, L., Khreishah, A., Zhang, Y., Yan, M. (eds.) WASA 2017. LNCS, vol. 10251, pp. 47–59. Springer, Cham (2017). https://doi.org/10.1007/978-3-319-60033-8_5

21. Wu, C., Yang, Z., Zhou, Z., Qian, K., Liu, Y., Liu, M.: PhaseU: real-time LOS identification with WiFi. In: 2015 IEEE Conference on Computer Communications (INFOCOM), pp. 2038–2046 (2015). https://doi.org/10.1109/INFOCOM.2015.7218588

22. Pu, Q., Gupta, S., Gollakota, S., Patel, S.: Whole-home gesture recognition using wireless signals. In: Proceedings of the 19th Annual International Conference on Mobile Computing and Networking (MobiCom 2013), pp. 27–38 (2013). https://doi.org/10.1145/2500423.2500436

23. Liu, J., Wang, L., Guo, L., Fang, J., Lu, B., Zhou, W.: Research on CSI-based human motion detection in complex scenarios. In: 2017 IEEE 19th International Conference on e-Health Networking, Applications and Services (Healthcom), pp. 1–6 (2017). https://doi.org/10.1109/HealthCom.2017.8210800

24. Wang, X., Yang, C., Mao, S.: PhaseBeat: exploiting CSI phase data for vital sign monitoring with commodity WiFi devices. In: 2017 IEEE 37th International Conference on Distributed Computing Systems (ICDCS), pp. 1230–1239 (2017). https://doi.org/10.1109/ICDCS.2017.206

25. Arshad, S., et al.: Wi-Chase: a WiFi based human activity recognition system for sensorless environments. In: 2017 IEEE 18th International Symposium on a World of Wireless, Mobile and Multimedia Networks (WoWMoM), pp. 1–6 (2017). https://doi.org/10.1109/WoWMoM.2017.797431541 (Sept. 2017), 27 pages. https://doi.org/10.1145/3130906

26. Liu, X., Cao, J., Tang, S., Wen, J., Guo, P.: Contactless respiration monitoring via off-the-shelf WiFi devices. IEEE Trans. Mob. Comput. 15(10), 2466–2479 (2016). https://doi.org/10.1109/TMC.2015.2504935

27. Palipana, S., Rojas, D., Agrawal, P., Pesch, D.: FallDeFi: ubiquitous fall detection using commodity wi-fi devices. Proc. ACM Interact. Mob. Wearable Ubiquit. Technol. 1(4), 25 p. (2018). Article 155

28. Fang, B., Lane, N.D., Zhang, M., Kawsar, F.: HeadScan: a wearable system for radio-based sensing of head and mouth-related activities. In: 2016 15th ACM/IEEE International Conference on Information Processing in Sensor Networks (IPSN), pp. 1–12 (2016). https://doi.org/10.1109/IPSN.2016.7460677

29. Zhou, Z., Yang, Z., Wu, C., Sun, W., Liu, Y.: LiFi: line-of-sight identification with WiFi. In: 2014 IEEE Conference on Computer Communications (INFOCOM), pp. 2688–2696 (2014). https://doi.org/10.1109/INFOCOM.2014.6848217

# Internet of Things

# Analysis and Implementation of Multidimensional Data Visualization Methods in Large-Scale Power Internet of Things

Zhoubin Liu[1], Zixiang Wang[1], Boyang Wei[2], and Xiaolu Yuan[3(✉)]

[1] State Grid Zhejiang Electric Power Research Institute, Hangzhou 310014, Zhejiang, China
{liuzhoubin,wangzixiang}@zj.sgcc.com.cn
[2] Georgetown University, Washington DC 20007, USA
bw558@georgetown.edu
[3] RUN Corporation, Wuxi 214000, Jiangsu, China
rubin0513@gmail.com

**Abstract.** In the large-scale power Internet of things, a large amount of data is generated due to its diversity. Data visualization technology is very important for people to capture the mathematical characteristics, rules and knowledge of data. People tend to get limited and less valuable information directly form large data when rely only on human-being's cognition. Therefore, people need new means and technologies to help display these data more intuitively and effectively. Data visualization mainly aims at conveying and communicating information clearly and effectively in term of graphical display, which can make data more human-readable and intuitive. Multidimensional data visualization refers to the methods to project multidimensional data to two-dimensional plane. It has important applications in exploratory data analysis, and verification of clustering or classification problems. This paper mainly studies the data visualization algorithm and technology in large-scale power Internet of things. Specifically, the traditional Radviz algorithm is selected and improved. The improved radviz-t algorithm is designed and implemented, and the unknown information of data transmission is obtained by analyzing its visualization effect. Finally, the methods used to study fault detection ability of radviz-t algorithm are discussed in detail.

**Keywords:** Data visualization · Radviz algorithm · Power Internet of Things

## 1 Introduction

Visualization is used to facilitate people's understanding of data in term of graphics display. With the advent of the computer age, the concept of visualization is more enriched, and the process of data visualization is more complex [1, 2]. Visualization technology is the most effective way to interpret data information by manipulating huge data processing in computer into a form that can be intuitively understood. The development of visualization has mainly gone through three stages: scientific

Q. Li et al. (Eds.): BROADNETS 2019, LNICST 303, pp. 135–143, 2019.
https://doi.org/10.1007/978-3-030-36442-7_9

visualization, information visualization and data visualization. Visualization technology is becoming more and more mature, as more tools emerged. In programming field, visualization software can directly translate algorithms into dynamic graphics. Visualization covers a broad range, not limited to data visuals, visuals from daily electrical power production such as electrical substation monitoring videos, furnace flame monitoring videos, and security monitoring videos can also be means of visualization.

The foundation of visualization builds upon data. Exploring the meaning of data, acquiring the information transmitted by data, and expressing data with appropriate visualization methods have also become the core of visualization inquiry [4]. Data are categorized and data processing methods are categorized. Exact same data could focus on dramatically different perspectives. Understanding the data itself, determining the center of gravity of the data, choosing the appropriate method of data transmission, and viewing the effect of data transmission are particularly important in each step of data visualization. Only by accomplishing the steps above could guarantee "deep meaning" contained in the data be displayed, and the data can be better managed, transmitted and saved, and more "secrets" about unknown data be explored.

This paper focuses on the research of visual clustering analysis (Radviz algorithm). Through the recognition of the principle and design implementation, Radviz algorithm still has room for improvement. Two classical large data visualization tools, D3 and R, are used to implement the algorithm, and some indexes are used to analyze the algorithm and explore the significance of improvement. Meanwhile, through the application of multi-dimensional data in the Internet of Things, the strength of Radviz and improved algorithm for fault detection are explored in depth.

## 2  Visualized Background and Radviz Algorithms

Data visualization can be divided into four stages: the era of poor computing, the era of first computing, the era of later computing and the era of big data at present. The era of poor computing refers to the era before the invention of calculators. The era of computing first refers to the era from calculator to modern computer. Post-computing era refers to the era of rapid development of the Internet. The era of big data is the era in which massive data are acquired and processed. In the earliest era of poor computing, people often used numbers and simple graphics to label data; first, in the computing age, formal formats gradually formed; then, in the computing age, Internet computers developed rapidly and charts became very popular; then, in the era of big data, the means of visualization are more varied [6].

With the popularity of computers and the development of the Internet, increasing attentions are paid to data visualization as the system has improved and technology is more cutting-edge. Visualization has become a complete subject and field of study. In different fields, it has related applications [7]. The annual IEEE VIS conference held in the UK is a summary and discussion conference on the field of visualization research. At present, the core technology of the conference is virtual reality technology, which has the characteristics of three-dimensional interaction and provides a good medium environment for the study of more complex structures.

Large and complex dimensions of data limit the amount of information people can directly acquire, thus application or technology to display multi-dimensional data visuals is needed [8]. Because the static state cannot meet the requirements of multi-dimensional data, we use dynamic visuals that can change or display processing through graphic changes. The core of such application is to reduce data dimensionality to dynamically display the information.

Radviz, a radial coordinate visualization method, is a multi-dimensional data visualization technology. It reduces the dimensionality of multi-dimensional data and projects it in low-dimensional space. Through certain characteristics, its clusters and analyses the similar data points. It has a circle arranged in the order of dimensions. Different rankings have a great impact on the final performance. The method also applies the idea of projecting the dimensions into several new dimensions, expanding the sorting space, to obtain better visual clustering effect.

The important concept is called dimension anchor, which fixes dimension to a circle. There visually serves as a spring-like pulling force between the anchors of each dimension. There is a special case when the resultant force becomes zero. At this time, the pulling force of non-neutral dimension is proportional to the value of dimension. When the values of dimensions between data points are similar, their positions in the circle are similar. By describing the clustering structure of data points, the visual clustering effect is displayed. The number of anchors on the ring is high scalable, so that the number of data points in the garden can be increased. The radial projection mechanism of data points using spring force can also maintain the original characteristics of multi-dimensional data in low-dimensional space. At the same time, it can keep the process simple and beautiful.

The prototype of Radviz algorithm was first proposed by Huffman et al. Radviz technology was proposed by Ankerst in 1996. Since then, various methods of multi-dimensional data analysis related to Radviz have been elvolved. Radviz is widely used in biology and medicine because of its outstanding ability to express genetic data. Radviz algorithm also has certain strength for fault detection, making it wide-applicable, such as sensor networks [10].

However, there are many constraints on the Radviz algorithm itself. Firstly, as the core content of the algorithm, dimension anchor cannot be operated at will, whether to add, delete, move or rearrange. This feature facilitates the exploration and application of Radviz in the field of interaction, but at the same time, the order of finding dimension anchors also increases the workload of users. It is a NP problem to find the best ranking method of dimension anchors. Some researchers point out that the quality of dimension anchors ranking can be guaranteed by heuristic strategy. They have creatively designed a vectorized Radviz called VRV. VRV is the number of dimensions of the extended algorithm, which makes the space of dimension anchor sorting become larger, and the way to obtain it become more flexible, from which we can find a more optimized result [11].

## 3 Radviz's Principle and Performance Evaluation

The dimensionality reduction of multidimensional data can be classified into linear and non-linear categories. Radviz is a non-linear visualization technology based on geometric technology. The dimension of multi-dimensional data is m. The results of these data are mapped to a circle through $R_m \sim R_2$. The number of springs obtained is n and the distances between springs are qual. Each dimension of data has its own characteristic variables, which are the observed values after mapping each point. Each spring is fixed at both ends, with one end at the observation point of the circumference and the other end $\{S_1 \sim S_m\}$ on the variables mapped by the data points. As shown in the figure (m = 6) (Fig. 1):

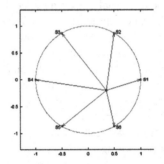

**Fig. 1.** Traditional Radviz model

The m-dimensional data mapping has a total of M data features, which are mapped to he circumference of two-dimensional space. The circumference is evenly divided into m parts. The data points are mapped into variables $\{S_1 \sim S_m\}$ on the circumference, that is, the equivalence points of the circumference. After the m-dimension mapping, every equal point will produce a spring. There are m springs in total, and the elastic coefficients of each spring are different. Assuming that the coordinates of equilibrium point are $(X_i, Y_i)$ in the circle, the point is the projection point of the m-dimensional data space point $\{S_1, S_2, S_3..., S_m\}$ mapped to the two-dimensional space, and the equilibrium position of the point is $Ai = \{X_i, Y_i\}\, T_j$. The tension relationship between the observation point of the equilibrium position and the point on the circumference of the other end of the spring is expressed as $u_{ij} = S_j - A_i$. Among them, $S_j = (d_{1j}, d_{2j})$ represents the $jth$ variable mapped to the circumference of two-dimensional space, and $A_i = (X_{1i}, Y_{2i})$.

According to the nature of spring, the spring force is determined by the stretching length and coefficient of spring. $K_{ij}$ represent stiffness constant of a scalar spring. When the spring is in a static state at the equilibrium point, the resultant force of the spring is 0.

$$\sum_{j=1}^m f_{ij} = 0, \text{ that is } \sum_{j=1}^m (S_j - A_i)k_{ij} = 0$$

According to the above formula, the position of equilibrium point can be obtained.

$$A_i = \frac{\sum_{j=1,2,\ldots,m} k_{ij} S_j}{\sum_{j=1,2,\ldots,m} k_{ij}}$$

Suppose there is a number w:

$$w_{ij} = \left(\sum_{j=1}^{m} k_{ij}\right)^{-1} k_{ij}$$

Then the formula of the value of the observation point is obtained.

$$A_i = \sum_{j=1}^{m} w_{ij} S_j$$

In order to avoid the possible negative value of the coordinate axis, the initial data should be processed. According to Radviz visual measurement model, the evaluation indexes of the algorithm are mainly divided into three aspects: data scale, visual effect and feature preservation.

## 4   Improved Radviz-t Algorithm and System Implementation

Due to the impact of the era of big data, the application of traditional Radviz algorithm has become extremely limited, which is unable to deal with big data [12]. For this reason, Radviz algorithm is improved.

To avoid the disadvantage of the original Radviz algorithm, which maps all data points directly to two-dimensional space, occlusion coincidence occurs between the points. Although the data dimension points correspond to the circumference, the arrangement order on the circumference also has a great impact on the final rendering. The criteria to select the RVM model indicators to improve the visualization effect mainly involve three indicators of the visualization effect: density, chaos and coverage. According to the previous model, in order to optimize the visualization effect mapped by Radviz, the chaos must be increased, the density must be reduced, and the coverage must be low. The new algorithm is called Radviz-t algorithm.

Indicator formulas for visual effects: Assuming the data set is $\{B_1, B_2, B_3, \ldots, B_m\}$ and the projection point is $A_i = (X_i, Y_i)$, the total coverage of $B_i$ for $B_j$ is as follows:

$$OCC(B) = \sum_{i=1}^{m} \left[ \sum_{j=1,i \neq j}^{m} OCC(B_i, B_j) \right] i \neq j = 1, 2, \ldots, m$$

$B_i$ density: _defaults to 2_:

$$CLO(B_i) = \begin{cases} 1, \exists A_j \in \{A_1, A_2, \ldots, A_m\} \text{ and } \sqrt{(A_{i1} - A_{j1})^2 + (A_{i2} - A_{j2})^2} < \eta, i \neq j \\ 0, \ other \end{cases}$$

Total density:

$$CLO(B) = \sum_{i=1}^{m} CLO(B_i)$$

Visual effects are expressed by ß, data coverage, density and confusion are expressed by c, g and l, respectively. According to the research, the overall relationship between them is as follows:

$$\beta = 1 - (0.228c + 0.403g + 0.369l)$$

From the formula above that there are three main indicators for visualization effect, namely, density, coverage and confusion. There is a certain relationship among these three indicators. When the density, coverage and chaos are given a certain set of values, the relationship is the most balanced and the visualization effect is the best. This balance is called "view effect value" and is expressed in Q.

$$Q_{c,g,l} = Q\{c, g, l\}$$

Given the multi-dimensional data is visualized, in order to present the best results, $m$ times of random experiments are carried out and Monte Carlo algorithm is used to find the optimal case. Suppose that in the first experiment, there is a mode $a_i$, which keeps the optimal projection performance within a certain number of times.

$$\exists a = \sum_{i=1}^{m} x_{max}$$

The final visual effect of the algorithm is determined by remove the visual effect value in this mode, in which the optimization is realized. Assuming that in AI mode, the probability of optimal projection effect is x, the visual effect value in this mode is $Q_{max}$, and the data coverage, data density and data confusion of the three index values are expressed by c\ g\ l, respectively, the formula is derived:

$$Q_{max} = \sum_{i=1}^{m} a_i\{x_{max}\}$$

From this, we can see that the visualization effect of this algorithm reaches the best state, which satisfies the modification of the visualization effect of Radviz algorithm. Because the improved algorithm needs to be designed for the optimal selection, the probability is selected to a certain extent, and the maximum probability is selected, from the probability distribution histogram. Through the visualization design of probability distribution histogram, we can directly get the function of the algorithm for data exploration and selection. According to the principle, probability distribution histogram is the necessary content for the improvement of Radviz-t algorithm. We use probability distribution histogram to calculate the index of visual view effect, and then select the index value that can present the optimal visualization effect for implementation.

The improved algorithm adds random experiment module and view effect evaluation module to the original algorithm. Random experiment module randomizes the order of array; first it randomizes the order of dimension ranking to ensure the randomness of the experiment, then randomly carries out multiple groups of experiments, and chooses the best visual effect map according to the view effect value. After starting the random experiment, in order to disrupt the dimension ranking, the original Radviz Implementation dimension ranking is chosen to numbering the variables of each dimension from 1 to m to form an array, and the random algorithm is used to disrupt the processing, so that the ranking of arrays is different in every second, and the visual effect of the output is also different to ensure the random experiment to follow. The visual effect changes when the order of dimension changes every second. To describe this effect, the view effect value is used. In the current second dimension order, the coordinates of the observation points in the current view are established and printed. Then, according to the formula principle, four modules are set up to calculate the data density, coverage, confusion, and the final calculation of the view effect value.

The data are processed by random module and then entered the optimal selection module. The module uses Monte Carlo algorithm to call the calculated visual effect value according to the characteristics of smaller view effect value and better visual effect. The idea of bubble sorting is used to ensure the minimum value of view effect, and the dimension corresponding to the visual effect map and the current visual effect map is sorted. In each second of random experiment, if a smaller view effect value appears, the original view effect value and its visual effect map and dimension ranking will be saved. Until the end of the experiment, if there is no smaller view effect value, it will output the view effect value saved in the previous second and the corresponding dimension ranking and visualization effect of the value, which is called optimal visualization, the whole process served as optimal selection.

For the start and end of the random experiment, the number and time of the experiment can be controlled manually by the button. In order to better explore data, when improving, the default choice of dimensions is changed to all choices, and the exploration of dimensions is abandoned. The advantage is faster exploration speed. The overall design page diagram is shown in the following Fig. 2:

Characteristic analysis:

(1) There is no need to select dimensions;
(2) Random ranking of dimensions is changed;
(3) The beginning and end of the experiment can be controlled at any time;
(4) Retaining color selectivity;
(5) Record the current optimal view effect value;
(6) Output the number of random experiments and the optimal visualization effect.

# RadViz-t Implementation

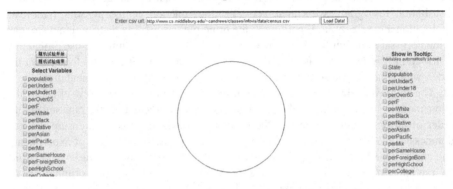

**Fig. 2.** The improved page is shown as Radviz-t Implementation

Radviz Implementation proposed that the smaller the view effect value, the better the view effect is. When number of random trials increases, the view effect value gradually decreases. Therefore, increasing the number of random experiments can improve the visualization effect. This means it can also increase the probability of finding the optimal visualization effect map of the whole data. Radviz algorithm itself has the ability of fault diagnosis. When identifying data points, it chooses to mark outliers with purple color. The ID of all fault points is 0. However, when dealing with high-dimensional data, the fault state of outliers cannot be observed intuitively, so Radviz-t is generated based on the original algorithm. Radviz-t Implementation based on Radviz-t increases the probability of choosing the optimal visualization effect map with the increase of random experiment numbers, by which the display of fault state becomes more and more obvious. The fault point begins to separate from the convergent normal area and produces an independent area, that is, the coverage between the fault point and the observation point in the normal area is less than 0. Finally, user can directly observe the purple point to get the information of the point in the current failure state (Fig. 3).

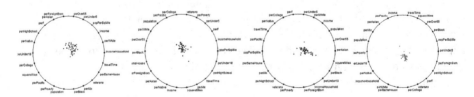

**Fig. 3.** Radviz-t dimension sorting change per second (uncolored and colored)

# 5   Conclusion

This paper first introduces the importance of data visualization in the power Internet of Things. Then, the principle and design of original Radviz algorithm and improved Radviz-t algorithm are discussed. An algorithm module is added to the original model to improve the original algorithm. When presenting the overall design, in addition to improving the original algorithm, the function of the original algorithm has been adjusted accordingly, and the Radviz-t implementation corresponding to the improved algorithm has been realized. This paper also gives a brief introduction to the new functions implemented by Radviz-t Implementation and analyses its properties. Finally, through the application of actual power data, the implementation of the algorithm and the improved algorithm is tested, and the correctness of the algorithm is verified.

**Acknowledgements.** This work was supported by State Grid Zhejiang Electric Power Corporation Technology Project (Grant No. 5211DS16001R).

# References

1. Sharko, J., Grinstein, G., Marx, K.A.: Vectorized Radviz and its application to multiple cluster datasets. IEEE Trans. Vis. Comput. Graph. **14**(6) (2008)
2. Novikova, E., Kotenko, I.: Visual analytics for detecting anomalous activity in mobile money transfer services. In: Teufel, S., Min, T.A., You, I., Weippl, E. (eds.) CD-ARES 2014. LNCS, vol. 8708, pp. 63–78. Springer, Cham (2014). https://doi.org/10.1007/978-3-319-10975-6_5
3. Lehmann, D.J., Theisel, H.: General projective maps for multidimensional data projection. In: Computer Graphics Forum, vol. 35, no. 2 (2016)
4. Daniels, K., Grinstein, G., Russell, A., Glidden, M.: Properties of normalized radial visualizations. Inform. Vis. **11**(4), 273–300 (2012)
5. Ravichandran, S., Chandrasekar, R.K., Uluagac, A.S., Beyah, R.: A simple visualization and programming framework for wireless sensor networks: PROVIZ. Ad Hoc Netw. **53**, 1–16 (2016)
6. Orsi, R.: Use of multiple cluster analysis methods to explore the validity of a community outcomes concept map. Eval. Program Plann. **60**, 277–283 (2016)
7. Keim, D.A., Mansmann, F., Schneidewind, J., Ziegler, H.: Challenges in visual data analysis. In: Tenth International Conference on Information Visualization, IV 2006, pp. 9–16. IEEE (2006)
8. Rao, T. R., Mitra, P., Bhatt, R., Goswami, A.: The big data system, components, tools, and technologies: a survey. Knowl. Inform. Syst. 1–81 (2018)
9. Li, P., Chen, Z., Yang, L.T., Zhang, Q., Deen, M.J.: Deep convolutional computation model for feature learning on big data in internet of things. IEEE Trans. Ind. Inform. **14**(2), 790–798 (2018)
10. Senaratne, H., et al.: Urban mobility analysis with mobile network data: a visual analytics approach. IEEE Trans. Intell. Transp. Syst. **19**(5), 1537–1546 (2018)
11. Zhou, F., Huang, W., Li, J., Huang, Y., Shi, Y., Zhao, Y.: Extending dimensions in radviz based on mean shift. In: 2015 IEEE Pacific Visualization Symposium (PacificVis), pp. 111–115. IEEE (2015)
12. Shi, L., Liao, Q., He, Y., Li, R., Striegel, A., Su, Z.: SAVE: Sensor anomaly visualization engine. In: 2011 IEEE Conference on Visual Analytics Science and Technology (VAST), pp. 201–210. IEEE (2011)

# Device-Free Gesture Recognition Using Time Series RFID Signals

Han Ding[1]($\boxtimes$), Lei Guo[2], Cui Zhao[1], Xiao Li[2], Wei Shi[1], and Jizhong Zhao[1]

[1] School of Computer Science and Technology, Xi'an Jiaotong University,
Xi'an, China
{dinghan,zjz}@xjtu.edu.cn, zhaocui@stu.xjtu.edu.cn, weishi0103@sina.com
[2] School of Software and Engineering, Xi'an Jiaotong University, Xi'an, China
{gl0103,lixiao0906}@stu.xjtu.edu.cn

**Abstract.** A wide range of applications can benefit from the human motion recognition techniques that utilize the fluctuation of time series wireless signals to infer human gestures. Among which, device-free gesture recognition becomes more attractive because it does not need human to carry or wear sensing devices. Existing device-free solutions, though yielding good performance, require heavy crafting on data preprocessing and feature extraction. In this paper, we propose RF-Mnet, a deep-learning based device-free gesture recognition framework, which explores the possibility of directly utilizing time series RFID tag signal to recognize static and dynamic gestures. We conduct extensive experiments in three different environments. The results demonstrate the superior effectiveness of the proposed RF-Mnet framework.

**Keywords:** Gesture recognition · RFID · Device free

## 1 Introduction

Human gesture recognition, which enables gesture-based Human-Computer Interaction (HCI), plays an important role in a wide range of applications, such as smart home, health care, especially the support for sign language (for example American Sign Language (ASL)) which can benefit the life of people who are deaf or hard of hearing. Traditional smart devices, say smart phones/pads, watches, and other wearable sensors, are widely used to recognize human gestures. However, such device-based approaches have the limitations that users need to carry or wear the devices, which are not convenient and feasible to real-world applications.

To overcome above limitations, a lot of effort have been made to explore device-free human gesture recognition techniques. Such methods usually require

This work is supported by NSFC Grants No. 61802299, 61772413, 61672424, Project funded by China Postdoctoral Science Foundation No. 2018M643663.

Q. Li et al. (Eds.): BROADNETS 2019, LNICST 303, pp. 144–155, 2019.
https://doi.org/10.1007/978-3-030-36442-7_10

users to perform hand or body motion and recognize these motions to be the HCI operations. One possible approach is to use imagery-based devices, *e.g.*, Kinect and LeapMotion [2,3], to track human motions in a natural way. However, these systems require line-of-sight and raise concerns on user privacy. Another approach, which is our focus in this paper, is to use radio frequency (RF) based sensing techniques of wireless devices. Generally, such wireless signal based solutions identify the gesture based on the rationale that there will be changes (*i.e.*, amplitude or phase) of time series RF signals when human performs specific gesture before the sensing system. By extracting certain features [7,8], the system can recognize the gesture through pre-defined similarity measures, such as Dynamic Time Warping (DTW) or distance-based classifiers. However these approaches all need heavy crafting on data preprocessing and feature engineering. And the performance is highly dependent on the selection of feature extraction algorithms.

In recent years, convolutional neural networks (CNN) have led to impressive success on objection recognition, audio classification, *etc.* [12,13]. A key superiority of CNN is its ability to automatically learn complex feature representations using its convolutional layers. Inspired by this, it is natural to ask a question: is it possible to automatically learn the feature representation from time series and realize the gesture recognition? In this paper, we design a multi-branch CNN network, namely RF-Mnet, for profiling time series and classifying gestures.

Specifically, with the rapid development of RFID techniques, RFID tag is no longer only an identity of certain product, it serves as wireless sensor for various applications [10,24,27]. The passive tag has the property of low cost and battery-free access. Inspired by this, we explore the possibility of recognizing human gestures via RFID systems. Our prototype of RF-Mnet is shown in Fig. 3(a). We deploy an array of 49 passive tags ($7 \times 7$ tag array) as the sensing plane. RF-Mnet do not need the user to carry any devices. The user just performs the gestures (including static gestures and dynamic gestures) in the air before the plane. The induced variations of RF signals (*i.e.*, RSS and phase) can be collected and correlated to the gestures. We implement a prototype of RF-Mnet using Commercial-off-the-shelf (COTS) RFID devices, and extensive experiments prove the effectiveness of our solution.

## 2    RF Signal Properties

Our RF-based motion estimation relies on transmitting the RF signal and receiving its replies (*i.e.*, reflections). In our system, we adopt the widely-used wireless technology: UHF RFID.

To capture the reflections from human, we deploy an RFID tag array. The human body is made up primarily of water from the RF point of view. Water is strongly absorptive around 900 MHz (with the dielectric constant of around 80 at room temperature), and radio waves are reflected by body parts. That is, the human body is both a reflector and an absorber of RF energy.

In ideal circumstances (like the *anechoic chamber*), the RF wave leaves from the reader antenna and strikes the tag. However, in real scenarios in which most

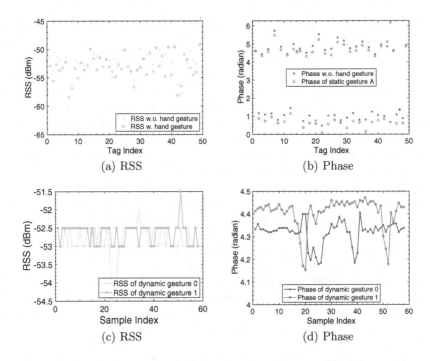

**Fig. 1.** RSS and phase distribution (a)(b) w./w.o. hand gesture. (c)(d) dynamic gesture 0 and 1.

RFID systems are used, the wave emitted from the reader antenna will interact with many other objects besides the tag itself. For example, in a typical office building, the integrated backscattered signal of a tag shall be the addition of the *direct* beam along the path between the reader and the tag, and those that are *reflected* (*i.e.*, from the floor, a distant wall, and nearby furniture). We can write the resulting signal as:

$$S_{total} = S_{dir} + S_{ref_0} \qquad (1)$$

Similarly, let us consider, when the human body (*i.e.*, wonderful reflector) exists, the newly resulting signal of a tag is the interaction of $S_{total}$ and the reflection ($S_{ref_h}$) from the human:

$$\hat{S}_{total} = S_{total} + S_{ref_h} \qquad (2)$$

Human reflected wave ($S_{ref_h}$) will add to (*i.e.*, in phase) or subtract from (*i.e.*, out of phase) $S_{total}$, causing the received signal vary. Specifically, this interaction happens even when the human body is far from the direct beam from a tag.

***Preliminary Experiment:*** Typical commercial off-the-shelf (COTS) RFID reader (*e.g.*, Imping R420) can report the channel parameters, *i.e.*, received signal strength (RSS) and phase, of each interrogated tag. To investigate the influence of human to backscattered tag signals, we conduct a group of proof-of-concept experiments. We first collect and calculate the average RSS and phase value of each tag in a $7 \times 7$ tag array in the static environment. Then the volunteer performs a static hand gesture (i.e., letter 'A' as shown in Fig. 3(b)) before the tag array. Figure 1(a) and (b) compare the RSS and phase of each tag with and without the hand gesture. We can observe that almost all tag signals vary (e.g., increase or decrease) with a human hand nearby the array, which demonstrates that the reflected wave from human body has essential importance. In addition, we also let the volunteer perform two dynamic gestures, e.g., moving the hand to write the number 0 and 1 in the air. Figure 1(c) and (d) illustrate the RSS and phase of tag #1, which tell that different gestures have different waveform profiles. In a nutshell, the received signals of tags contain the information of human reflections, inspiring us to infer the human gesture using RF time series signals.

## 3   Method

In this section, we define the RFID time series classification problem. Then we introduce the RF-Mnet framework.

### 3.1   Problem Definition

A time series is a sequence of data points with timestamps. In this paper, we use $7 \times 7$ tags in the implementation. We denote the time series of tag $i$ as $T_i = t_{i1}, t_{i2}, \ldots, t_{in}$, where $t_{ij}$ is the value at timestamp $j$ and there are $n$ timestamps. Thus, a real time series of RF-Mnet is

$$\mathcal{T} = \begin{bmatrix} T_1 \\ T_2 \\ \vdots \\ T_m \end{bmatrix} = \begin{bmatrix} t_{11}, t_{12}, \ldots, t_{1n} \\ t_{21}, t_{22}, \ldots, t_{2n} \\ \vdots, \vdots, \ldots, \vdots \\ t_{m1}, t_{m2}, \ldots, t_{mn} \end{bmatrix} \tag{3}$$

where $m$ is the number of tags ($m = 49$).

A labeled time series dataset is denoted as $D = (\mathcal{T}^k, y^k)_{k=1}^{N}$, which contains $N$ time series and $y^k$ is the associated label. $y^k$ is a real value and $y^k \in [1, C]$, where $C$ is the number of distinguishing labels (i.e., classes). Thus the problem we solve in this paper is to establish a model that can predict an unlabeled time series $\mathcal{T}^k$.

### 3.2   RF-Mnet Framework

The overall architecture is illustrated in Fig. 2. The RF-Mnet framework has three stages: multi-branch input stage, feature extraction stage, and gesture

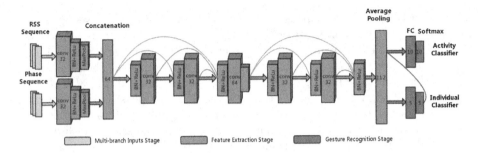

**Fig. 2.** Overall architecture of RF-Mnet.

recognition stage, in which the input is the time series data, and the output is its label.

***Multi-branch Inputs Stage:*** Different gestures might have different influences on RF channels, such as the change of energy attenuation and propagation path, which are reflected in RSS and phase variations of RF signals. Thus, we take RSS series and phase series as multi-branch inputs, which will provide us a bigger picture of the human motion. Each time series in the multi-branch has the same length.

***Feature Extraction Stage:*** We employ two steps for feature extraction: local convolution and global convolution. The multi-branch inputs $(\boldsymbol{X}^0)$ are first fed into the local convolution block which includes a batch normalization (BN) layer, a 2-D convolutional layer (Conv) and a rectified linear unit (ReLu), then the output feature map of local convolution is:

$$\boldsymbol{X} = BN(ReLu(\boldsymbol{W}\boldsymbol{X}^0 + \boldsymbol{b})) \tag{4}$$

where $\boldsymbol{W}$ represents the convolution filters and $\boldsymbol{b}$ is the bias. In particular, the filter size of the Conv layer is 3. The number of filters for both Conv layers is 32. The output of two local convolutional layers will capture a different dimension of features from original signals.

After extracting feature maps from each branch, we then concatenate all features and feed them into the global convolutional stage. Deep convolutional neural networks are proved to be capable of capturing the hierarchy of features [25], where the lower layers respond to primitive features, and the higher layers extract more complex feature informations. Such low and high-level features are both important and complementary in estimating human gestures, which motivates us to incorporate multi-layer information together. Hence, we employ a two-layer DenseNet [12] architecture in global convolutional layers, as shown in Fig. 2. Each layer is a stack of two dense blocks. Each dense block is constructed with two basic blocks. In each basic block, the input of $l$-th layer is the concatenation of the feature maps produced in all preceding layers $0, 1 \ldots, l-1$. If we denote the sequential operations of BN, ReLU, and Conv as $H$, the feature map of $l$-th basic block as $\boldsymbol{X}_l$, then $\boldsymbol{X}_l$ can be calculated is:

$$\boldsymbol{X}_l = H([\boldsymbol{X}_0, \boldsymbol{X}_1, \ldots, \boldsymbol{X}_{l-1}]) \tag{5}$$

Between two dense blocks, there is a transition layer which is composed of a BN layer, a ReLU layer, and an $1 \times 1$ Conv layer followed by a $3 \times 3$ average pooling layer. The transition layer reduces network parameters by converting the numbers of filter to half.

DenseNet achieves better performance by mitigating vanishing-gradient and enhancing the delivery of features. However, it is possible that the net will overfit the training data since time series data always lacks of complex structures compared with 3-D images that DenseNet is proved to be effective for object detection tasks. Hence, a global average pooling layer (kernel size of 3) is adopted to minimize overfitting and reduce the parameters.

*Gesture Recognition Stage:* During the gesture recognition stage, we propose two tasks which involve gesture classification and individual classification. The latter is able to authenticate the user identity when s/he conducts a gesture, which can be applied in applications which have privacy and security concerns. The input of individual classifier is the outputs of feature extraction stage $(\hat{\boldsymbol{X}}_i)$, we then feed the features into a fully connected layer followed by a Softmax activation function. The output of individual classifier can be denoted as:

$$\hat{\boldsymbol{Y}}_i = Softmax(\hat{\boldsymbol{W}}\hat{\boldsymbol{X}}_i + \hat{\boldsymbol{b}}) \tag{6}$$

where $\hat{\boldsymbol{W}}$ and $\hat{\boldsymbol{b}}$ are parameters. The output $\hat{\boldsymbol{Y}}_i$ is the predicted possibility of each label for $i$th input series data. Since gesture classifier contains individual related features, the input of gesture classifier is the concatenation of the output of feature extraction stage $(\hat{\boldsymbol{X}}_i)$ and the output of individual classifier $Y_i$. Then, the output of gesture classifier can be described as:

$$\hat{\boldsymbol{Z}}_i = Softmax(\hat{\boldsymbol{W}}[\hat{\boldsymbol{X}}_i, \hat{\boldsymbol{Y}}_i] + \hat{\boldsymbol{b}}) \tag{7}$$

In addition, to train the network, we use cross entropy function to calculate the loss between predictions and the real labels for gesture classification and individual classification. We define $L_y$ as loss function of gesture classifier, and $L_z$ as loss function of individual classifier.

$$\boldsymbol{L_y} = -\sum_{c=1}^{N} \boldsymbol{y}_c \log(\hat{\boldsymbol{y}}_c) \tag{8}$$

$$\boldsymbol{L_z} = -\sum_{c=1}^{M} \boldsymbol{y}_c \log(\hat{\boldsymbol{y}}_c) \tag{9}$$

where $M, N$ is the number of gesture and individual classes respectively. Then, the composite loss can be denoted as:

$$\boldsymbol{L} = \alpha * \boldsymbol{L_y} + \beta * \boldsymbol{L_z} \tag{10}$$

where $\alpha, \beta$ are hyper-parameters.

## 4    Implementation and Evaluation

### 4.1    Implementation

As shown in Fig. 3(a), RF-Mnet consists of an Impinj reader (Speedway R420) and a $7 \times 7$ Alien-9629 tag array. The whole system runs at the frequency of 922.375 MHz. In order to test the performance of RF-Mnet in multiple environments, we ask 5 volunteers to perform a large number of gestures in three indoor environments: Scene A (tag plane are placed in relatively open space), Scene B (tag plane are placed near walls and tables), Scene C (multiple objects are placed around the RFID tag plane). In each scene, there are people walking around during experiments occasionally.

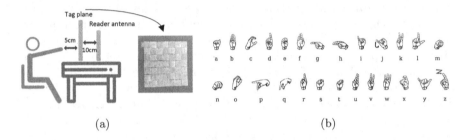

(a)                                                    (b)

**Fig. 3.** (a) Experimental setups of RF-Mnet. (b) Static gestures: the ASL fingerspelling alphabet [1].

**Dataset:** In each scene, we collect two kinds of gestures for each person, including static and dynamic gestures. As illustrated in Fig. 3(b), the static gestures are gestures corresponding to 26 English letters specified by American Sign Language (ASL). Dynamic gestures are a handwritten number (0–9,) in the air. The experimental dataset includes 39000 static gestures (5 users × 3 positions × 26 gestures × 100 instances), 15000 dynamic gestures (5 users × 3 positions × 10 gestures × 100 instances).

**Parameter Setting:** We implement our network with Pytorch 0.4.0. The Adam optimizer with $\beta_1 = 0.9$ and $\beta_2 = 0.999$ was used to train the network. The initial learning rate is 0.001, and it decreases by 50% every 3 epochs. We train the network 40 epochs in total.

### 4.2    Performance of Gesture Recognition

**Overall Accuracy:** We first test the overall gesture recognition accuracy of RF-Mnet. In this trail of experiments, 70% and 30% of the data collected in three environments are used for training and testing. The accuracy is shown in

Fig. 4. We can observe that the accuracy of static gesture recognition is higher (say, average accuracy 99.5%). The reason lies in that the human hand is moving during writing the dynamic numbers (*i.e.*, 0–9) in the air. The phase and amplitude of tag signals change dynamically due to the reflections of the moving hand. Thus, dynamic gestures are more susceptible to multipath interference, yielding lower accuracy. However, the average accuracy can still reach 92.5%. The results prove the effectiveness of our framework.

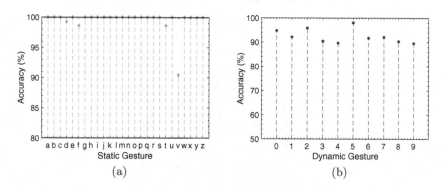

**Fig. 4.** Overall gesture recognition accuracy. (a) Static gestures. (b) Dynamic gestures.

**Impact of Human Diversity:** Next, we examine the usability of the system. We invite five volunteers to perform both static and dynamic gestures, 100 times for each. We balance the diversity of the volunteers in terms of their gender (3 males and 2 females), age (ranging from 22 to 28 years old), and other physical conditions (*e.g.*, 158–185 cm in height, 45–70 kg in weight, *etc.*). Note that when performing the dynamic gestures, they are naturally moving their fingers before the tag plane according to their writing habits. Figure 5 compares the average accuracy of static and dynamic gesture recognition. In particular, the accuracy of each volunteer for static gesture recognition is above 95%.

**Impact of Distance:** We then check the impact of distance between the user hand and the tag plane. We vary the distance from 10 cm to 30 cm. Other settings are consistent as default. We choose four representative static gestures and five dynamic gestures. Specifically, the static gestures involved in this experiment are *a*, *h*, *o*, *v*, and the dynamic gestures are 1, 3, 5, 7 and 9. The average recognition accuracy are plotted in Fig. 6. The average recognition accuracy over five distances is 99.8% for static gestures 90.1% for dynamic gestures. As expected, when enlarging the distance, the accuracy becomes lower. We envision the reason is that a larger distance to may weaken the direct interference from human hand, and involve extra influence from ambient factors, which introduces irregular variations to tag signals and induces lower accuracy.

**Impact of Environments:** Since the experiments involve three environments, we also compare the accuracy in different scenarios. The accuracy is shown in

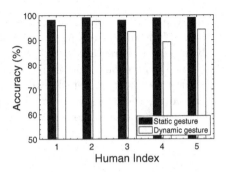

**Fig. 5.** Accuracy v.s. human diversity.

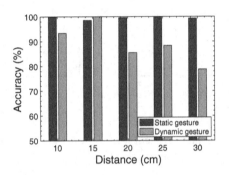

**Fig. 6.** Accuracy v.s. distance.

Fig. 7. The overall accuracy of three environments reaches 92.4%. In particular, the accuracy of Scene C is lowest because of its rich multipath property.

## 5    Related Work

Existing studies on the gesture/posture recognition can be classified into following categories:

**Computer Vision Based Gesture Recognition:** Vision based gesture recognition systems capture fine-grained gesture movements using cameras or light sensors [15,19,20,23,26]. For example, Okuli [26] adopts LED and light sensors to locate user's finger. In-air [19] uses built-in RGB camera of off-the-shelf mobile devices to recognize a wide range of gestures. However, these systems are susceptible to lighting condition changes, which are not suitable for applications where occlusions are everywhere. Most importantly, it will expose user privacy. In contrast, RF-Mnet has no requirement for line-of-sight and is lightweight and scalable.

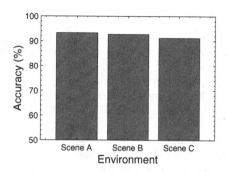

**Fig. 7.** Accuracy v.s. environment.

**RF-Based Gesture Recognition:** Prior works on RF-based gesture recognition span a wide spectrum, which can be divided into two categories: device-based and device-free approaches. Device-based methods use wearable devices, such as sensors or tags, for tracking finger or body movements [6,9,11,14,17,18, 22]. For example, RF-IDraw [22] leverages beamsteering capability of multiple antennas to detect the direction of the tagged finger and then track the tag by computing the location of intersected beams, which requires a large number of antennas that incurs heavy cost. FitCoach [9] perform fine-grained exercise recognition including exercise types, the number of sets and repetitions by using inertial sensors from wearable devices. These systems rely on wearable devices which are not friendly for users. Another appealing solution is device-free gesture recognition. There exists many systems that track the motion of object by receiving RF signals reflected by objects [4,5,21,24]. For example, WiZ [4] and WiTrack [5] combine frequency modulated continuous wave (FMCW) and multiple antennas technologies for motion tracking. However, both methods require dedicated devices that incur high host for daily gesture monitoring. Tadar [24] arranges a group of tags as an antenna array which receives reflections from surrounding objects and tracks human movements, while it cannot perform fine-grained gesture recognition. Rio [16] detects gestures by touching the surface of the tag with a finger which limits the position of the finger to some extent. In contrast, RF-Mnet is built on COTS RFID devices. Our system enables fine-grained gesture recognition without the need for users to carry the sensing devices.

## 6    Conclusions

In this paper, we propose an effective deep-learning based framework, namely RF-Mnet, to recognize device-free human gestures. RF-Mnet leverages a COTS RFID tag array as the sensing plane, which allows a user to perform in-air gestures, to capture the time series signals for gesture analysis. Extensive experiments from three environments demonstrate the effectiveness of proposed framework. In particular, RF-Mnet can achieve 99.5% and 92.3% average accuracy for static and dynamic gesture recognition respectively.

# References

1. American Sign Language (2019). https://www.nidcd.nih.gov/health/american-sign-language
2. Leap Motion (2017). https://www.vicon.com
3. X-Box Kinect (2017). https://www.xbox.com
4. Adib, F., Kabelac, Z., Katabi, D.: Multi-person motion tracking via RF body reflections (2014)
5. Adib, F., Kabelac, Z., Katabi, D., Miller, R.C.: 3D tracking via body radio reflections. In: Proceedings of USENIX NSDI (2014)
6. Bu, Y., et al.: RF-Dial: an RFID-based 2D human-computer interaction via tag array. In: Proceedings of IEEE INFOCOM (2018)
7. Ding, H., et al.: A platform for free-weight exercise monitoring with RFIDs. IEEE Trans. Mob. Comput. **16**(12), 3279–3293 (2017)
8. Ding, H., et al.: Close-proximity detection for hand approaching using backscatter communication. IEEE Trans. Mob. Comput. **18**(10), 2285–2297 (2019)
9. Guo, X., Liu, J., Chen, Y.: FitCoach: virtual fitness coach empowered by wearable mobile devices. In: Proceedings of IEEE INFOCOM (2017)
10. Han, J., et al.: CBID: a customer behavior identification system using passive tags. IEEE/ACM Trans. Network. **24**(5), 2885–2898 (2016)
11. Hao, T., Xing, G., Zhou, G.: RunBuddy: a smartphone system for running rhythm monitoring. In: Proceedings of ACM UbiComp (2015)
12. Huang, G., Liu, Z., Van Der Maaten, L., Weinberger, K.Q.: Densely connected convolutional networks. In: Proceedings of IEEE CVPR (2017)
13. Krizhevsky, A., Sutskever, I., Hinton, G.E.: ImageNet classification with deep convolutional neural networks. In: Proceedings of IEEE ICONIP (2012)
14. Mokaya, F., Lucas, R., Noh, H.Y., Zhang, P.: MyoVibe: vibration based wearable muscle activation detection in high mobility exercises. In: Proceedings of ACM UbiComp (2015)
15. Plotz, T., Chen, C., Hammerla, N.Y., Abowd, G.D.: Automatic synchronization of wearable sensors and video-cameras for ground truth annotation-a practical approach. In: Proceedings of IEEE ISWC (2012)
16. Pradhan, S., Chai, E., Sundaresan, K., Qiu, L., Khojastepour, M.A., Rangarajan, S.: RIO: a pervasive RFID-based touch gesture interface. In: Proceedings of ACM MobiCom (2017)
17. Ren, Y., Chen, Y., Chuah, M.C., Yang, J.: Smartphone based user verification leveraging gait recognition for mobile healthcare systems. In: Proceedings of IEEE SECON (2013)
18. Shangguan, L., Zhou, Z., Jamieson, K.: Enabling gesture-based interactions with objects. In: Proceedings of ACM MobiSys (2017)
19. Song, J., et al.: In-air gestures around unmodified mobile devices. In: Proceedings of ACM UIST (2014)
20. Taylor, J., et al.: Efficient and precise interactive hand tracking through joint, continuous optimization of pose and correspondences. ACM Trans. Graph. **35**(4), 143 (2016)
21. Wang, C., et al.: Multi-touch in the air: device-free finger tracking and gesture recognition via COTS RFID. In: Proceedings of IEEE INFOCOM (2018)
22. Wang, J., Vasisht, D., Katabi, D.: RF-IDraw: virtual touch screen in the air using RF signals. In: Proceedings of ACM SIGCOMM (2014)

23. Xiao, R., Harrison, C., Willis, K.D., Poupyrev, I., Hudson, S.E.: Lumitrack: low cost, high precision, high speed tracking with projected M-sequences. In: Proceedings of ACM UIST (2013)
24. Yang, L., Lin, Q., Li, X., Liu, T., Liu, Y.: See through walls with COTS RFID system! In: Proceedings of ACM MobiCom (2015)
25. Zeiler, M.D., Fergus, R.: Visualizing and understanding convolutional networks. In: Fleet, D., Pajdla, T., Schiele, B., Tuytelaars, T. (eds.) ECCV 2014. LNCS, vol. 8689, pp. 818–833. Springer, Cham (2014). https://doi.org/10.1007/978-3-319-10590-1_53
26. Zhang, C., Tabor, J., Zhang, J., Zhang, X.: Extending mobile interaction through near-field visible light sensing. In: Proceedings of ACM Mobicom (2015)
27. Zhao, C., et al.: RF-Mehndi: a fingertip profiled RF identifier. In: Proceedings of IEEE INFOCOM (2019)

# Dynamic IFFSM Modeling Using IFHMM-Based Bayesian Non-parametric Learning for Energy Disaggregation in Smart Solar Home System

Kalthoum Zaouali[1]([✉]), Mohamed Lassaad Ammari[2,3], Amine Chouaieb[3,4], and Ridha Bouallegue[4]

[1] Ecole Nationale d'Ingénieurs de Tunis, Innov'Com Laboratory, Université Tunis El Manar, Tunis, Tunisia
zaoualikalthoum@gmail.com
[2] Department of Electrical and Computer Engineering, Laval University, Quebec, Canada
mlammari@gel.ulaval.ca
[3] Chifco Company, Tunis, Tunisia
amine.chouaieb@chifco.com
[4] Sup'Com, Innov'Com Laboratory, Carthage University, Tunis, Tunisia
ridha.bouallegue@supcom.tn

**Abstract.** Recently, the analysis and recognition of each appliance's energy consumption are fundamental in smart homes and smart buildings systems. Our paper presents a novel Non-Intrusive Load Monitoring (NILM) recognition method based on Bayesian Non-Parametric (BNP) learning approach to solve the problem of energy disaggregation for smart Solar Home System (SHS). Several researches assumed that there is prior information about the household appliances in order to restrict those that do not hold the maximum expectation for inference. Therefore, to deal with the unknown number of electrical appliances in a SHS, we have adapted a dynamic Infinite Factorial Hidden Markov Model (IFHMM) -based Infinite Factorial Finite State Machine (IFFSM) to our NILM times-series modeling as an unsupervised BNP learning method. Our suggested method can grip with few or nappropriate learning data as well as to standardize electrical appliance modeling. Our proposed method outperforms FHMM-based FSM modeling results illustrated in literature.

**Keywords:** Bayesian Non-Parametric (BNP) · Energy disaggregation · Infinite Factorial Finite State Machine (IFFSM) · Infinite Factorial Hidden Markov Model (IFHMM) · Non-Intrusive Load Monitoring (NILM) · Solar Home System (SHS)

## 1 Introduction

Nowadays, many countries have widely adopted renewable energy sources to reduce the adverse effects of traditional fossil-fueled electricity generation on the

Q. Li et al. (Eds.): BROADNETS 2019, LNICST 303, pp. 156–175, 2019.
https://doi.org/10.1007/978-3-030-36442-7_11

environment and cut down on their bills. The social and technical developments, as well as economic profits can be ensured by energy saving and Internet of Things (IoT) technologies improvements especially for renewable energy.

In fact, understanding daily activities and energy consumption behaviors can contributes to energy bills reducing, a fine management and an optimal usage of electrical appliances in a Solar Home System (SHS). IoT and smart home technologies are able to guarantee an efficient management of energy and can satisfy the SHS implementation requirements. Furthermore, sensor-based technologies and data science permit the solar field automation for electricity saving and participate for green energy awards. Thanks to IoT technology, smart metering infrastructure, sensors and analytics tools, it is possible to connect solar panels into one system and manage and remote the smart home via mobile or Web applications. Our SHS described in Fig. 1 is extensively considered to lead to energy efficiency goals that require the ability to monitor electric energy appliances via smart meter technology. This smart device promotes aggregated electric energy signal acquisition and ingestion before analysis and data mining.

Currently, there has been a wide interest in the field of Non-Intrusive Load Monitoring (NILM), which involves several methods for monitoring electric appliances and providing appropriate notifications on usage patterns to home-owners. NILM approach gives rise to energy disaggregation by discovering the energy behavior of each household appliance using a single smart meter.

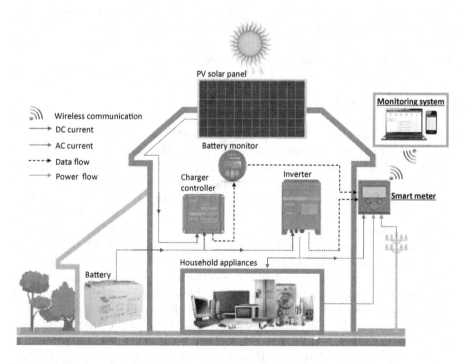

**Fig. 1.** Solar home system based on smart metering for an efficient green energy management

Since the number and the states of electrical appliances differ from one household to another, several researches assumed that there is prior information about the household appliances in order to restrict those that do not hold the maximum expectation for inference. One of the key areas of the NILM study is the use of unsupervised learning for energy disaggregation which can be an efficient solution to the problem of dependence on a set of training data. Applying unsupervised learning in NILM can reduce the IT complexity of smart homes solution deployment.

In this paper, we suggest a BNP model as an unsupervised learning method to deal with the problem of energy disaggregation with unknown number of electrical appliances in a SHS. We may apply this unsupervised model to solve the problem of a few or an inappropriate learning data as well as to standardize the electrical appliance modeling. To address the NILM challenge for energy disaggregation based on unsupervised learning, we have implemented our suggested Infinite Factorial Finite State Machine (IFFSM) building an Infinite Factorial Hidden Markov Model (IFHMM). This practical method contributes to inform homeowners about each household appliances energy consumption. The results of our proposed BNP model are compared with the Factorial Hidden Markov Model (FHMM) model results illustrated in [32].

The rest of this paper is organized as follows. Section 2 discuss some related work. Section 3 exposes our proposed method focusing on the adaptation of time series modeling by IFFSM to the energy disaggregation problem. The source separation technique using IFHMM-based BNP model, the inference algorithms as well as their different related process and models are detailed in followed subsections. In Sect. 4, we discuss the experimental results of the suggested method. The conclusions and perspective work are given in Sect. 5.

## 2    Related Work

According to Hart [20], indirect monitoring methods can measure the non-electrical characteristics, from which the power demand of each device is deduced. Device labeling is one of those indirect methods reported in the literature. The signals are detected by a central hub which estimates the energy consumption of each device. Nevertheless, this approach requires the characterization of each device as well as the installation of a central signature detector. This method is expensive and takes a long time to install. For this reason, researchers have been thinking of Wireless Sensor Network (WSN) to identify the power consumed by each device. These sensors make it possible to monitor the human behavior and to control the functioning of the device according to the indicators brought on the temperature, the luminosity, the movements of the inhabitants [28]. This approach also requires the installation of several sensors which reflects the same high cost constraint. In addition, the Conditional Demand Analysis (CDA) technique is emphasized since it does not require the installation of additional meters. However, the CDA requires a large participant base, in which each participant must complete a detailed questionnaire and unusual cases will not be examined. In order to overcome the high cost and intrusive complexity of the WSN and its installation, some emphasis has shifted to NILM methods which automatically deduce the energy signals from each device by a single sensor at a single measurement point.

Since then, several research work has been suggested with improvements to the initial design and different approaches. Taking advantage of NILM approach to infer individual appliance's consumption through a smart meter, several researches have demonstrated its utility for appliance classification, demand response, energy feedback, as well as activity recognition [45,49,54]. Other work that incorporates consumer behavior information has begun to explore learning algorithms that operate on a large number of samples obtained from many homes over extended periods of time. The real-time identification of faulty appliance behavior is a desired technique to save energy by analyzing wasted energy reasons as well as servicing appliance in failure [47]. Much work has been developed for use with NILM systems based on electrical signature detection and classification utilizing different machine learning as shown in Table 1 that summarizes the state of art of some NILM techniques and their characterizations.

As a non-parametric machine learning, the Hidden Markov Model (HMM) was used to describe and identify electrical uses by modeling the combination of stationary stochastic processes that translate the steady-state power level of the different combined waveforms into the total load curve [55]. As an extended version of HMM model, FHMM model was widely used for NILM approach. Kolter et al. [32] proposed a device-level power sub-meter learning model using the prediction maximization technique and then performed rough identification by Gibbs sampling. Such model does not tolerate non-stationary noise, and therefore requires training data to be collected on all devices in the home. For a Conditional FHMM (CFHMM), the state of each hidden variable is dependent on the state of each variable of all other Markov strings in the previous time interval [28].

The Finite State Machine (FSM) is an extension of the HMM model that has a set of input and output vectors, a transition matrix that considers the current input and previous state and returns the future state, as well as a send matrix that treats the input and the current state and returns the output. The FSM has been used as an unsupervised modelling framework for NILM in multiple research [24,39]. Under this model, the future hidden state depends only on a finite number of previous entries.

The source separation problem was closely related to independent component analysis (ICA) based on Discriminative Disaggregation Sparse Coding (DDSC) or Source-separation via Tensor and Matrix Factorizations (STMF) methods. Kolter [30] suggested a NILM based on unsupervised machine learning structured by the DDSC algorithm that builds a base matrix dictionary corresponding to a small subset of device types that contains devices with similar functionality. In order to overcome the limitations of the DDSC learning model which does not take into account the dependence between the devices, Figueiredo [16] presented a new model based on the separation of the global electrical signal sources by the

**Table 1.** State of art of different NILM based on non-parametric techniques via supervised and unsupervised machine learning

| Acronyms | References | NILM features | Network | Sampling frequency | Simpling interval | Event-based | Scal-ability | Disaggregation accuracy | Applications |
|---|---|---|---|---|---|---|---|---|---|
| DDSC | [30] | Source separation | PLC | 0.1 Hz | 15 min | No | No | + | The vast majority of appliances |
| STMF | [16,46] | Source separation | PLC | 0.1 Hz | 15 min | No | No | ++ | The vast majority of appliances |
| CO | [20] | Steady states | HAN | 0.1 KHz | 10 s–1 s | No | No | ++ | Appliances with power > 50 W |
| HMM | [29,33] | Steady states | HAN | 0.1 KHz | 10 s–1 s | No | No | +++ | Appliances with power > 50 W |
| NN | [7,49] | Steady states | HAN | 0.1 KHz | 10 s–1 s | – | Yes | +++ | Appliances with power > 50 W |
| DL | [26,27,34] | Steady states | HAN | 0.1 KHz | 10 s–1 s | – | Yes | +++++ | Appliances with power > 50 W |
| FHMM | [31] | Steady states | HAN | 0.1 KHz | 10 s–1 s | No | No | +++ | Appliances with power > 50 W |
| AFHMM | [31] | Steady states | HAN | 0.1 KHz | 10 s–1 s | No | No | +++ | Appliances with power > 50 W |
| DTW | [23,40] | Steady states Transient states | HAN | 0.1 KHz | 10 s–1 s | Yes | No | +++++ | Appliances with power > 50 W |
| CFHMM | [29] | Steady states Transient states | WSN | 0.1 KHz–1 KHz | 1 s–1 ms | No | No | ++++ | Appliances with power > 50 W |
| RF | [35,42] | Steady states Transient states | WSN | 1 KHz–20 KHz | 1 ms–50 µs | No | Yes | +++++ | Appliances with power > 50 W |
| DT | [19] | Steady states Transient states | WSN | 1 KHz–20 KHz | 1 ms–50 µs | No | Yes | +++++ | Appliances with power > 50 W |
| CDM | [4] | Steady states Transient states | WSN | 1 KHz–100 KHz | 1 ms–10 ms | Yes | Yes | ++++++ | Appliances with power > 50 W |
| SVM | [10] | Steady states Transient states | WSN | 1 KHz–100 KHz | 1 ms–10 ms | Yes | Yes | +++++++ | Appliances with power > 50 W |
| NFL | [11,12] | Steady states Transient states | WSN | 1 KHz–100 KHz | 1 ms–10 ms | Yes | Yes | +++++++ | Appliances with power > 50 W |
| MLC | [50] | Steady states Transient states | WSN | 1 KHz–100 KHz | 1 ms–10 ms | Yes | Yes | +++++++ | Appliances with power > 50 W |
| GSP | [21,56] | Steady states Transient states | WSN | 1 KHz–100 KHz | 1 ms–10 ms | Yes | Yes | +++++++ | Appliances with power > 50 W |
| EMI | [1] | | WSN | 100 KHz | 10 ms | Yes | No | ++++++++ | Some appliance models |
| GMM | [3] | Steady states Transient states | WSN | 1 KHz–100 KHz | 1 ms–10 ms | Yes | Yes | +++++++ | Appliances with power > 50 W |

factorization in non-negative tensors. This method has generated great interest in the concept of blind separation of sources. However, these tow methods assume a fixed and known number of latent sources.

Many other unsupervised machine learning have been introduced in recent research work to address NILM requirements [22,41]. NILM based on Bayesian Non-Parametrics (BNP) dynamical system is introduced to solve the problem of inferring the operational state of individual electrical appliances though aggregate measurements. An unbiased algorithm for neural variational identification and filtering was investigated in [36]. BNP techniques are also used in different fields, such as driving-styles analysis and recognition for smart transportation and vehicle calibration [52], intelligent dynamic spectrum access notification in cognitive radio environments [1,2,17,51,53], detection and estimation of sparse acoustic channels [9,25], video and image segmentation, reconstruction and recognition [13,15,43,44], data clustering and classification [6,14,37], etc. Numerous surveys have been carried out to describe the different learning models and the extractions features methods for NILM and home monitoring systems [3,8]. The advantages of the BNP models are that they can take a complex problem and create a model describing the appropriate solution. Due to its non-parametric nature with infinite memory chains length bonus, BNP models are able to deal with an unbounded number of states. Our proposed NILM method focuses on BNP models which are used most often in source separation problem.

## 3     Adaptation of Time Series Modeling by IFFSM to the Energy Disaggregation Problem

The approach adopted in our energy disaggregation problem is summarized in Fig. 2 which introduces the NILM approach based on an automatic unsupervised learning. In particular, the sources separation method allowing the processing of data from the smart meter, as well as the use of BNP model in conjunction with the adopted automatic learning method. Our proposed approach for energy disaggregation problem is described in detail in the following sections and subsections.

### 3.1     Source Separation Using IFHMM-based BNP Model

In our study, the FSM relies on a finite memory denoted **L** and a finite set denoted $\mathcal{X}$. Each new entry $x_t$ is able to modify the FSM state as well as the observed output. The future state and the output depend only on the current state and the input. The FSM can be modeled as single HMM where the vector containing the last $n$ inputs can characterize each state. Making an inference on this model has a $\mathcal{O}(T|\mathbf{S}|^{2L})$ complexity, but it can be reduced to $\mathcal{O}(T|\mathbf{S}|^{L+1})$ by exploiting the suggested IFFSM machine learning based on IFHMM model that requires approximation inference methods to avoid dependence on memory length $L$.

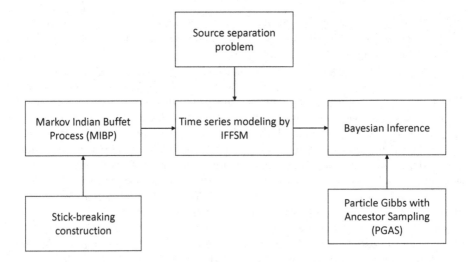

**Fig. 2.** Adopted NILM approach based on source separation

In the proposed IFHMM model, we assume binary input variables $x_{tn} = 0$ for $t \leq 0$. This model can be generalized to obtain additional properties concerning the fundamental structure of the model. We consider two methods for generalization that our inference algorithm can handle with little or no modifications. First, we assume that the input vectors $x_{tk}$ belong to a finite set $X$, so that we can give $|X|^L$ possible states in each parallel Markov chain. However, we can also consider that the set $X$ is a countable infinite set, which implies that the input vectors do not necessarily contain discrete values. The resulting model is no longer an FSM model, but an infinite factorial model in which the hidden variables affect present observations, as well as future observations [18].

To solve the energy disaggregation problem using the suggested unsupervised learning method, we assume that the observation $y_t$ represents a sample of the general smart meter signal, which depends on the signals generated by the different active electrical appliances. We suppose that there is theoretically an infinite number of sources that display the observed sequence $\{\mathbf{y}_t\}_{t=i}^T$, where $T$ is the number of time steps. Every source is modeled by a dynamic system model wherein the input symbol correspond to the $n'th$ source at time $t$ is denoted by $x_{tn} \in \mathcal{X}$ setting the first-order Markov chain, where $\mathcal{X}$ may be a discrete or continuous state space.

Each element of the auxiliary binary matrix $\mathbf{S}$, denoted by $s_{tn} \in \{0, 1\}$, reveals the source state (active or inactive) at time instant $t$, and can be expressed as:

$$x_{tn}|s_{tn} \sim \begin{cases} \delta_0(x_{tn}) & \text{if } s_{tn} = 0 \text{ ;} \\ \mathcal{U}(\mathcal{X}) & \text{if } s_{tn} = 1 \end{cases} \tag{1}$$

where $delta_0(.)$ designates the Dirac measure at $t = 0$, and $\mathcal{U}(\mathcal{X})$ is the uniform law on the set $\mathcal{X}$. The entries $x_{tn}$ are independent and identically distributed conditionally on the auxiliary variables $s_{tn}$.

Therefore, the proposed dynamic model depends basically on this conditional probability $p(x_{tn}|s_{tn}, x_{(t-1)n}) >$ with $s_{tn} = 0$ for $T \leq 0$. This transition model performs the active states $x_{tn}$ evolving over-time as dynamics of the global smart meter signal. During the multi-channel propagation of the individual signals, the electric waves can be reflected and that may cause reception delays. Considering this memory effect, we can note that the hidden state $x_{tn}$ affects the observations $y_t$ as well as the last future observations $y_{t+1} \ldots y_{t+L-1}$ if $s_{tn} \neq 0$, where $L$ denotes the last states of overall Markov chains.

The expression of $y_t$ likelihood is as follows:

$$p(y_t|\mathcal{X}, \mathbf{S}) = p(y_t|\{x_{tn}, s_{tn}, x_{(t-1)n}, s_{(t-1)n}, \cdots x_{(t-L+1)n}, s_{(t-L+1)n}\}_{n=1}^N) \quad (2)$$

The considered IFFSM graphic model is shown in Fig. 3.

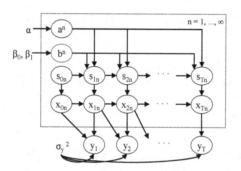

**Fig. 3.** IFFSM modeling with memory length $L = 2$.

**Markov Indian Buffet Process.** Among the BNP models, the Markov Indian Buffet Process (MIBP) is the fundamental building block of the Infinite Factorial Hidden Markov Model (IFHMM) as an IFFSM. To process an infinite number of sources, the binary matrix $\mathbf{S}$ is distributed as MIBP priors with distribution parameters $\alpha, \beta_0, \beta_1$ [48] as:

$$\mathbf{S} \sim lBP(\alpha, \beta_0, \beta_1) \quad (3)$$

This MIBP prior distribution ensures that, for any finite number of time instant $T$, only a finite number of Markov chain $N$ become active, while the rest of them remain in the zero state without affecting the observations.

We consider the total consumption of the household, the energy disaggregation concept is to predict the number of the active electrical appliances as well as their corresponding consumption. We treat a 24-hour segment for 6 different houses. Each electrical appliance holds 4 different states: one inactive state and 3 active states pointing different values of energy consumption.

A symmetric Dirichlet a prior distribution is placed on the vectors of the transition probabilities as $a_i^n$ $Dirichlet(1)$, where $a_{ij}^n = p(x_{tn} = j|s_{tn} = 1, x_{(t-1)n} = i, s_{(t-1)n}$. The energy consumption of the electrical appliance $n$ at the time instant $t$ is null when $x_{tn} = 0$ ($s_{tn} = 0$), and the total energy consumption is given by:

$$y_t = \sum_{n=1}^{N} P_{x_{tn}}^n + \varepsilon_t \tag{4}$$

where $\varepsilon_t \sim \mathcal{N}(0, \sigma_y^2 I)$ represents additive Gaussian noise with the hyperparameter $\sigma_y^2$ is the noise variance and $I$ is the identity matrix.

To get a maximum accuracy, we associate to each electrical appliance an estimated chain. Thus, the accuracy of each estimated appliance consumption evaluating the performance of our method is determined by:

$$accuracy = 1 - \frac{\sum_{t=1}^{T} \sum_{n=1}^{N} |x_t^{(n)} - \hat{x}_t^{(n)}|}{2 \sum_{t=1}^{T} \sum_{n=1}^{N} x_t^{(n)}} \tag{5}$$

where $\hat{x}_t^{(n)} = P_{x_{tn}}^n$ is the estimated consumption of each $n$ electrical appliance at time instant $t$. In the case where the inferred number of electrical appliances is less than the available electrical appliances number, the additional chains will be grouped in a category of unknown electrical appliances $x_t^{(unknown)}$.

**Stick-Breaking Construction.** The Stick-Breaking construction can be adapted to the MlBP to efficiency improve inference algorithms by introducing the transition probability from inactive state to active one denoted by $a^n$, as well as the self-transition probability of the $n$-th Markov chain active state denoted by $b^n$. These two hidden variables $a^n$ and $b^n$ are described as follows:

$$a^n = p(s_{tn} = 0|s_{(t-1)n} = 0) \tag{6}$$

$$b^n = p(s_{tn} = 0|s_{(t-1)n} = 1) \tag{7}$$

The matrix of the transition probabilities of the $n$-th Markov chain can be written as follows:

$$\mathbf{A}^n = \begin{pmatrix} 1 - a^n & a^n \\ 1 - b^n & b^n \end{pmatrix} \tag{8}$$

The distribution on the variables $a^n$ is given by:

$$a^1 \sim Beta(\alpha, 1) \tag{9}$$

and the distribution probability on the variables $a^n$ can be given by:

$$p(a^{(n)}|a^{(n-1)}) \propto (a^{(n-1)})^{(-\alpha)} (a^{(n)})^{(\alpha-1)} I(0 \le a^{(n)} \le a^{(n-1)}) \tag{10}$$

with $I(.)$ characterizes the indicator function which is worth 1 if its argument is true and 0 if its argument is false, $\alpha$ is the concentration parameter that controls the number of active Markov chains. Independently of $n$, the MIBP prior over variables $b^n$, distributed according to a Beta process, is defined by:

$$b^n \sim Beta(\beta_0, \beta_1) \tag{11}$$

## 3.2  Inference Algorithms

Each Bayesian model looks for a posterior inference calculated according to the posterior distribution of hidden variables. Several time series BNP models take a proximate inference algorithm if the posterior distribution cannot be obtained directly. Such inference algorithm can be based on Markov Chain Monte Carlo (MCMC) methods. In particular, we have used a Gibbs Sampling inference algorithm that combines MCMC and Sequential Monte Carlo (SMC) standard tools. The suggested IFHMM model is incorporated with the Gibbs sampling-based inference algorithm to treat the parallel chains number and transition variables. The Inference algorithm based on MIBP prior starts with introducing new inactive chains $N_{new}$ utilizing a slice sampling method and an auxiliary slice variable to provide a finite factorial model. Thus, the number of parallel chains is increased from $N_+$ to $N^{\ddagger} = N_+ + N_{news}$ and consequently the $N_+$ chains number cannot be updated. The first sampled auxiliary slice variable $\mathcal{V}$ is distributed as:

$$\mathcal{V}|S, \{a^n\} \sim Uniform(0, a_{minimale}) \tag{12}$$

where $a_{minimal} = \min_{n:\exists t, s_{tn} \neq 0} a^n$ can be a Beta distribution.

The next new variables $a^n$ make the following sampling iterations until $a^n < \mathcal{V}$:

$$p(a^n|a^n) \propto \exp\left(\beta_0 \sum_{t-1}^{T} \frac{1}{t}(1-a^n)^t\right) \times (a^n)^{-\beta_0-1}(1-a^n)^T, (0 \leq a^n \leq a^{n-1}) \tag{13}$$

Moreover, the next step of the inference algorithm is focused on sampling the states $s_{tn}$ and the input symbols $x_{tn}$ of all chains of the IFFSM model. This compacted sampling eliminates the chains that remain inactive throughout the observation time, which allows the update of $N_+$ chains number. We put forward the use of Particle Gibbs algorithm for inference in non-Markovian latent variable models.

Despite the efficiency and the simplicity of the Gibbs sampling model of each element $x_{tn}|s_{tn}$, this technique cannot provide good mixing properties owning to the high coupling of successive steps. The Gibbs sampling model can utilize the Forward Filtering Backward Sampling (FFBS) method to treat the successive sampling of chains according a complexity of $\mathcal{O}(TN^{\ddagger}|\mathcal{X}|^{L+1})$ to our suggested IFFSM model while $L > 1$. However, the exponential dependence on $L$ can prevent convergence of the FFBS calculation. Thus, to deal with this problem, a Particle Gibbs with Ancestor Sampling (PGAS) algorithm is combined with the inference algorithm to jointly sample the matrices $\mathcal{X}$ and $\mathbf{S}$. If $P$ particles are used for the PGAS kernel, the complexity of our adopted algorithm is about $\mathcal{O}(PTN^{\ddagger}L^2)$.

Finally, inference algorithm will sample the global variables joining the transition and the emission probabilities depending on their posterior distribution so as to evaluate the likelihood $p(y_t|\mathcal{X}, \mathbf{S})$.

This final step of the inference algorithm is focused on the sampling of the global variables of the model from their full conditional distributions under the Stick-Breaking construction is expressed as follows:

$$p(a^n|\mathbf{S}) = Beta(1 + tr_{00}^k, tr_{01}^k) \qquad (14)$$

with $tr$ the number of transitions between states in the $N$ column of $\mathbf{S}$.

With regard to the transition probabilities of the active state to the inactive state $b^n$, we have:

$$p(b^n|\mathbf{S}) = Beta(\beta_1 + tr_{00}^k, \beta_2 + tr_{01}^n) \qquad (15)$$

We also define the extended matrix $\mathcal{X}^{extended}$ of size $TxLN_+$ with:

$$\mathcal{X}^{(n)} = \begin{pmatrix} x_{1n} & 0 & \cdots & 0 \\ x_{2n} & x_{1n} & \cdots & 0 \\ \vdots & \vdots & \ddots & \vdots \\ x_{Tn} & x_{(T-1)n} & \cdots & x_{(T-L+1)n} \end{pmatrix} \qquad (16)$$

**Particle Gibbs with Ancestor Sampling.** Compared to the FFBS method, the PGAS algorithm offers several benefits in addition to its non-exponential complexity. Notably, it can be applied independently of $\mathbf{X}$ proprieties as finite or infinite set. Furthermore, its inference properties are better than those of FFBS algorithm which cannot contribute to the $N^{\ddagger} - 1$ propriety of the observation chains. For energy disaggregation problems, the FFBS is often limited to later local modes in which several Markov chains correspond to a single hidden source. However, for each simultaneous instant $t$, the PGAS algorithm can treat and sample all chains in parallel. This characterization does not exist in FFBS and FHMM models [5]. The integration of such algorithm involves the elimination of certain $a^n$ and $b^n$ variable as well as the updated chains $N_+$.

In order to better adapt the PGAS algorithm to our energy disaggregation problem based on source separation approach, we can refer to [38] describing the appropriate steps as well as the theoretical justification of the PGAS algorithm.

We assume that the suggested PGAS model is composed of a set of $P$ particles which represent the hidden states $\{x_{tn}\}_{n=1}^{N^{\ddagger}}$ at an instant $t$. Let defining the $i$-th particle state at time $t$ by $x_t^i$ of length $N^{\ddagger}$ as well as its ancestor indexes $a_t^i \in 1, \ldots, P$. Given $\mathbf{x}_{1:t}^i$ as the ancestral trajectory of the particle $\mathbf{x}_t^i$, the recursive form of the particle trajectory is described as follows:

$$\mathbf{x}_{1:t}^i = \left(\mathbf{x}_{1:t-1}^{a_t^i}, \mathbf{x}_t^i\right) \tag{17}$$

A fixed reference particle denotes by $\mathbf{x}_t^*$ is generated by previous iteration outputs is required for the PGAS algorithm extension in order to introduce novel inactive chains. Thus, the corresponding ancestor indexes $a_t^P$ of the fixed particle $\mathbf{x}_t^P$ are variables and randomly picked. Hence, a distribution form $q_t\left(\mathbf{x}_t|\mathbf{x}_{1:t-1}^{a_t}\right)$ is required to specify the propagate way of the particles, so that, we assume the following expression:

$$q_t\left(\mathbf{x}_t|\mathbf{x}_{1:t-1}^{a_t}\right) = p\left(\mathbf{x}_t|\mathbf{x}_{t-1}^{a_t}\right) \tag{18}$$

$$= \prod_{n=1}^{N^{\ddagger}} p\left(x_{tn}|s_{tn}, x_{(t-1)n}^{a_t}\right) p\left(s_{tn}|s_{(t-1)n}^{a_t}\right) \tag{19}$$

For each instant $t$, the PGAS inference algorithm samples the particles at instant $t-1$ taking into account their weight specification $w_{t-1}^i$ and their distributed propagation $q_t\left(\mathbf{x}_t|\mathbf{x}_{t-1}^{a_t}\right)$.

The following equations develop the weights expression as well as the ancestor weights:

$$W_t(\mathbf{x}_{1:t}) = \frac{p\left(\mathbf{x}_{1:t}|\mathbf{y}_{1:t}\right)}{p\left(\mathbf{x}_{1:t-1}|\mathbf{y}_{1:t-1}\right) q_t\left(\mathbf{x}_t|\mathbf{x}_{1:t-1}\right)} \tag{20}$$

$$\propto \frac{p\left(\mathbf{y}_{1:t}|\mathbf{x}_{1:t}\right) p\left(\mathbf{x}_{1:t}\right)}{p\left(\mathbf{y}_{1:t-1}|\mathbf{x}_{1:t-1}\right) p\left(\mathbf{x}_{1:t-1}\right) p\left(\mathbf{x}_t|\mathbf{x}_{t-1}\right)} \tag{21}$$

$$\propto p\left(\mathbf{y}_t|\mathbf{x}_{t-L+1:t}\right) \tag{22}$$

with $\mathbf{y}_{\tau_1:\tau_2}$ denotes the set of observations $\{\mathbf{y}_t\}_{t=\tau_1}^{\tau_2}$. The $W_t(\mathbf{x}_{1:t})$ equation involves that obtained weights $w_{t-1}^i$ depending essentially on the likelihood evaluation at time $t$. The estimated ancestor weights $\tilde{w}_{t-1|T}^i$ of the reference particle are given by:

$$\tilde{w}_{t-1|T}^i = w_{t-1}^i \frac{p\left(\mathbf{x}_{1:t-1}^i, \mathbf{x}_{t:T}^*|\mathbf{y}_{1:T}\right)}{p\left(\mathbf{x}_{1:t}^i|\mathbf{y}_{1:t-1}\right)} \tag{23}$$

$$\propto w_{t-1}^i \frac{p\left(\mathbf{y}_{1:T}|\mathbf{x}_{1:t}^i, \mathbf{x}_{t:T}^*\right) p\left(\mathbf{x}_{1:t-1}^i, \mathbf{x}_{t:T}^*\right)}{p\left(\mathbf{y}_{1:t-1}|\mathbf{x}_{1:t-1}^i\right) p\left(\mathbf{x}_{1:t-1}^i\right)} \tag{24}$$

$$\propto w_{t-1}^i p\left(\mathbf{x}_t^*|\mathbf{x}_{t-1}^i\right) \prod_{T=t}^{t+L-2} p\left(\mathbf{y}_T|\mathbf{x}_{1:t}^i, \mathbf{x}_{t:T}^*\right) \tag{25}$$

Calculating the weights $\tilde{w}_{t-1|T}^i$ for $L > 1$ and $i = l, \ldots, P$ increasing the model complexity by $\mathcal{O}(PN^{\ddagger}L^2)$.

To deal with the last equation, we have to take only the IFHMM factors which depending on the particles index $i$, without memory ($L = 1$) and with a transition probability $p(\mathbf{x}_t|\mathbf{x}_{t-1})$ factored in parallel through the IFHMM model. The weights expression can be written as follows:

$$\tilde{w}_{t-1|T}^i \propto w_{t-1}^i p\left(\mathbf{x}_t^*|\mathbf{x}_{t-1}^i\right) \tag{26}$$

## 4  Simulations and Results

We have implemented the proposed IFFSM based on IFHMM model with the following parameters to obtain the desired results:

- Length of the observation sequence $T$.
- Number of possible states $Q = 4$
- Iteration number $i = 10000$
- Hyper-parameters of our IFFSM model:
    - $\lambda_1 = 15$ and $\lambda_2 = 10$ are the parameters of the Gaussian distribution on consumption, i.e. $P_q^n \sim \mathcal{N}(15.10)$
    - $\gamma_1 = 0$ and $\gamma_2 = 0.5$ are the hyper-parameters of the Gaussian distribution on the white noise denoted as $\varepsilon_t \sim \mathcal{N}(0, 0.5)$.
    - $\beta_0 = 1$ is the hyper-parameter of the distribution on the self-transition of the inactive state $a_i^n \sim Dirichlet(1)$.
    - $\beta_1 = 0.1$ and $\beta_2 = 2$ are the Beta distribution hyper-parameters involving the transition from an active state to an inactive state $b^n$, i.e. $b^n \sim Beta(0.1, 2)$.
    - MIBP distribution on the activation binary matrix $\mathbf{S}$ is defined as $\mathbf{S} \sim MIBP(l, 0.1, 2)$.

The visualization of the general load curve which represents the total consumption of the electrical appliances is given by the Fig. 4.

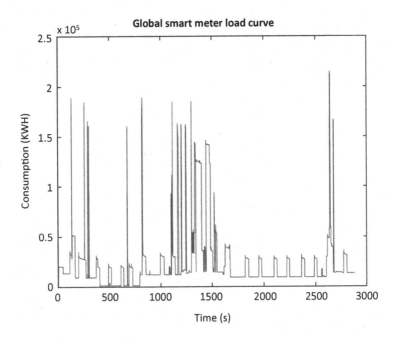

**Fig. 4.** The global smart meter load curve

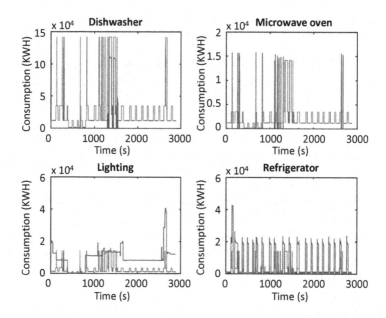

**Fig. 5.** The estimated load curve of each electrical appliance

After implementing our suggested IFFSM model adapted to the energy disaggregation problem, we obtain the load curves for a specific electrical appliance. Each load curve represents the evolution of the consumption of each electrical appliance over time. A visualization on Matlab allows to obtain the results presented by the Fig. 5.

## 5    Performance Evaluation and Discussion

To evaluate the performance of our model, we apply the formula of the equation:

$$accuracy = 1 - \frac{\sum_{t=1}^{T} \sum_{n=1}^{N} |x_t^n - \hat{x}_t^n|}{2 \sum_{t=1}^{T} \sum_{n=1}^{N} x_t^n} \qquad (27)$$

This formula allows to test the performance of our IFFSM model to estimate the individual energy consumption of each appliance installed in smart homes. The performance of our model is calculated for the 4 appliances that consume the most energy. Table 2 shows the different performances of our applied model.

We have to test the performance of our IFFSM model and compare the accuracy with an existing Markov model. Thus, we have used the Reference Energy Disaggregation Dataset (REDD) in order to be able to evaluate our model. We have compared the obtained results with the FHMM model results implemented in [32]. Table 3 visualizes the performance of each model for 6 smart homes described by the REDD database.

**Table 2.** Performances evaluation: the accuracy of the estimated energy consumption for 4 household appliances situated in a smart home environment

|  | Smart home electrical appliances | | | |
|---|---|---|---|---|
|  | Dishwasher | Microwave oven | Refrigerator | Lighting |
| Accuracy (%) | 66.3 | 67.2 | 70.1 | 61.4 |
| Average | 66.75 | | | |

**Table 3.** Comparison of our IFFSM model performances to the FHMM model

|  | Accuracy (%) | |
|---|---|---|
|  | Proposed IFFSM model (IFHMM) | Compared model (FHMM) |
| Smart home 1 | 66.7 | 71.5 |
| Smart home 2 | 66.3 | 59.6 |
| Smart home 3 | 66.2 | 59.6 |
| Smart home 4 | 66.7 | 69.0 |
| Smart home 5 | 66.4 | 62.9 |
| Smart home 6 | 66.7 | 54.4 |
| Average | 66.5 | 64.5 |

We demonstrate that the proposed model slightly exceeds the FHMM model in the energy disaggregation problem. Thus, the accuracy of the model does not change substantially in different houses, this is because the model is completely unsupervised and does not require learning data but can determine consumption from prior observations and distributions.

Thus, the proposed model is more practical because it is not realistic to believe that learning data can be obtained from all the homes where we will implement our solution and we should not expect to have a model for each electrical appliance installed in each home.

## 6   Conclusion

In this paper we have proposed a NILM approach based on an unsupervised learning to solve the energy disaggregation problem. In fact, we have adapted an IFFSM- based time series modeling to the source separation problem with an unknown number of sources. We have developed a BNP model which aims to disaggregate the general load curve at the end of a smart meter. We have implemented the suggested IFFSM model building an IFHMM model and we have obtained satisfactory results compared to the FHMM model which requires learning data and test data. The proposed model builds on the MIBP to recognize an infinite number of hidden Markov chains with either discrete or continuous states. We have put forward a PGAS algorithm for posterior inference to deal with the FFBS complexity. We have successfully implemented our proposed model and we have visualized the experimental results using the REDD data-set.

However, our proposed approach is inappropriate if we want to use it on a large number of SHS deployed in smart cities or smart grids. Being a blind method, it is difficult to recognize each estimated chain of a specific device with the lack of sufficient individual information from each appliance for every SHS. Being a blind method, it is difficult to recognize each estimated chain of a specific device for lack of sufficient individual information from each appliance for every SHS. To deal will this limitation, we can improve our suggested IFHMM model by dividing the chains between the houses in a hierarchical way while calculating each activation function individually for every house and inferring the common features between different houses. In further research, we can improve our inference algorithm scalability to contribute for both a larger number of appliances and larger observation sequences. Furthermore, we can transform our model from a time-invariant load model using an offline static database to an on-line unsupervised model for autonomous household database construction in order to recognize the real-time behavior of the power consumption [22].

**Abbreviations**

The following abbreviations are used in this manuscript:

| | |
|---|---|
| AFHMM | Additive Factorial Hidden Markov Models |
| ANN | Artificial Neural Network |
| BNP | Bayesian Non-Parametric |
| CDA | Conditional Demand Analysis |
| CFHMM | Conditional Factorial Hidden Markov Models |
| CO | Combinatorial Optimization |
| DDSC | Discriminative Disaggregation Sparse Coding |
| DL | Deep Learning |
| DT | Decision Tree |
| DTW | Dynamic Time Warping |
| CDM | Committee Decision Mechanism |
| EMI | Electromagnetic Interference |
| FFBS | Forward Filtering Backward Sampling |
| FHMM | Factorial Hidden Markov Models |
| FSM | Finite State Machine |
| GMM | Gaussian Mixture Model |
| GSP | Graph Signal Processing techniques |
| HMM | Hidden Markov Model |
| IFFSM | Infinite Factorial Finite State Machine |
| IFHMM | Infinite Factorial Hidden Markov Model |
| IoT | Internet of Things |
| KNN | k-nearest neighbor |
| MCMC | Markov Chain Monte Carlo |
| MIBP | Markov Indian Buffet Process |
| MLC | Multi-Label Classification |
| NFL | Neuro-Fuzzy Logic algorithm |

| NILM | Non Intrusive Load Monitoring |
| PGAS | Particle Gibbs with Ancestor Sampling |
| REDD | Reference Energy Disaggregation Dataset |
| RF | Random Forest |
| SHS | Solar Home System |
| SMC | Sequential Monte Carlo |
| STMF | Source-separation via Tensor and Matrix Factorizations |
| SVM | Support Vector Machine |
| WSN | Wireless Sensor Network |

# References

1. Aboulian, A., Donnal, J.S., Leeb, S.B.: Autonomous calibration of non-contact power monitors. IEEE Sens. J. **18**(13), 5376–5385 (2018)
2. Ahmed, M.E., Kim, D.I., Kim, J.Y., Shin, Y.: Energy-arrival-aware detection threshold in wireless-powered cognitive radio networks. IEEE Trans. Veh. Technol. **66**(10), 9201–9213 (2017)
3. Alcalá, J., Ureña, J., Hernández, Á., Gualda, D.: Event-based energy disaggregation algorithm for activity monitoring from a single-point sensor. IEEE Trans. Instrum. Meas. **66**(10), 2615–2626 (2017)
4. Andrean, V., Zhao, X., Teshome, D.F., Huang, T., Lian, K.: A hybrid method of cascade-filtering and committee decision mechanism for non-intrusive load monitoring. IEEE Access **6**, 41212–41223 (2018)
5. Aueb, M.T.R., Yau, C.: Hamming ball auxiliary sampling for factorial hidden Markov models. In: Advances in Neural Information Processing Systems, pp. 2960–2968 (2014)
6. Campbell, T., Kulis, B., How, J.: Dynamic clustering algorithms via small-variance analysis of Markov chain mixture models. IEEE Trans. Pattern Anal. Mach. Intell. **41**(6), 1338–1352 (2019)
7. Chang, H., Lee, M., Lee, W., Chien, C., Chen, N.: Feature extraction-based hellinger distance algorithm for nonintrusive aging load identification in residential buildings. IEEE Trans. Ind. Appl. **52**(3), 2031–2039 (2016)
8. Dan, W., Li, H.X., Ce, Y.S.: Review of non-intrusive load appliance monitoring. In: 2018 Proceedings of the IEEE 3rd Advanced Information Technology Electronic and Automation Control Conference (IAEAC), pp. 18–23, October 2018
9. Diamantis, K., Dermitzakis, A., Hopgood, J.R., Sboros, V.: Super-resolved ultrasound echo spectra with simultaneous localization using parametric statistical estimation. IEEE Access **6**, 14188–14203 (2018)
10. Duarte, C., Delmar, P., Goossen, K.W., Barner, K., Gomez-Luna, E.: Non-intrusive load monitoring based on switching voltage transients and wavelet transforms. In: 2012 Future of Instrumentation International Workshop (FIIW), pp. 1–4. IEEE (2012)
11. Ducange, P., Marcelloni, F., Antonelli, M.: A novel approach based on finite-state machines with fuzzy transitions for nonintrusive home appliance monitoring. IEEE Trans. Ind. Inform. **10**(2), 1185–1197 (2014)
12. Egarter, D., Bhuvana, V.P., Elmenreich, W.: PALDi: online load disaggregation via particle filtering. IEEE Trans. Instrum. Meas. **64**(2), 467–477 (2015)

13. Erdil, E., Ghani, M.U., Rada, L., Argunsah, A.O., Unay, D., Tasdizen, T., Cetin, M.: Nonparametric joint shape and feature priors for image segmentation. IEEE Trans. Image Process. **26**(11), 5312–5323 (2017)
14. Fan, W., Bouguila, N., Du, J., Liu, X.: Axially symmetric data clustering through dirichlet process mixture models of Watson distributions. IEEE Trans. Neural Netw. Learn. Syst. **30**(6), 1683–1694 (2019)
15. Fan, W., Sallay, H., Bouguila, N.: Online learning of hierarchical Pitman-Yor process mixture of generalized dirichlet distributions with feature selection. IEEE Trans. Neural Netw. Learn. Syst. **28**(9), 2048–2061 (2017)
16. Figueiredo, M., Ribeiro, B., de Almeida, A.: Electrical signal source separation via nonnegative tensor factorization using on site measurements in a smart home. IEEE Trans. Instrum. Meas. **63**(2), 364–373 (2014)
17. Ford, G., et al.: Wireless network traffic disaggregation using Bayesian nonparametric techniques. In: 52nd Annual Conference on Information Sciences and Systems, CISS 2018, Princeton, NJ, USA, 21–23 March 2018, pp. 1–6. IEEE (2018)
18. Gael, J.V., Teh, Y.W., Ghahramani, Z.: The infinite factorial hidden Markov model. In: Advances in Neural Information Processing Systems, pp. 1697–1704 (2009)
19. Gillis, J.M., Alshareef, S.M., Morsi, W.G.: Nonintrusive load monitoring using wavelet design and machine learning. IEEE Trans. Smart Grid **7**(1), 320–328 (2016)
20. Hart, G.W.: Nonintrusive appliance load monitoring. Proc. IEEE **80**(12), 1870–1891 (1992)
21. He, K., Stankovic, L., Liao, J., Stankovic, V.: Non-intrusive load disaggregation using graph signal processing. IEEE Trans. Smart Grid **PP**(99), 1 (2017)
22. Hosseini, S.S., Kelouwani, S., Agbossou, K., Cardenas, A., Henao, N.: Adaptive on-line unsupervised appliance modeling for autonomous household database construction. Electr. Power Energy Syst. J. **112**, 156–168 (2019)
23. Iwayemi, A., Zhou, C.: SARAA: semi-supervised learning for automated residential appliance annotation. IEEE Trans. Smart Grid **8**(2), 779–786 (2017)
24. Jia, R., Gao, Y., Spanos, C.J.: A fully unsupervised non-intrusive load monitoring framework. In: Proceedings of the IEEE International Conference on Smart Grid Communications (SmartGridComm), pp. 872–878, November 2015
25. Jing, L., He, C., Huang, J., Ding, Z.: Joint channel estimation and detection using Markov chain Monte Carlo method over sparse underwater acoustic channels. IET Commun. **11**(11), 1789–1796 (2017)
26. Kaselimi, M., Protopapadakis, E., Voulodimos, A., Doulamis, N., Doulamis, A.: Multi-channel recurrent convolutional neural networks for energy disaggregation. IEEE Access **7**, 81047–81056 (2019)
27. Khodayar, M., Mohammadi, S., Khodayar, M.E., Wang, J., Liu, G.: Convolutional graph autoencoder: a generative deep neural network for probabilistic spatio-temporal solar irradiance forecasting. IEEE Trans. Sustain. Energy 1 (2019)
28. Kim, H., Marwah, M., Arlitt, M., Lyon, G., Han, J.: Unsupervised disaggregation of low frequency power measurements. In: Proceedings of the 2011 SIAM International Conference on Data Mining, pp. 747–758. SIAM (2011)
29. Kim, M.S., Kim, S.R., Kim, J., Yoo, Y.: Design and implementation of MAC protocol for SmartGrid HAN environment. In: Proceedings of the IEEE 11th International Conference on Computer and Information Technology, pp. 212–217, August 2011
30. Kolter, J.Z., Batra, S., Ng, A.Y.: Energy disaggregation via discriminative sparse coding. In: Advances in Neural Information Processing Systems, pp. 1153–1161 (2010)

31. Kolter, J.Z., Jaakkola, T.: Approximate inference in additive factorial HMMs with application to energy disaggregation. In: Artificial Intelligence and Statistics, pp. 1472–1482 (2012)

32. Kolter, J.Z., Johnson, M.J.: REDD: a public data set for energy disaggregation research. In: Workshop on Data Mining Applications in Sustainability (SIGKDD), San Diego, CA, vol. 25, pp. 59–62 (2011)

33. Kong, W., Dong, Z.Y., Ma, J., Hill, D.J., Zhao, J., Luo, F.: An extensible approach for non-intrusive load disaggregation with smart meter data. IEEE Trans. Smart Grid **9**(4), 3362–3372 (2018)

34. Kong, W., Dong, Z.Y., Wang, B., Zhao, J., Huang, J.: A practical solution for non-intrusive type II load monitoring based on deep learning and post-processing. IEEE Trans. Smart Grid 1 (2019)

35. Kramer, O.: Non-intrusive appliance load monitoring with bagging classifiers (2015)

36. Lange, H., Bergés, M., Kolter, Z.: Neural variational identification and filtering for stochastic non-linear dynamical systems with application to non-intrusive load monitoring. In: 2019 IEEE International Conference on Acoustics, Speech and Signal Processing (ICASSP), ICASSP 2019, pp. 8340–8344. IEEE (2019)

37. Lhéritier, A., Cazals, F.: A sequential non-parametric multivariate two-sample test. IEEE Trans. Inf. Theory **64**(5), 3361–3370 (2018)

38. Lindsten, F., Jordan, M.I., Schön, T.B.: Particle Gibbs with ancestor sampling. J. Mach. Learn. Res. **15**(1), 2145–2184 (2014)

39. Liu, B., Yu, Y., Luan, W., Zeng, B.: An unsupervised electrical appliance modeling framework for non-intrusive load monitoring. In: Proceedings of the IEEE Power Energy Society General Meeting, pp. 1–5, July 2017

40. Liu, B., Luan, W., Yu, Y.: Dynamic time warping based non-intrusive load transient identification. Appl. Energy **195**, 634–645 (2017)

41. Liu, Q., Kamoto, K.M., Liu, X., Sun, M., Linge, N.: Low-complexity non-intrusive load monitoring using unsupervised learning and generalized appliance models. IEEE Trans. Consum. Electron. **65**(1), 28–37 (2019)

42. Mei, J., He, D., Harley, R.G., Habetler, T.G.: Random forest based adaptive non-intrusive load identification. In: Proceedings of the International Joint Conference on Neural Networks (IJCNN), pp. 1978–1983, July 2014

43. Meng, N., Sun, X., So, H.K., Lam, E.Y.: Computational light field generation using deep nonparametric Bayesian learning. IEEE Access **7**, 24990–25000 (2019)

44. Mesadi, F., Erdil, E., Cetin, M., Tasdizen, T.: Image segmentation using disjunctive normal Bayesian shape and appearance models. IEEE Trans. Med. Imaging **37**(1), 293–305 (2018)

45. Nalmpantis, C., Vrakas, D.: Machine learning approaches for non-intrusive load monitoring: from qualitative to quantitative comparation. Artif. Intell. Rev. **52**(1), 217–243 (2019)

46. Rahimpour, A., Qi, H., Fugate, D., Kuruganti, T.: Non-intrusive energy disaggregation using non-negative matrix factorization with sum-to-k constraint. IEEE Trans. Power Syst. **32**(6), 4430–4441 (2017)

47. Rashid, H., Singh, P., Stankovic, V., Stankovic, L.: Can non-intrusive load monitoring be used for identifying an appliance's anomalous behaviour? Appl. Energy **238**, 796–805 (2019)

48. Ruiz, F.J.R., Valera, I., Svensson, L., Perez-Cruz, F.: Infinite factorial finite state machine for blind multiuser channel estimation. IEEE Trans. Cogn. Commun. Netw. **4**(2), 177–191 (2018)

49. Shin, C., Rho, S., Lee, H., Rhee, W.: Data requirements for applying machine learning to energy disaggregation. Energies **12**(9), 1696 (2019)
50. Tabatabaei, S.M., Dick, S., Xu, W.: Toward non-intrusive load monitoring via multi-label classification. IEEE Trans. Smart Grid **8**(1), 26–40 (2017)
51. Varela, P.M., Hong, J., Ohtsuki, T., Qin, X.: IGMM-based co-localization of mobile users with ambient radio signals. IEEE Internet Things J. **4**(2), 308–319 (2017)
52. Wang, W., Xi, J., Zhao, D.: Driving style analysis using primitive driving patterns with Bayesian nonparametric approaches. IEEE Trans. Intell. Transp. Syst. 1–13 (2018)
53. Xu, Y., Cheng, P., Chen, Z., Li, Y., Vucetic, B.: Mobile collaborative spectrum sensing for heterogeneous networks: a Bayesian machine learning approach. IEEE Trans. Sig. Process. **66**(21), 5634–5647 (2018)
54. Yan, D., et al.: Household appliance recognition through a Bayes classification model. Sustain. Cities Soc. **46**, 101393 (2019)
55. Zaidi, A.A., Kupzog, F., Zia, T., Palensky, P.: Load recognition for automated demand response in microgrids. In: Proceedings of the IECON 2010–36th Annual Conference on IEEE Industrial Electronics Society, pp. 2442–2447, November 2010
56. Zhao, B., Stankovic, L., Stankovic, V.: On a training-less solution for non-intrusive appliance load monitoring using graph signal processing. IEEE Access **4**, 1784–1799 (2016)

# Distributed Integrated Modular Avionics Resource Allocation and Scheduling Algorithm Supporting Task Migration

Qing Zhou[1]([⊠]), Kui Li[1], Guoquan Zhang[1], and Liang Liu[2]

[1] National Key Laboratory of Science and Technology on Avionics Integration,
China Aeronautical Radio Electronics Research Institute,
Shanghai 200233, China
zhouqingavic@163.com
[2] College of Computer Science and Technology,
Nanjing University of Aeronautics and Astronautics,
29 Jiangjun Avenue, Nanjing 210016, China

**Abstract.** At present, the avionics system tends to be modularized and integrated, and the distributed integrated modular avionics system (DIMA) is proposed as the development direction of the next generation avionics system. In order to support the operation of complex tasks, DIMA needs to have an effective resource allocation and scheduling algorithm for task migration and reorganization to achieve reconstruction. However, many current resource allocation and scheduling algorithms, used in traditional avionics systems, are not available for DIMA. In view of the above problems, the paper analyzes the characteristics of the DIMA avionics system architecture model and builds abstract models of the computing resources, computing platforms and tasks. Based on the established model, an efficient task scheduling algorithm, resource allocation algorithm and task migration algorithm for DIMA avionics architecture are designed. And we do simulation experiments to establish the model, and compare the designed EWSA algorithm with the mainstream algorithm JIT-C. The results show better performance in terms of workflow average completion time, successful scheduling completion rate and optimization rate. In addition, considering the failure in the process of executing the mission, we proposed a mission migration and reorganization algorithm WMA and set different time and number of fault resources of the aircraft in the simulation experiments to evaluate the performance of WMA algorithm.

**Keywords:** Distributed integrated modular avionics systems · Resource allocation · Scheduling algorithm · Task migration

## 1 Introduction

The avionics system is the critical system to the mission for aircrafts. It covers various electronic systems such as communications, radar, surveillance, and flight control. If the engine is the heart of the aircraft, then the avionics system is the brain [1]. It can be said that there is no advanced aircraft without advanced avionics systems [2]. Recently, the avionics system tends to be modular and integrated, subsystems such as flight management, navigation, display control, radar, photoelectric detection, fire control, and mission control, are all implemented with universal function module [3]. Distributed Integrated Modular Avionics (DIMA) system has been proposed as the direction of the development of next-generation avionics system [4].

The DIMA is essentially a parallel distributed computer system that integrates heterogeneous hardware resources, which uses switched network to interconnect hardware facilities of different physical areas of the aircraft and applies application software system to share underlying hardware devices, to realize highly generalized hardware devices and lower degree of software and hardware binding coupling [5].

However, compared to other distributed system architectures, DIMA architecture has many special features owing to its requirements of strong real time, high security and reliability [6]. First of all, the energy consumption of the computing platform service is very small compared to the braking system, thus the energy consumption caused by the algorithm is basically negligible [7]. Secondly, the aeronautical system has a very high real-time demand for task scheduling, thus the cost factor can be ignored while sharing internal resources of the aircraft [8]. In addition, in order to ensure the high safety and reliability of the avionics system, it is necessary to consider the emergency measures of the aircraft in the case of computational resource failure [9]. For aircraft clusters, we need to consider the differences between different types of aircrafts for the different sensor information they carried. Moreover, location information also needs to be considered when assigning tasks in the process of dynamic flight of aircrafts [10].

Therefore, in the face of complex DIMA avionics architecture, how to redistribute integrated heterogeneous computing resources, how to efficiently implement task scheduling on shared resources, and how to improve the reliability and security of application systems under DIMA system are the urgent problems to be solved in the current aviation research. However, with the strong real-time, high security and reliability requirements of DIMA avionics system, related scheduling strategies of traditional aeronautical systems applying on DIMA show poor performance [11].

In view of the above problems, the paper analyzes the characteristics of the DIMA avionics system architecture model and abstracts the computing resources, computing platforms and tasks. Based on the established model, an efficient task scheduling algorithm, resource allocation algorithm and task migration algorithm for DIMA avionics architecture are designed.

## 2   Related Works

At present, although few scholars have studied task scheduling and migration algorithms for DIMA avionics systems, many excellent task scheduling algorithms have been proposed in other distributed fields such as cloud computing.

Gupta et al. [12] proposed an online-aware job scheduling algorithm named MPA2OJS for multi-platform applications. To improve job execution time and optimize throughput, MPA2OJS uses dynamic heuristic job scheduling mapping schemes and efficiently schedules new high-performance cloud computing (HPC) workflows and load fluctuations for heterogeneous computing resource platforms based on internal application characteristics, job requirements, platform capabilities and dynamic requirements. The algorithm takes the availability of platform resources, the applicability and sensitivity of jobs on the platform into consideration, which balances the load of multiple computing platforms and distributes job streams to the most profitable computing platform efficiently.

Dong et al. [13] proposed a task scheduling algorithm with the most efficient priority named MESF to reduce energy consumption by limiting the number of active servers and the time of response. It takes advantage of the integer programming optimization solution to trade off active server and the time of response. Experimental simulation results show that MESF can save about 70% of energy consumption compared with the random task scheduling scheme.

Panigrahy et al. [14] proposed a geometric heuristic algorithm by sorting VM request queues of virtual machines. Li et al. [15] proposed a method based on multidimensional knapsack problem to solve the virtual machine VM placement. By using the virtual machine VM placement schedule, the total working time requested by the virtual machine on the same physical machine was reduced.

Khanna et al. [16] proposed a dynamic host management algorithm, which is activated when the physical machine becomes underloaded or overloaded. At the same time, the algorithm reduces SLA violations, minimizes migration costs and optimizes the number of physical servers. Beloglazov and Buyya et al. [17] analyzed historical data of resource utilization of virtual machines and proposed dynamic virtual machine integration to achieve the effect of power saving.

Taheri and Zamanifar et al. [18] introduced a two-stage virtual machine integration mechanism to deal with incomplete VM migration. Perplex VMs are virtual machines that should be integrated but have no suitable host placement. As a result, the system terminates the migration and VMs is replaced in its previous location. This problem will lead to a waste of CPU capacity and power consumption, increasing network overhead. According to the proposed framework, VMs are migrated from the overused host to other hosts in the first phase, and VMs from the low-load host are sent to other hosts in the second phase.

Sahni and Vidyarthi [19] etc. proposed a workflow scheduling algorithm JIT-C with deadline constraint. JIT-C algorithm is the core technology in the workflow task until the execution and ready to make a proper scheduling decisions, in order to make proper scheduling decisions/supply, this algorithm considering the performance of the virtual machine VM cloud computing platform change, loop controller monitors the task execution progress monitoring, and according to the latest information resources allocation/scheduling decisions. JIT-C algorithm scheduling workflow tasks has good performance, but because of the avionics system's strong real-time, high security and reliability, JIT-C algorithm can not meet the requirements of avionics system.

JIT-C has some shortcomings used in DIMA system platform. We model the computing resources, the computing platform and tasks by analyzing the characteristics of DIMA avionics system. Based on the established model, an efficient workflow-based task scheduling algorithm EWSA is proposed and compared with JIT-C algorithm, which focused on DIMA avionics system architecture. Through the simulation experiments, we evaluate these algorithms in terms of the average completion time and the optimization rate. In addition, considering the failure in the process of executing the mission, we proposed a mission migration and reorganization algorithm WMA and set different time and number of fault resources of the aircraft in the simulation experiment to evaluate the performance of WMA algorithm.

# 3   DIMA Avionics System Resource and Task Model

Based on the requirements of strong real-time performance, high security and reliability of DIMA avionics system, we modeled the computing resources, and proposed the computing platform and workflow task model for the computing resources model.

## 3.1   Computing Resource Model

A computing resource has a variety of parameter information such as processor, memory, bandwidth, etc., and each processor has one or more cores, each core has a fixed computing capacity. Each computing resource can run one or more subsystem tasks that share the underlying hardware facilities such as memory, processors, and so on. In order to ensure that the tasks of each subsystem are not interfered with each other when sharing resources, the concept of partition is proposed in the standard specification of ARINC653. By dividing different tasks into different partition systems, the task systems running on the same computing resources are not interfered with each other.

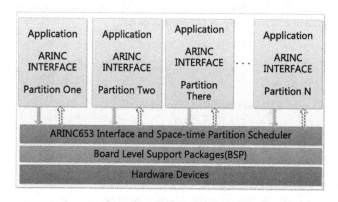

**Fig. 1.** ARINC653 partition software diagram

Figure 1 shows ARINC653 partition software, the core concept is to propose time and space partition isolation [20]. On the same computing resource, time-division or space-division scheduling schemes are implemented between partitions, and partitions are isolated from each other, which improve fault tolerance and security of the system. In the partition, multiple task processes share the hardware resource information obtained by the partition system, and the scheduling between task processes can be specified by the partition system designer, such as FCFS, MFS, etc. Therefore, in order to comply with the ARINC653 standard specification, the task process application is run by creating a VmPartition during the abstract modeling of computing resources of DIMA avionics system.

Figure 2 shows the model diagram of computing resources. It is assumed that an aircraft has multiple computing resources, and different computing resources have different parameters such as the number of processors, processing speed and memory configuration, etc., and the external service is composed of virtualized resources, including computing services and storage services. All computing resources inside the aircraft are interconnected through a strong real-time communication network to share sensor interfaces. Assuming that all computing and storage services are provided by the same aircraft, the average bandwidth between computing services is roughly equal. In addition, the storage service is implemented by the unified allocation of the local storage service of the aircraft, and the computing service is provided in the form of different types of VmPartition. Virtual partition has different CPU types, memory RAM and other configuration information resources, and virtual partition is dynamically submitted by the user to create and destroy. Different process task types are running between virtual partitions. If two tasks $t_i$ and $t_j$ are running on different virtual partitions and there is data transmission association between tasks, the calculation formula of data transmission time $TT(e_{ij})$ is as follows: $\frac{d_{t_i}}{\beta}$, where $d_{t_i}$ is the data file size output from task $t_i$ to task $t_j$, and $\beta$ is the average bandwidth of data transmission within the aircraft. If two tasks $t_i$ and $t_j$ are scheduled to run on the same virtual partition $t_j$, the data transfer time between them is zero. In the research work of this paper, the scheduling scheme between virtual partitions on the same computing resource is Time Shared or Space Shared strategy, and the intra-partition tasks uses the first-come-first-served (FCFS) strategy while scheduling between processes.

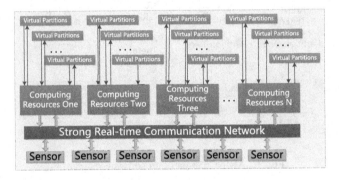

**Fig. 2.** The model diagram of calculation resource

## 3.2  Computing Platform Model

The computing platform model in this paper is similar to the one used in literature [20], as shown in Fig. 3. The resources of the aircraft provide services externally with virtualized service resources, which mainly involve three stages in planning workflow task execution. The first stage completes resource provisioning, which involves identifying and mapping the number of computational resources required to perform this task and the configuration of the VmPartition. The second stage dynamically creates the VmPartition onto the appropriate physical computing resources and generates an appropriate schedule for task execution in that partition. The third stage is a process that exists from the start of operation of the aircraft to the end of the aircraft, that is, to scan whether the physical computing resources of the aircraft fail. In case of failure, the failure resource processing module is called to implement the emergency plan to ensure the execution of subsequent tasks.

Workflow tasks submitted to the computing platform contain associated high-quality service (QOS) requirements, such as deadline constraints and resource specification requests. The deadline limit refers to the latest completion time for executing the workflow, and the resource specification required to perform the task refers to the computing resource requirements (such as memory, computing power, I/O, etc.). Based on these input requests, the Workflow Management System (WMS) automatically identifies the required resources and maps the tasks to the corresponding VmPartition.

WMS is mainly composed of three modules: resource allocation mapping module, workflow scheduling module and execution management module. Resource allocation mapping module consists of two sub-modules: resource capability assessment module and resource mapping management module. Resource capacity assessment module analyzes workflow structure to determine the number of resource requests, and resource mapping management module maps the corresponding resource requests to the corresponding VmPartition. Workflow scheduling management module and execution management module work together to dynamically create VmPartition on aircraft physical computing resources. Workflow scheduling management module mainly involves the actual scheduling of any workflow node to the VmPartition execution. The execution management module is used to track the execution status of the task node to record the Real Start Time (RST) and Real Finish Time (RFT) of the task node and update the status of resource information. This computing model is task-driven and the size of the fetched resources may vary at runtime.

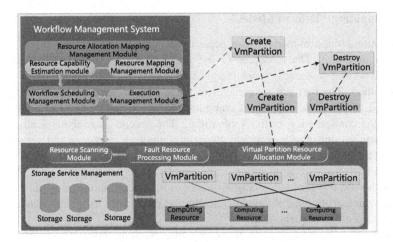

**Fig. 3.** Computing platform model diagram

The aircraft is mainly composed of resource scanning module, fault resource processing module and virtual partition resource allocation module to manage various computing resource information of the aircraft. The resource scanning module works in cooperation with the fault resource processing module. When the fault resource is scanned, the fault resource processing module executes the corresponding emergency plan. The virtual partition resource allocation module actually creates/destroys VmPartition on the physical computing resources.

### 3.3 Task Model

In the field of real aviation, the combat mission of aircraft is realized step by step. As shown in Fig. 4, when the aircraft strikes the target, it must first discover the target, identify and track the target, and then carry out anti-reconnaissance to prevent its exposure in the process of tracking. When a series of pre-task execution is completed, the target is hit. In the execution of a series of predecessor tasks, there are data transfer association requests between the subtasks, and subsequent tasks cannot begin execution until all prior tasks have been completed. In addition, task nodes that are simultaneously executable can be executed concurrently.

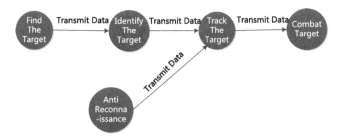

**Fig. 4.** Operational task execution diagram of the aircraft.

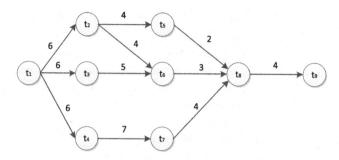

**Fig. 5.** Workflow task model diagram

According to the description above, we abstract similar aviation missions into a workflow task model, as shown in Fig. 5, a workflow application W = (T, E) be modeled as a directed acyclic graph (DAG), where $T = \{t_1, t_2, \ldots, t_n\}$ means task node set, E is a set of tasks with data transmission between or control set of dependencies to the edge. Dependency $e_{ij}$ is a set of priority constraints of task $(t_i, t_j)$, where $t_i, t_j \in T$ and $t_i \neq t_j$, indicate that task $t_j$ (subtask) cannot start execution before the end of task $t_i$ (parent task) execution, because there are relevant data transfer dependencies between task $t_i$ and task $t_j$. Subtask nodes are not able to start executing until all parent task nodes have finished executing and all dependencies (including data transfer and control) have been preprocessed. The workflow deadline D is defined as the latest execution time limit for executing the workflow. In this paper, each task node is modeled as a node with a fixed number of instructions, and each node has a corresponding output data file size. As shown in the figure, each directed edge represents the time required to transfer data files between tasks.

## 4    Efficient Task Scheduling Algorithm Under DIMA Avionics System

Based on the computing resources, computing platform and task model of DIMA avionics system in Sect. 4, this section introduces an efficient task scheduling algorithm. Firstly, several variables used in the algorithm and their meanings are introduced, as shown in Table 1.

**Table 1.** Definition of basic mathematical symbols

| Symbol | Meaning |
|---|---|
| D | Workflow W deadline |
| $T(t_i, VmPartition_v)$ | Task $t_i$ execution time on virtual partition type $VmPartition_v$ |
| $TT(e_{ik})$ | Data transfer time between task $t_i$ and task $t_k$ |
| $MET(t_i)$ | The minimum execution time for task $t_i$ |
| $\{t_{entry}\}$ | A task without a parent node |
| $\{t_{exit}\}$ | A task without child nodes |
| $EST(t_i)$ | The earliest execution time for task $t_i$ |
| $EFT(t_i)$ | The earliest time task $t_i$ ends |
| $RST(t_i)$ | Task $t_i$ actually starts execution time |
| $RFT(t_i)$ | Task $t_i$ actual end time |
| $EXST(t_i)$ | Task $t_i$ is expected to start time |
| $EXFT(t_i)$ | Task $t_i$ expected end time |
| $LST(t_i)$ | Task $t_i$ starts execution at the latest |
| $LFT(t_i)$ | Task $t_i$ end at the latest |
| $MET\_W$ | Workflow W minimum execution time |
| $RATE$ | Workflow W completion time optimization rate |

The operational equations of the above basic mathematical symbols are shown as follows, some of which are similar to literature [20]:

(1) Firstly, the proposed algorithm should satisfy the real-time requirement of avionics system. It is necessary to find a scheduling scheme for workflow $W$ and optimize the total completion time $TCT$ (Total Completed Time) of workflow $W$ under the limitation of deadline $D$. The optimization objective is shown in Eq. (1).

$$Optimize\, TCT = \mathop{max}_{t_i \in W}\{(RFT(t_i))\}, TCT \leq D \qquad (1)$$

Where $TCT$ is the total completion time of workflow $W$, $t_i$ is the task node in $W$, and $RFT(t_i$ is the actual completion time of task node $t_i$.

(2) $MET(t_i)$: the minimum execution time of task $t_i$, assuming that task $t_i$ can be executed on $VmPartition_{type}$ and has a minimum execution time $ET(t_i, VmPartition_v)$ on a certain VmPartition type. The calculation formula is shown in (2):

$$MET(t_i) = \mathop{min}_{VmPartition_v \in VmPartition_{type}}\{ET(t_i, VmPartition_v)\} \qquad (2)$$

(3) $TT(e_{ij})$: the time for transferring data between task $t_i$ and task $t_j$, assuming that task $t_i$ and task $t_j$ are assigned to execute on different VmPartition. The average bandwidth between the VmPartition of the aircraft is $\beta$, the data size from task $t_i$

to task $t_j$ is $d_{t_i}$, and the calculation formula of data transmission time is shown in (3):

$$TT\left(e_{ij}\right) = \frac{d_{t_i}}{\beta} \tag{3}$$

(4)  $EST(t_i)$: the earliest start time of task $t_i$, when the tasks of the predecessor nodes of task $t_i$ are all executed with the minimum execution time, and after the relevant data dependence transferring to $t_i$, $t_i$ can start execution. The calculation formula is shown in (4):

$$\begin{cases} EST\left(t_{entry}\right) = 0 \\ EST(t_i) = \underset{t_p \in t_{i'sparent}}{max} \left\{EST(t_p) + MET\left((t_p) + TT\left(e_{pi}\right)\right\} \end{cases} \tag{4}$$

(5)  $EFT(t_i)$: the earliest end time of task $t_i$, the calculation formula is shown in (5):

$$EFT(t_i) = EST(t_i) + MET(t_i) \tag{5}$$

(6)  $EXFT(t_i)$: the expected end time of task $t_i$, assuming that the number of scheduled tasks at the same level is N, the minimum execution time of all task nodes are at the same level. The calculation formula is shown in (6):

$$EXFT(t_i) = EXST(t_i) + \frac{\sum_{j=1}^{N}\left(MET\left(t_j\right)\right)}{N} \tag{6}$$

(7)  $EXST(t_i)$: the expected start time of task $t_i$, which is defined as the expected start time that $t_i$ can start execution after all previous task nodes of task $t_i$ have been scheduled and executed. The calculation formula is shown in (7):

$$\begin{cases} EXST\left(t_{entry}\right) = 0 \\ EXST(t_i) = \underset{t_p \in t_{i'sparent}}{max} \left\{EXFT(t_p) + TT\left(e_{pi}\right)\right\} \end{cases} \tag{7}$$

(8)  $LFT(t_i)$: the latest completion time of task $t_i$ with the limit of workflow deadline D. The calculation formula is shown in (8):

$$\begin{cases} LFT\left(t_{exit}\right) = D \\ LFT(t_i) = \underset{t_c \in t_{i'schildren}}{min} \left\{LFT(t_c) - MET(t_c) - TT(e_{ic})\right\} \end{cases} \tag{8}$$

(9)  $LST(t_i)$: the latest start time of task $t_i$ with the limit of workflow deadline D. The calculation formula is shown in (9):

$$LST(t_i) = LFT(t_i) - MET(t_i) \tag{9}$$

(10)  *MET_W*: The minimum execution time of workflow W, which is defined as the total time required when all task nodes on the critical path (the longest execution path) complete execution on the fastest virtual machine. The calculation formula is shown in (10):

$$MET\_W = \max_{t_i \in W} \{EFT(t_i)\} \tag{10}$$

(11)  *RATE*: the optimization rate of completion time with the constraint of deadline D. The calculation formula is shown in Eq. (11):

$$RATE = \frac{TET - D}{D} \tag{11}$$

## 4.1  Preprocessing Algorithm

In order to save the time of data transfer between tasks, the serial task nodes in workflow W need to be combined into one task node in advance. The pre-processing shown in Algorithm 1, is implemented by tksqueue, which firstly queues the entry task, queues all its sub-task nodes when it exits the queue, merges the existing serial task nodes, and changes the direction of the merged task node and data transmission time.

---

**Algorithm 1.** *Pre-processing (W)*
Input: A workflow of N tasks DAG W(T, E)
1. Begin
2.    $tksqueue \leftarrow \{t_{entry}\}$
3.    While *tksqueue* not null then
4.        $t_p \leftarrow tksqueue(front)$ // Retrieves the queue header element
5.        $S_c \leftarrow \{t_c | t_c \in W \ \&\& \ t_c \text{is a subtask node of } t_p\}$
6.        If $|S_c| == 1 \ \&\& \ t_c$ has only one parent node $t_p$
7.           Combine task $t_p$ and $t_c$ into task $t_{p+c}$ and replace $_p$ nodes
8.           Set the parent of the $t_c{'}$ s node to $t_{p+c}$
9.           Change the $t_{p+c}$ transfer time to $t_c$ to $t_c{'}$ s
10.          Update $MET(t_{p+c})$
11.          Add $t_{p+c}$ to *tksqueue*
12.       Else
13.          Add child node task $t_p{'}$ s to *tksqueue*
14.       End if
15.    End While
16. End

---

## 4.2   Resource Allocation Mapping Algorithm

The resource allocation mapping algorithm is used to allocate VmPartition for task mapping at the same level in workflow W. In addition, the corresponding task node binding is assigned to VmPartition.

Specific resource allocation algorithm OptimalVmPartitionMapping, shown in Algorithm 2, firstly calculates the average expected with hierarchical task execution time and start time by using Eqs. (5) and (6) with AverageExcutionTime method (line 2). The allocateResourceType method (line 8) is then used to allocate appropriate computing resources to each task using the expected end time and expected start time difference of the task. When mapping VmPartition resource configuration, it may occur that the expected end time of the task node is greater than the expected start time, that is, EXFT $\geq$ EXST. In this case, it is unreasonable to continue scheduling the task at the current expected end time. Therefore, we need to appropriately extend the expected end time of the task node to appropriately shrink the VmPartition resource configuration of the task node.

---

**Algorithm 2.** *OptimalVmPartitionMapping* (*taskList*)

1. Begin
2. *averageExcutionTime* ←*caculateAverageExcutionTime(taskList)*
3.    For each $t_i \in$ *taskList*
4.        *taskVmPartitionMap* = $\emptyset$
5.    Sets the expected end time*EXFT($t_i$)* ← *averageExcutionTime* - TT($t_i$)
6.        If *EXFT($t_i$)* $\leq$ *LFT($t_i$)* then
7.            *tempTime* ← *EXFT($t_i$)* - *EXST($t_i$)*
8.            *taskVmPartitionMap* ← *allocateResourceType(tempTime)*
9.        Else
10.        *taskVmPartitionMap* ← Assigns a virtual partition configuration 11. of a standard configuration type
11.        Bind the task Id of taskVmPartitionMap
12.        Set the creation time of taskVmPartitionMap to *EXFT($t_i$)*
13.        Add VmPartitionMa to VmPartitionList*p*
14.    End For
15. End

---

## 4.3   Ewsa Algorithm

When workflow W is in the existing vehicle scheduling on computing resources platform, first initializes the workflow W with the Initialize method, then respectively using Eqs. (2), (3) and (4) to calculate the minimum execution time MET in the workflow W task node, the data transmission time TT and the earliest start time EST, and identify critical path task node in workflow W. Equation (10) is used to calculate the minimum execution time in the workflow W, and judge the workflow scheduling can be. The Algorithm is shown in Algorithm 3.

---

**Algorithm 3.** *Efficient Workflow Schedule Algorithm* (EWSA)
Input:
Workflow DAG W(T, E) composed of N task nodes
The deadline D defined by workflow W
1. Begin
2.   *Initialize*(W)
3.   lculate the minimum execution time of workflow W MET_W
4.   utilization (11)
5.   If $D \geq MET\_W$ then
6.   Invoke the preprocessor pre-processing (W) to preprocess the workflow
8.       Calculate the workflow task nodes MET, LFT, LST using equations (2), (8), and (9)
9.       *analysisLevel*(W)
10.     $level \leftarrow max\{level(W)\}$
11.     $for(i \leftarrow 1; i \leq level; i \leftarrow i + 1)$
12.     $taskList \leftarrow \{t_i | t_i \in W \,\&\&\, leval(t_i) == level\}$
13.     *OptimaVmPartitionMapping*(taskList)
14.     end for
15.   {tentry}←All entry nodes in workflow W
16.   *SheduleAviationCloudlet*$(\{t_{entry}\})$
17.   While $t_i \in W \,\&\&\, t_i$ unscheduled execution then
18.     $t_i$ is sent to the scheduling manager to perform the scheduling
19.       Update task $t_i's$ time start execution time RST and actual end time RFT
to_be_scheduled← $\{t_i \in W \,|\, \forall t_p \in t_i'sparent \&\& t_p$ 24. completed $\}$
20.     *SheduleAviationCloudlet* (to_be_scheduled)
21.   end while
22.   Else
23.     It is recommended to modify the
24. deadline D && D ≥MET_W
25.   end If
26. End

---

When it is judged that the workflow W can be scheduled within the deadline D, the Pre-processing algorithm is used to pre-process the workflow. Equations (2), (8) and (9) were respectively used to update the task nodes MET, LFT and LST in workflow

W. The scheduling hierarchy of each task node in workflow task W and corresponding task nodes at the same level are analyzed by *analysis Level* method. The resource allocation mapping algorithm OptimaVmPartitionMappin maps one type of *VmPartition* separately to generate the corresponding expected scheduling plan. The *ScheduleAviationCloudlet* method (lines 14–20) is in the task scheduling management module, which performs the actual task scheduling and works with the execution manager to dynamically create the corresponding task's VmPartition on the aircraft's computing resources and execute the task scheduling on VmPartition. During this process, the execution manager monitors the execution status of the scheduling task, and records the actual start time RST and the actual end time RFT of the task, which are used by the scheduling manager to decide the scheduling plan for the next task. When the execution manager finds that the task has finished executing, it works with the scheduling manager to destroy the task's VmPartition and release system resources.

When the task scheduling management module dynamically creates the *VmPartition* where the task resides, there may be insufficient computing resources available for the aircraft to meet VmPartition requirements. At this point, we need to dynamically adjust the configuration type of VmPartition based on the available computing resources of the current aircraft. In the process of adjusting the virtual partition configuration VmPartition, the following conditions must be met:

(1) Task $t_i$ executing on VmPartition is expected to finish no later than task $t_i$, that is, $EXFT(t_i) \leq LFT(t_i)$.
(2) Task $t_i$ binded on *VmPartition* executes on virtual partitions, there is no such node $t_c$ subtasks, $t_c \in t_i'$ s children, making the expected end time of $t_c$ is greater than the start time at the latest.

### 4.4    Task Migration Wma Algorithm

The fault resource processing module works together with the resource scanning module. When the resource scanning module finds that there is a fault computing resource, *VmPartition* on the fault resource is sent to the fault resource processing module. The fault resource processing module calls the migration algorithm to realize task migration and reorganization, as shown in **Algorithm 4.**

---

**Algorithm 4.** *WorkflowMigrationAlgorithm* (*VmPartitionList*)

1. Begin
2. If *VmPartitionList* not null Then
3.   $R \leftarrow \{R_k | R_k$ is the computing resource available in the current aircraft$\}$
5.   For each VmPartition$_i$ $\in$ *VmPartitioonList*
6.     For each $R_k \in R$
7.       If The remaining compute resources available in $R_i$ satisfy the VmPartition$_i$request Then
8.         Move VmPartition$_i$ to compute resource $R_i$
9.         Restart the task node in Restart the task node in 11.VmPartition$_i$and Break
10.        End If
11.      End For
12.    If VmPartition$_i$ can't find  moving computing 15.resources && (The task on VmPartition$_i$ belongs to the 16.critical path task || VmPartition$_i$ has special priority tasks) Then
13.      The maximum number of remaining computing 19.resources available in $R_j \leftarrow R$
14 *selectVmPartition* $\leftarrow$ 21. *selectVmPartitionMigrateChoice*(VmPartition$_i$, $R_j$, choice)
15.                    Suspend selectVmPartition and join the wait queue 23.*VmPartitionWaitingList*
16. Free the system resources occupied by select VmPartition
17. Moving VmPartition$_i$ to compute resource $R_j$ and restart 26.the task
18.      Else
19.      Add VmPartition$_i$ to the wait queue 29.*VmPartitionWaitingList*
20.      End If
21.    End For
22. End

---

This algorithm realizes *VmPartition* movement by traversing (lines 5–9) other computing resources available to the aircraft. If it finds the computing resource $R_i$ available to accommodate VmPartition's request, move VmPartition directly to the computing resource $R_i$ and restart the task on *VmPartition*.

If it traverses all available computing resources and find that no computing resources are suitable for VmPartition's resource requirements, we select part of the virtual partition in a preemptive manner to temporarily suspend and join the waiting queue. As shown in 12–16 lines, we select a virtual partitions with selectVmPartitionMigrateChoice method, which realizes two selection strategy as follows:

(1)  Select the selectVmPartition in the compute resource $R_j$ that occupies the least computing resources and is applicable.
(2)  select the task node in the compute resource $R_j$ with the longest execution end time and is applicable for selectVmPartition.

During processing the task migration, whenever the task node finishes execution and releases the system computing resources, the fault resource processing module will give priority to the tasks pending execution queue to ensure that the suspended tasks can be executed as soon as possible. Although the migration and reorganization of tasks in a preemptive way will lead to the delay of the entire workflow W, it can ensure that the aircraft can still perform the workflow tasks normally in the case of resource failure. Simulation results show that in workflow with relatively simple relational complexity, the probability of delayed task caused by failure of fault resource is obvious. Moreover, delay rates are relatively low or even better in complex workflows.

# 5    Experimental Simulation and Result Analysis

In the paper, we use the simulation tool CloudSim [11] to build the resource model, task model and computing platform model for the proposed distributed integrated modular avionics DIMA system. the proposed algorithm EWSA is compared with the algorithm JIT-C. Moreover, in order to evaluate the performance of WMA algorithm, we set different time and number of fault resources of the aircraft in the simulation experiment.

## 5.1    Simulation Experiment Configuration Ewsa Algorithm

In the simulation environment of CloudSim, we assume that the computing resources of the aircraft have both isomorphic and heterogeneous types. The isomorphic computing resources are consistent with all kinds of computing resources processing capacity and memory configuration within the aircraft. For heterogeneous computing resources, the internal computing resources of the aircraft are inconsistent and the processing capabilities are different. The isomorphic computing resource information is shown in Table 2, and the other is shown in Table 3. We assume that there are 10 physical computing resources inside an aircraft, and the average network bandwidth between internal virtual partitions VmPartition is 200 MBps. In addition, as shown in Table 4, this paper sets up virtual partition VmPartition of three different reference configuration types in the simulation process. At the same time, this paper sets the physical computing resources to scan time interval of 200 s.

**Table 2.** Isomorphic computing resource information table

| Quantity | Cores | Capacity (GHz) | RAM (GB) |
|----------|-------|----------------|----------|
| 10       | 2     | 4              | 8        |

**Table 3.** Heterogeneous computing resource information table

| Serial number | Cores | Capacity (GHz) | RAM (GB) |
|---|---|---|---|
| #1 | 1 | 2 | 4 |
| #2 | 1 | 4 | 8 |
| #3 | 1 | 6 | 16 |
| #4 | 2 | 4 | 8 |
| #5 | 2 | 8 | 16 |
| #6 | 2 | 12 | 32 |
| #7 | 4 | 8 | 16 |
| #8 | 4 | 16 | 32 |
| #9 | 8 | 16 | 32 |
| #10 | 8 | 32 | 64 |

**Table 4.** VmPartition baseline configuration type

| VmPartition type | Cores | Capacity (GHz) | RAM (GB) | Storage (GB) |
|---|---|---|---|---|
| Small | 1 | 1 | 1.7 | 160 |
| Medium | 1 | 2 | 3.75 | 410 |
| Large | 2 | 4 | 7.5 | 840 |

When evaluating algorithm performance, you need to define an expiration date for each workflow W. If the deadline is very loose, there is enough time to schedule the workflow in real time under the constraints of the deadline. Therefore, it is necessary to comprehensively evaluate the performance analysis of the algorithm for all possible periods: emergency, moderate and loose deadlines.

Thus, we use the rules in Eq. (12) to set the deadline [20]:

$$Deadline \ D = (1 + \mu) \times MET\_W \qquad (12)$$

Among them, MET_W is the shortest execution time of the workflow, and $\mu$ is the deadline date factor, which is defined as follows:

*Strict DeadLines:* $0 \leq \mu < 1.5$
*Moderate DeadLines:* $1.5 \leq \mu < 3$
*Relaxed DeadLines:* $3 \leq \mu < 4.5$

During the experiment, the workflow is generated by random values. The number of task node instructions is among the range of $[2.5 \times 10^5, 5 \times 10^5]$, and the data size between task nodes is among the range of $[2 \times 10^2 \text{ MB}, 5 \times 10^2 \text{ MB}]$. The workflow W is randomly generated, which has one or more ingress nodes and one egress node. The workflow W is incremented with the number of its task nodes among [5, 10, 15, 20, 25, 30, 35], and its workflow relationship is more and more complex. The workflow algorithm with the same task node performs 100 times each time, and we take the average to evaluate the performance of the algorithm.

## 5.2    Comparison of Experimental Results

In this paper, we compare EWSA with JIT-C in terms of the average completion time of workflows, the rate of successful scheduling completion and the rate of optimization. In addition, by setting different time and number of fault resources of the aircraft, we evaluate the performance of the WMA algorithm in terms of the rate of the task migration delay.

(1)  Comparison of average completion time of workflows of different algorithms

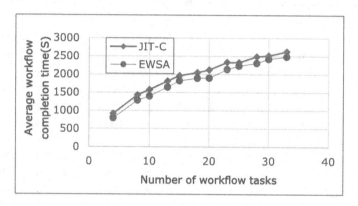

**Fig. 6.** Workflow completion time

Figure 6 is the comparison of the average completion time of different algorithms under different task numbers. As the number of workflow task nodes increases, the average completion time also increases, and the results show that the EWSA algorithm is always slightly better than the JIT-C algorithm. The reason is that when the EWSA algorithm schedules the workflow task nodes, the task nodes on the critical path acquire more computing resources by evaluating the amount of resources required for each level of the task node. At the same time, tasks on non-critical paths get longer execution time without affecting the execution of the next-level task nodes. The EWSA algorithm optimizes the rational allocation of resources for the entire workflow under the distributed integrated modular DIMA avionics system, thereby reducing the execution time of critical paths.

(2)  Comparison of successful scheduling completion rate and optimization rate of different algorithms

We set the deadline limit of workflow tasks under different constraints as STRICT, MODERATE, RELAXED, to evaluate the performance of the algorithm. As shown in Table 5, JIT-C algorithm meets real-time requirements of workflow tasks under STRICT constraints, and the successful scheduling rate is about 75%, while EWSA algorithm is as high as 90%. This is because EWSA algorithm dynamically adjusts the resource requests of workflow task nodes based on the number of resources available on the current platform, and macroscopically enables workflow task nodes to meet the requirements every time they request resources. Both JIT-C algorithm and EWSA algorithm can achieve 100% under MODERATE and RELAXED deadline constraint, caused by the extended deadline limit that leads workflow tasks to having more relaxed time to implement scheduling on limited resources without delay.

**Table 5.** Completion rate and optimization rate of workflow successful scheduling

| Tasks deadline limit | | | Workflow | | | | | |
|---|---|---|---|---|---|---|---|---|
| | | | 5 | 10 | 15 | 20 | 25 | 30 |
| STRICT | JIT-C | Success rate | 78% | 76% | 74% | 73% | 74% | 75% |
| | | Optimize rate | 5% | 5% | 5% | 5% | 4% | 5% |
| | EWSA | Success rate | 90% | 89% | 91% | 86% | 88% | 87% |
| | | Optimize rate | 17% | 14% | 15% | 15% | 12% | 12% |
| MODERATE | JIT-C | Success rate | 100% | 100% | 100% | 100% | 100% | 100% |
| | | Optimize rate | 41% | 42% | 42% | 42% | 41% | 42% |
| | EWSA | Success rate | 100% | 100% | 100% | 100% | 100% | 100% |
| | | Optimize rate | 44% | 45% | 44% | 45% | 46% | 44% |
| RELAXED | JIT-C | Success rate | 100% | 100% | 100% | 100% | 100% | 100% |
| | | Optimize rate | 70% | 70% | 70% | 69% | 67% | 68% |
| | EWSA | Success rate | 100% | 100% | 100% | 100% | 100% | 100% |
| | | Optimize rate | 70% | 72% | 70% | 70% | 70% | 70% |

The rate of optimization is calculated according to the deadline defined by the workflow, and the calculation formula is shown in Eq. (11). In order to see the optimization effect of workflow completion time more intuitively, we present the optimization results in the form of bar chart, which is shown in Fig. 7. The optimization rate of EWSA algorithm is higher about 10% than that of JIT-C algorithm under STRICT constraints. Under MODERATE and RELAXED constraints, they are nearly stable and the optimization rate of EWSA algorithm is about 3% higher than that of JIT-C algorithm. As the deadline constraint is more relaxed, JIT-C algorithm can always find the appropriate resources while EWSA algorithm optimization degree is close to saturation. In addition, with the increase of the number of workflow task nodes, the complexity of workflow relationship becomes higher and higher. Many tasks are close to serial execution, and the concurrent execution rate is lower, resulting in the lower degree of optimization.

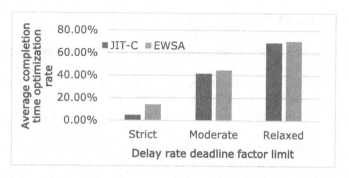

**Fig. 7.** Completion time optimization rate

**(3) Migration task delay rate of wma algorithm**

We set different time and numbers of failure resources of the aircraft in our simulation to verify the efficiency of task migration and reorganization, and to evaluate the performance of WMA algorithm. The computational resource failure time is set as 400 s after the simulation is started, and the number of computational resource failures is randomly selected among [1, 8]. Experimental results, in Fig. 8, show that when the number of workflow task nodes is small and their relationship is simple, the rate of workflow delay is high at about −11%. As the workflow task node number increases, the relationship is more and more complex, thus the delay rate is gradually decreasing, the completion time optimized slightly at around 2%. Because the more complex the relationship between task nodes within the workflow, the number of concurrent execution of task nodes within the same hierarchy decreases, making the execution of the entire workflow task nodes close to serial.

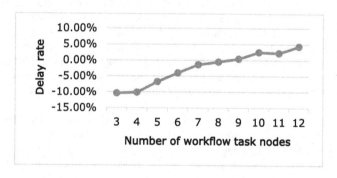

**Fig. 8.** Migration task delay rate

## 6  Conclusion and Future Work

Under the current trend of integrated modularization of avionics systems, by analyzing the characteristics of the future distributed avionics system DIMA architecture model, the paper focuses on the task scheduling and resource allocation of DIMA and uses simulation tool CloudSim to model its computing resources, tasks and computing platforms. Based on the established model, an efficient workflow-based task scheduling algorithm EWSA is proposed and compared with JIT-C algorithm, which shows better performance in terms of average workflow completion time and the rate of optimization. In addition, considering the failure in the process of executing the mission, we proposed a mission migration and reorganization algorithm WMA and set different time and number of fault resources of the aircraft in the simulation experiment to evaluate the performance of WMA algorithm.

The related scheduling algorithms and models proposed in the paper are currently applicable to single aircraft platforms. Considering multi-aircrafts cluster is the next research direction.

**Acknowledgments.** This work was supported in part by the Aeronautical Science Foundation of China under Grant 20165515001.

## References

1. Wang, T., Qingfan, G.: Research on distributed integrated modular avionics system architecture design and implementation. In: IEEE AIAA Digital Avionics Systems Conference, pp. 1–53 (2013)
2. Annighofer, B., Thielecke, F.: A systems architecting framework for optimal distributed integrated modular avionics architectures. CEAS Aeronaut. J. **6**(3), 485–496 (2015)
3. Swanson, D.L.: Evolving avionics systems from federated to distributed architectures. In: Proceedings of the 17th DASC Digital Avionics Systems Conference 1998. The AIAA/IEEE/SAE, 1: D26/1-D26/8, vol. 1. IEEE (1998)
4. Han, P., Zhai, Z., Nielsen, B., et al.: A modeling framework for schedulability analysis of distributed avionics systems. arXiv: Software Engineering, pp. 150–168 (2018)
5. Li, X., Xiong, H.: Modeling and analysis of integrated avionics processing systems. In: 2010 IEEE/AIAA 29th Digital Avionics Systems Conference (DASC), pp. 6.E.4-1–6.E.4-8. IEEE (2010)
6. Bao, L., Bois, G., Boland, J., et al.: Model-based method to automate the design of IMA avionics system based on cosimulation. SAE Int. J. Aerosp. **8**(2), 234–242 (2015)
7. Yunsheng, W., Savage, S., Hang, L., et al.: The architecture of airborne datalink system in distributed integrated modular avionics. In: Integrated Communications, Navigation and Surveillance Conference (2016)
8. Robati, T., Gherbi, A., Mullins, J., et al.: A modeling and verification approach to the design of distributed IMA architectures using TTEthernet. Procedia Comput. Sci. **83**, 229–236 (2016)
9. Wang, H., Niu, W.: A review on key technologies of the distributed integrated modular avionics system. Int. J. Wirel. Inf. Netw. **25**(3), 358–369 (2018)

10. Zhou, Q., Xiong, Z., Zhan, Z., et al.: The mapping mechanism between distributed integrated modular avionics and data distribution service. In: Fuzzy Systems and Knowledge Discovery, pp. 2502–2507 (2015)
11. Calheiros, R.N., Ranjan, R., Beloglazov, A., et al.: CloudSim: a toolkit for modeling and simulation of cloud computing environments and evaluation of resource provisioning algorithms. Softw.: Pract. Exp. **41**(1), 23–50 (2011)
12. Gupta, A., Faraboschi, P., Gioachin, F., et al.: Evaluating and improving the performance and scheduling of HPC applications in cloud. IEEE Trans. Cloud Comput. **4**(3), 307–321 (2016)
13. Dong, Z., Liu, N., Rojas-Cessa, R.: Greedy scheduling of tasks with time constraints for energy-efficient cloud-computing data centers. J. Cloud Comput. **4**(1), 5 (2015)
14. Panigrahy, R., Talwar, K., Uyeda, L., et al.: Heuristics for vector bin packing. research. microsoft.com (2011)
15. Li, K., Zheng, H., Wu, J.: Migration-based virtual machine placement in cloud systems. In: 2013 IEEE 2nd International Conference on Cloud Networking (CloudNet), pp. 83–90. IEEE (2013)
16. Khanna, G., Beaty, K., Kar, G.: Application performance management in virtualized server environments. In: 10th IEEE/IFIP, IEEE 2006 Network Operations and Management Symposium, 2006. NOMS 2006, pp. 373–381 (2006)
17. Beloglazov, A., Buyya, R.: Optimal online deterministic algorithms and adaptive heuristics for energy and performance efficient dynamic consolidation of virtual machines in cloud data centers. Concurr. Comput.: Pract. Exp. **24**(13), 1397–1420 (2012)
18. Taheri, M.M., Zamanifar, K.: 2-phase optimization method for energy aware scheduling of virtual machines in cloud data centers. In: International Conference for Internet Technology and Secured Transactions, pp. 525–530 (2011)
19. Sahni, J., Vidyarthi, D.: A cost-effective deadline-constrained dynamic scheduling algorithm for scientific workflows in a cloud environment. IEEE Trans. Cloud Comput. **6**, 2–18 (2015)
20. Wang, Y., Cui, L., Wang, J., et al.: Spatial and temporal partitioning validation for ARINC635-based avionics software. In: International Conference on Electronics and Information Engineering (2015)

# Pervasive Computing

# An Intelligent Question and Answering System for Dental Healthcare

Yan Jiang, Yueshen Xu[✉], Jin Guo, Yaning Liu, and Rui Li

School of Computer Science and Technology,
Xidian University, Xi'an 710071, China
{yjiang_2, jguo_2, ynliul}@stu.xidian.edu.cn,
{ysxu, rli}@xidian.edu.cn

**Abstract.** The intelligent question and answering system is an artificial intelligence product that combines natural language processing technology and information retrieval technology. This paper designs and implements a retrieval-based intelligent question and answering system for closed domain, and focuses on researching and improving related algorithms. The intelligent question and answering system mainly includes three modules: classifier, Q&A system and Chatbots API. This paper focuses on the classifier module, and designs and implements a classifier based on neural network technology, mainly involving word vector, bidirectional long short-term memory (Bi-LSTM), and attention mechanism. The word vector technology is derived from the word2vec tool proposed by Google in 2013. This paper uses the skip-gram model in word2vec. The Q&A system mainly consists of two modules: semantic analysis and retrieval. The semantic analysis mainly includes techniques such as part-of-speech tagging and dependency parsing. The retrieval mainly relates to technologies such as indexing and search. The Chatbots API calls the API provided by Turing Robotics. The intelligent question and answering system designed and implemented in this paper has been put into use, and the user experience is very good.

**Keywords:** Question and answering · Word2vec · Skip-gram · Bi-LSTM · Attention mechanism · Part-of-speech

## 1 Introduction

The intelligent question and answering system is a new type of information service system that combines natural language processing and information retrieval technologies. The intelligent question and answering system can be divided into closed domain and open domain according to the knowledge field [1]. The closed domain focuses on answering specific domain questions. The questioner can only ask some domain related questions and get answers. The open domain system does not set the scope of the problem. The questioner can come up with any topic of interest to him, and can get the answer he wants from the system. According to the principle of answer generation, the question and answering system is divided into generation and retrieval.

© ICST Institute for Computer Sciences, Social Informatics and Telecommunications Engineering 2019
Published by Springer Nature Switzerland AG 2019. All Rights Reserved
Q. Li et al. (Eds.): BROADNETS 2019, LNICST 303, pp. 201–215, 2019.
https://doi.org/10.1007/978-3-030-36442-7_13

At present, Most of the popular question and answering robots are based on open domain intelligent question and answering systems, such as Microsoft Xiaobing based on Internet corpus and user click logs, and Baidu voice assistants based on Baidu search logs. These open intelligent robots cannot accurately answer questions about specific domains such as taxation, finance, oral, etc., while traditional artificial services have problems such as slow response and high labor costs. This paper implements a specific domain (Prosthodontics), search-based intelligent question and answering system.

The intelligent question and answering system consists of a classifier, Q&A system, and Chatbots API. The workflow of the intelligent question and answering system is that the user asks questions, and then the classifier classifies the problems. This paper has set up two categories, ordinary chat and Prosthodontics. If the problem is assigned to the Prosthodontics, the Q&A system designed by this paper provides answers to the questions. If the question is assigned to the ordinary chat, the answer provided by the Chatbots API.

The classifier is the focus of this paper. In order to classify such short questions, this paper adopts the popular neural network technology in recent years to design classifier. At first, the Bi-LSTM neural network model is used to construct the classifier. The accuracy of the classifier can only reach about 80%. Later, this paper employs the attention mechanism, which is emerging in the field of machine translation and is now widely applied in various fields of artificial intelligence, and has achieved good results [10–13, 16]. The accuracy of the classifier can reach about 87%.

Another focus of this paper is the design of the Q&A system. The Q&A system has two modules, semantic analysis and retrieval. The semantic analysis module includes Chinese word segmentation, part-of-speech tagging, and dependency parsing. They are also the three basic tasks in natural language processing (NLP). After the dependency paring, according to the characteristics of the Prosthodontics data, this paper designed some filter rules for dependency, which has obvious effects in practical applications. The retrieval module is implemented by Appach Lucene. Lucene has incremental indexing technology, which allows users to add, change, or remove documents in their index without the need to re-index all contents within the index. By using timestamps, the index will identify the documents that were newly created, modified, or deleted, and what will be re-indexed are only those contents related to those documents. Lucene has a world-class search technology, especially its field search functionality is very suitable for the application scenario of this paper.

The last module of this paper is Chatbots API. This paper calls the API provided by Turing Robotics. After testing the user experience is very good, the user can ask any domain of questions, it can give a suitable answer.

## 2   The Architecture of the Intelligent Question and Answering System

Figure 1 presents the whole architecture of the intelligent question and answering system, which contains four modules and each module is given a brief explanation as follows.

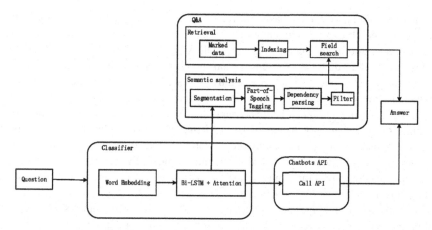

**Fig. 1.** The architecture of the intelligent question and answering system

**Classifier.** The classifier is an important module of the intelligent question answering system and the focus of this paper. The function of the classifier is to classify the questions asked by the user. This paper has set up two categories, ordinary chat and Prosthodontics. This paper designs and implements a classifier based on neural network technology.

**Q&A System.** The Q&A system is also the focus of this system. If the user's question is assigned to Prosthodontics, the Q&A system will answer the user's question. The Q&A system mainly involves two modules, semantic analysis and retrieval. After the semantic analysis of the user's question is completed, the search module will search with its result. If a similar problem is found, the search module will return the answer corresponding to the question.

**Chatbots API.** If the user's question is assigned to the ordinary chat category, this paper will call Turing Robot's API to answer the user's question.

## 3    The Classifier

This paper employs the 'jieba' Chinese segmentation tool to segment words for the Chinese Wikipedia dataset, and a 50-dimensional word vectors are trained using the skip-gram model in word2vec. For the professional dental dataset, this paper adopts the same tool. Through that process, the trained word vectors are obtained, which are used to encode the professional dental data. Then input those vectors into the classifier for training. The classifier model uses a Bi-LSTM model, incorporating the Attention mechanism.

The 'jieba' Chinese word segmentation supports three word segmentation modes: (1) the precise mode; it can cut the sentence most accurately and is suitable for text analysis; (2) the full mode; it can scan all the words that can be formed into words in sentences. It is very fast, but cannot solve ambiguity; (3) the search engine mode; it can split long words again on the basis of precise mode. It improves the recall rate and is suitable for search engine segmentation.

This paper adopts the search engine model. Although we are now in an era of data explosion, the data that can be used to train the classifier is still very small. The data used in this paper is labeled and processed by professional doctors. Search engine mode can increase the vocabulary and it has obvious effect in later practical applications. The 'jieba' word segmentation can be loaded into the developer's own defined dictionary to contain words that are not in the 'jieba' lexicon. Although 'jieba' has the ability of recognizing new words, self-adding new words on its own can guarantee higher accuracy. Moreover, the 'jieba' word segmentation not only can dynamically modify the dictionary in the program, but can also adjust the word frequency of a single word so that it can (or cannot) be separated.

Word2vec is a two-layer neural net that processes text [3]. Its input is a text corpus and its output is a set of vectors: feature vectors for words in that corpus. It does so in one of two ways, either using context to predict a target word (a method known as continuous bag of words, or CBOW), or using a word to predict a target context, which is called skip-gram. So we use the latter method because it produces more accurate results on large datasets (Fig. 2).

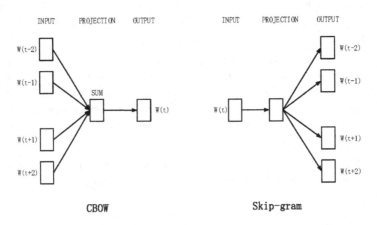

**Fig. 2.** Architecture of the CBOW and the Skip-gram.

The training objective of the Skip-gram model is to find word representations that are useful for predicting the surrounding words in a sentence or a document [4]. More formally, given a sequence of training words $w_1, w_2, w_3, \ldots, w_T$, the objective of the Skip-gram model is to maximize the average log probability

$$\frac{1}{T} \sum_{t=1}^{T} \sum_{-c \leq j \leq c, j \neq 0} \log p\left(w_{t+j} | w_t\right) \tag{1}$$

where c is the size of the training context (which can be a function of the center word $w_t$). The basic Skip-gram formulation defines $p\left(w_{t+j} | w_t\right)$ using the softmax function:

$$p(w_O|w_I) = \frac{\exp\left(v'_{w_O}{}^T v_{w_I}\right)}{\sum_{w=1}^{W} \exp\left(v'_{w_O}{}^T v_{w_I}\right)} \tag{2}$$

where $v_w$ and $v'_w$ are the "input" and "output" vector representations of w, and W is the number of words in the vocabulary. This formulation is impractical because the cost of computing $\nabla \log p(w_O + w_I)$ is proportional to W.

This paper uses the negative sampling [3] method in the skip-gram model to solve this problem. We define Negative sampling (NEG) by the objective

$$\log\sigma\left(v'_{w_O}{}^T v_{w_I}\right) + \sum_{i=1}^{k} E_{w_i \sim P_n(w)}\left[\log\sigma\left(-v'_{w_I}{}^T v_{w_I}\right)\right] \tag{3}$$

which is used to replace every $\log p(w_O|w_I)$ term in the Skip-gram objective. In this paper, the unigram distribution is used to select negative samples. The probability calculation formula for each word being selected as a negative sample is related to the frequency of its occurrence. The formula used for the implementation in the code is (Fig. 3):

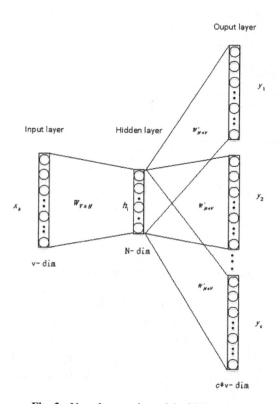

**Fig. 3.** Neural network model of Skip-gram

$$P(w_i) = \frac{f(w_i)^{3/4}}{\sum_{j=0}^{n} (f(w_i)^{3/4})} \tag{4}$$

Among them, v is the vocabulary size; N refers to the dimension of the word vector; c is twice the size of the word vector window; w refers to the central word vector matrix; $w'$ is the matrix formed by context words' vectors. In this paper, central word vector matrix w is used as the final word vector. This paper does not employ the traditional recurrent neural network model in the design of the classifier of neural network model. Because the traditional recurrent neural network model is prone to the problems of gradient disappearance and gradient explosion during training. In order to solve this problem, this paper adopts the LSTM network [4], a variant of the recurrent neural network.

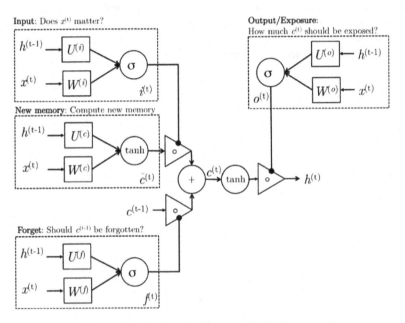

**Fig. 4.** Internal structure of LSTM

The Fig. 4 is described by the following formula:

$$
\begin{aligned}
i^{(t)} &= \sigma\big(w^{(i)}x^{(t)} + U^{(i)}h^{(t-1)}\big) && \text{(Input gate)} \\
f^{(t)} &= \sigma\big(w^{(f)}x^{(t)} + U^{(f)}h^{(t-1)}\big) && \text{(Forget gate)} \\
o^{(t)} &= \sigma\big(w^{(o)}x^{(t)} + U^{(o)}h^{(t-1)}\big) && \text{(Output gate)} \\
\tilde{c}^{(t)} &= \sigma\big(w^{(c)}x^{(t)} + U^{(o)}h^{(t-1)}\big) && \text{(New memory cell)} \\
c^{(t)} &= i^{(t)} \circ \tilde{c}^{(t)} + f^{(t)}c^{(t-1)} && \text{(Final memory cell)} \\
h^{(t)} &= o^{(t)} \circ \tan h\big(c^{(t)}\big)
\end{aligned}
\tag{5}
$$

The calculation process of LSTM is as follows:

1. The New memory cell uses the word $x^{(t)}$ at the current moment and the hidden state $h^{(t-1)}$ at the previous moment to generate a new memory $\tilde{c}^{(t)}$. So the new memory contains the attributes of the current word $x^{(t)}$.
2. The Input Gate uses the word $x^{(t)}$ at the current moment and the hidden state $h^{(t-1)}$ at the previous moment to determine how much the attribute of the word at the current moment should be kept. What is kept refers to $i^{(t)}$.
3. The Forget Gate uses the word $x^{(t)}$ at the current moment and the hidden state $h^{(t-1)}$ at the previous moment to determine how much the past memory $c^{(t-1)}$ should be forgotten. What is forgotten refers to $f^{(t)}$.
4. The Final memory cell adds up the new memory retained by Input Gate and the past memory forgotten by Forget Gate to generate the final memory $c^{(t)}$.
5. The Output Gate uses the word $x^{(t)}$ at the current moment and the hidden state $h^{(t-1)}$ at the previous moment to determine how much new memory $\tan h(c^{(t)})$ should be output, and the output amount is represented by $o^{(t)}$.

As the basic structure of the neural network model, the network structure diagram of the Bi-LSTM used in this paper is shown in Fig. 5 [14, 15]. In the built Bi-LSTM, the *jth* hidden state $h_j \rightarrow$ can only carry the *jth* word itself and a part of information in previous words. If the input is in reverse order, $h_j \leftarrow$ carries the *jth* word and a part of information in posterior words. Combining $h_j \rightarrow$ and $h_j \leftarrow$, $h_j[h_j \rightarrow, h_j \leftarrow]$ can contain the information before and after the *jth* word.

The built classifier aims to classify the questions raised by users, most of which are short sentences. Therefore, our system adds the attention mechanism [5] to the basic LSTM network. The Attention mechanism is to assign different weights to different words in a sentence, which is likely to improve the accuracy of the classifier, especially in the case of short sentences. Attention mechanism for the neural network in this paper

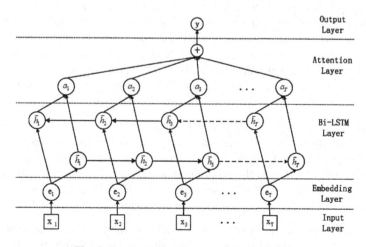

**Fig. 5.** Neural network model of the classifier

is to add a parameter to each moment of Bi-LSTM after the Bi-LSTM layer. This parameter is randomly given at the beginning, and then it is revised through continuous training. Figure 5 shows the neural network model of the classifier. The diagram contains Bi-LSTM and Attention mechanism.

# 4   Q&A System

In Sect. 3.1, we focus on how to classify the questions asked by users. In this section and the next section, we will focus on how to provide answers to questions asked by users. The answer comes from two modules: the self-designed search-based Q&A system; Call the more mature Chatbots API of the current technology. If the user's question is assigned to the oral prosthetics field, the Q&A system will provide the answer, otherwise the Chatbots API will provide the answer. The Q&A system consists of two modules, semantic analysis and retrieval.

## 4.1   Semantic Analysis

The semantic analysis module mainly performs two tasks, part-of-speech and dependency parsing. This paper uses the Stanford Corenlp tool to accomplish these tasks.

Syntactic parsing is one of the key techniques in natural language processing (NLP). Its basic task is to determine the syntactic structure of a sentence or the dependence of words in a sentence. In general, syntactic parsing is not the ultimate goal of NLP task, but it is often an important link or even a key link to achieve the ultimate goal.

Syntactic parsing is divided into two types: syntactic structure parsing and dependency parsing. Syntactic structure parsing refers to determining whether a composition of a word sequence (usually a sentence) conforms to a given grammar and analyzes a syntactic structure that conforms to grammar. Syntactic structure is generally represented by the tree data structure, typically referred to as syntactic parse trees. Dependency parsing refers to the analysis of the dependence between words and words for a sentence. In fact, we sometimes don't need or need to know the phrase structure tree of the whole sentence, and we must know the dependence between words and words in the sentence, so does this paper. "Dependence" refers to the dominant and dominated relationship between words. This relationship is not equal and has a direction. The dominant component is called the governor, and the dominated component is called the modifier. There are four rules for dependency parsing [6]:

(1)   A sentence has only one independent component;
(2)   The other components of a sentence are subordinate to a certain part;
(3)   No single component can be dependent on two or more components;
(4)   If component A is directly subordinate to component B, and component c is located between A and B in the sentence, component c is either subordinate to A, or subordinate to B, or subordinate to a component between A and B.

These four rules are equivalent to the formal constraints on the dependency graph and the dependency tree: single, headed, connective, acyclic and projective, and used to ensure that the dependency parsing result of the sentence is a tree with "root". This lays the foundation for the formal description of dependent grammar and its application in computer linguistics. Currently, there are two ways to do dependency parsing: rules based (rules designed by people), based on neural networks [2, 9]. With the rapid growth of marking datasets, neural network-based dependency parsing has become increasingly popular. Labeling datasets is time-consuming and labor-intensive, but with the development for many years, the number is growing. Everyone prefers to use an annotated dataset because its work for each process can be reused, such as part-of-speech tagging, named entity recognition, and so on. There are many shortcomings in rule-based dependency analysis: the rules are relatively rigid; moreover, there are many rules, which are difficult to unify. Therefore, this paper uses a neural network-based dependency parser to do dependency parsing analysis. This paper uses the dependency parsing analyzer of the Stanford Corenlp toolkit to implement dependency parsing. Figure 6 shows the dependency tree after dependency parsing of a sentence.

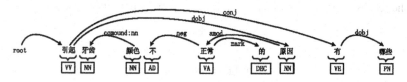

**Fig. 6.** Dependency tree

The dependencies contained in Fig. 6 are shown in Table 1:

**Table 1.** Dependency results

| Dependency |
| --- |
| (root,ROOT,引起#VV)(compound:nn,颜色#NN,牙齿#NN)(nsubj,正常#VA,颜色#NN)(neg, 正常 #VA, 不 #AD)(amod, 原因 #NN, 正常 #VA)(mark, 正常 #VA, 的 #DEC)(dobj,引起#VV,原因#NN)(conj,引起#VV,有#VE)(dobj,有#VE,哪些#PN) |

Analysis of the results of a large number of dependency parsing, combined with the analysis of the search results in next section, the paper finally determined that only eight dependencies were retained: amod (adjective modifier), compound:nn (noun modifies noun), advmod (adverb modifies adjective), nsubj (nounsubjective), dobj (direct objective), det (determiner), prnmod (parenthetical modifier), assmod (associative modifier). This paper also establishes filtering rules for each of these dependencies. For example, amod's filtering rules:

a. Both modified words and modifiers can only be Chinese characters at the same time, and cannot contain meaningless symbols such as punctuation.

b. Only retain the modified word part as NN (common noun), and modify the part of speech as JJ (adjective or numeral, ordinal).

Due to the limited length of the paper, the filtering rules for each dependency are not described here. The results of the dependency filtering for the example sentences in Fig. 6 are shown in Table 2.

**Table 2.** Selected dependency

| Selected dependency |
|---|
| (dobj, 引起#VV, 原因#NN)(compound:nn, 颜色#NN, 牙齿#NN)(nsubj, 正常#VA,颜色#NN)(amod,原因#NN,正常#VA)(dobj,有#VE,哪些#PN) |

The results after the combination of Table 2 dependency filter results are shown in Table 3.

**Table 3.** Sentence

| Sentence |
|---|
| 引起 牙齿 颜色 正常 原因 有 哪些 |

## 4.2 Retrieval

The retrieval module consists of two main steps, indexing and searching. Indexing means that the system uses the question and answer pairs marked by the professional doctor to establish an index; searching refers to the system using the results of the semantic analysis module to search for similar problems. If it is found, the answer corresponding to the question is returned. Implementation of the retrieval module employs Apache Lucene.

Apache Lucene is an ultra-fast, powerful, and full-featured text search engine library developed to empower mobile applications, websites, and solutions with search capabilities so they can deliver a seamless search experience. Written in Java, Apache Lucene functions as an indexing and search technology that allows the implementation of full-text search capabilities and efficient indexing processes on applications and websites that are running on multiple platforms.

As an indexing technology, Apache Lucene offers a feature called incremental indexing. This feature allows users to add, change, or remove documents in their index without the need to re-index all contents within the index. By using timestamps, the index will identify the documents that were newly created, modified, or deleted, and what will be re-indexed are only those contents related to those documents. This feature is very advantageous for users who are managing a large database and who need to change their data. The indexing capability of Apache Lucene is scalable and has a high performance. In fact, it lets users process a large number of documents and index them in less time. This is because the search engine library supports the indexing of over 50 GB of documents or data within an hour. In addition, as they perform indexing, they only need to allocate a minimal amount of RAM for such task which is 1 MB heap.

Apache Lucene makes it easy for searchers to find any information or content they need. Through the aid of its powerful, accurate, and efficient search algorithms, users will be able to precisely deliver whatever searchers are looking for. This makes the search engine library a truly world-class search technology that can handle sophisticated searches and generate search results in an advanced and modern way. For example, accurate searching can be implemented by enabling a fielded search functionality on websites and applications. Basically, searchers are looking for pieces of information that are stored in a database and are defined and labelled in the search index during the indexing process. These pieces of information are organized in specific areas called fields. In this paper, question, answer and disease fields are established when indexing. Using Lucene's fielded search functionality to search only in question field. In the process of searching, this paper also adds some rules, for example: import synonym dictionary to improve the hit rate; Repeated searches for some keywords help to improve the ranking of keyword-related questions [8].

Lucene enables users to rank their search results so that searchers will be able to see first the contents that are most relevant to what they are looking for. In addition, the search engine library allows users to use different ranking models as they rank their search results which include Vector Space Model and Okapi BM25 [3]. This paper uses Vector Space Model to rank.

## 5   Chatbots API

When the classifier in Sect. 3.1 divides the problem into the ordinary chat field, then the user's question is answered by the chatter robot. This paper calls Turing Robotics' Chatbots API to implement this function. The details of the technology are not covered here.

## 6   Experimental Results

In this section, this paper tests the performance of the classifier described in Sect. 3 and delay of the intelligent question and answering system. The data set used in this paper has two parts, ordinary chat data sets and Prosthodontics data sets. Prosthodontics data is some real patient problems crawled from the medical website, such as Chunyuyisheng, Haodaifu, etc., and then marked by a professional dentist. Ordinary chat data is crawled from the web. The two-part data format is shown in Fig. 7:

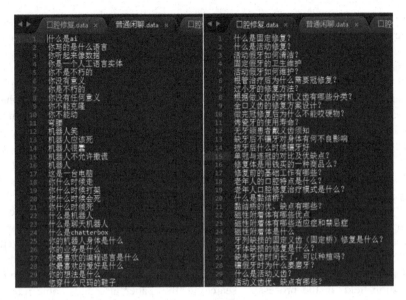

**Fig. 7.** Ordinary chat data sets and Prosthodontics data sets

## 6.1 Classifier Performance Test

The performance metric of the evaluation classifier is generally the classification accuracy, which is defined as the ratio of the number of samples correctly classified by the classifier to the total number of samples for a given test data set. The formula is as follows:

$n$ – the number of samples correctly classified
$N$ – the total number of samples

$$Accuracy = \frac{n}{N} \tag{6}$$

Evaluation indicators commonly used for the two-class classification are precision and recall [7]. Usually, the class of interest is positive, the other classes are negative, and the prediction of the classifier on the test data set is either correct or incorrect, the total number of occurrences of the four cases is recorded as:

$TP$ – Positive class is predicted as positive class
$FN$ – predict positive classes as negative classes
$FP$ – predict negative classes as positive classes
$TN$ – predict negative classes as negative classes.

The precision is defined as:

$$P = \frac{TP}{TP + FP} \tag{7}$$

The recall is defined as:

$$R = \frac{TP}{TP + FN} \tag{8}$$

In addition, there is an F1 value, which is the harmonic mean of the precision and recall. It is defined as:

$$\frac{2}{F_1} = \frac{1}{P} + \frac{1}{R} \tag{9}$$

$$F_1 = \frac{2TP}{2TP + FP + FN} \tag{10}$$

When both the precision and the recall are high, the F1 value will be high.

In this paper, the four performance indicators of the three classifiers described in Sect. 3.1 are tested separately. The results are shown in Table 4:

**Table 4.** Classifier performance test

|  | Precision | Recall score | F1 score | Accuracy |
|---|---|---|---|---|
| The classifier | 0.8923 | 0.9134 | 0.9027 | 0.8780 |

## 6.2 System Delay Test

The intelligent question and answering system provides a restful interface that can be easily embedded in any program. The intelligent question and answering system includes classifiers, Q&A, and chat API. These modules all involve certain delay problems. In this paper, the response time of the system is tested for different lengths of the user's question. The results are shown in Fig. 8. It can be seen from the Fig. 8 that

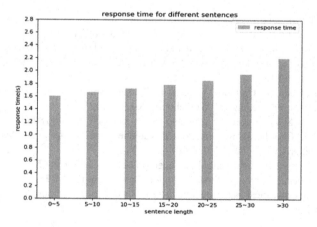

**Fig. 8.** System delay test

as the length of the user's problem increases, the response time of the system gradually increases. However, the overall response of the system is still very fast, which will give users a good experience.

## 7   Conclusion

This paper proposes three modules of classifier, Q&A and Chatbots API, which constitute a complete intelligent question and answering system. The paper focuses on the design process of the classifier. It is concluded that the effect of neural network technology on small and medium-sized data sets is not as good as the traditional statistical learning method. In the case of the data set used in this paper, AdaBoost is better than non-linear SVM. As for the dependency parsing in the semantic analysis module of Q&A, this paper does not use rule-based dependency parsing while adopts neural network technology-based dependency parsing, the effect is really better. According to the characteristics of users' questioning question and the search module, this paper designs some filtering rules of dependency relationship, so that the semantic analysis module analysis is very good.

In the future, this intelligent question and answering system will continue to collect corpus marking it by a professional dentist, and then import it into the database, so that the intelligent question and answering system can answer more questions. Will continue to work to improve the accuracy of the classifier on short problem data sets, to study dependency parsing to make it even better. The search algorithm will continue to be improved, so that users to get a more appropriate answer.

**Acknowledgement.** This paper is supported by Fundamental Research Fund for Central Universities (No. JBX171007), National Natural Science Fund of China (No. 61702391), and Shaanxi Provincial Natural Science Foundation (No. 2018JQ6050).

## References

1. Pundge, A.M., Khillare, S.A., Mahender, N.: Question answering system, approaches and techniques: a review. Int. J. Comput. Appl. **141**, 0975–8887 (2016)
2. Li, H., Zhang, Z., Ju, Y., Zhao, H.: Neural character-level dependency parsing for Chinese. In: Proceedings of the Thirty-Second AAAI Conference on Artificial Intelligence (AAAI), pp. 5205–5212 (2018)
3. Mikolov, T., Chen, K., Corrado, G., Dean, J.: Efficient estimation of word representations in vector space. In: Proceedings of Workshop at ICLR (2013)
4. Hochreiter, S., Schmidhuber, J.: Long short-term memory. 1735–1780 (1997)
5. Bahdanau, D., Cho, K., Bengio, Y.: Neural machine translation by jointly learning to align and translate. arXiv, pp. 1409–0473 (2014)
6. Chengqing Zong: Statistical natural language processing. 155–158 (2013)
7. Li, H.: Statistical Learning Methods. Tsinghua Press, Beijing (2012)
8. Jones, K.S.: A statistical interpretation of term specificity and its application in retrieval. J. Doc. (1972)

9. Chen, D., Manning, C.D.: A fast and accurate dependency parser using neural networks. In: Proceedings of the 2014 Conference on Empirical Methods in Natural Language Processing (EMNLP), pp. 740–750 (2014)

10. Vaswani, A., et al.: Attention is all you need. Neural Information Processing Systems (NIPS) (2017)

11. Mnih, V., Heess, N., Graves, A.: Recurrent model of visual attention. Advances in Neural Information Processing Systems 27 (NIPS) (2014)

12. Ba, J., Mnih, V., Kavukcuoglu, K.: Multiple object recognition with visual attention. arXiv (2014)

13. Xu, K., et al.: Show, attend and tell: neural image caption generation with visual attention. In: International Conference on Machine Learning (ICML) (2015)

14. Schuster, M., Paliwal, K.K.: Bidirectional recurrent neural networks. IEEE Trans. Signal Process. **45**(11), 2673–2681 (1997)

15. Graves, A., Schmidhuber, J.: Framewise phoneme classification with bidirectional LSTM and other neural network architectures. Neural Netw. **18**(5–6), 602–610 (2005)

16. Zhang, M., Wang, N., Li, Y., Gao, X.: Neural probabilistic graphical model for face sketch synthesis. TNNLS (2019)

# Classifier Fusion Method Based Emotion Recognition for Mobile Phone Users

Luobing Dong[1,2]([✉]) [iD], Yueshen Xu[1] [iD], Ping Wang[2] [iD], and Shijun He[1]

[1] Xidian University, Xi'an 710071, Shaanxi, China
{lbdong,ysxu}@xidian.edu.cn, 18700197078@163.com
[2] National Key Laboratory of Science and Technology on ATR,
College of Electronic Science National University of Defense Technology,
Changsha 410073, Hunan, China
760521407@qq.com

**Abstract.** With the development of modern society, people are paying more and more attention to their mental situation. An emotion is an external reaction of people's psychological state. Therefore, emotion recognition has attached widespread attention and become a hot research topic. Currently, researchers identify people's emotion mainly based on their facial expression, human behavior, physiological signals, etc. These traditional methods usually require some additional ancillary equipment to obtain information. This always inevitably makes trouble for users. At the same time, ordinary smart-phones are equipped with a lot of sensor devices nowadays. This enables researchers to collect emotion-related information of mobile users just using their mobile phones. In this paper, we track daily behavior data of 50 student volunteers using sensors on their smart-phones. Then a machine learning based classifier pool is constructed with considering diversity and complementary. Base classifiers with high inconsistent are combined using a dynamic adaptive fusion strategy. The weights of base classifiers are learned based on their prior probabilities and class-conditional probabilities. Finally, the emotion status of mobile phone users are predicted.

**Keywords:** Classifier fusion method · Emotion recognition · Dynamic adaptive fusion strategy · Machine learning

## 1 Introduction

Nowadays, people have greater material wealth than previous generations. We can access to an abundance of material resources. However, life has become more pressured and challenging than ever before. These pressures come from a wide range of sources, including our workplaces, the education system, our family and friends, etc. The constant pressure that they exert on us can have a profound effect on our mental health. This makes people pay more and more attention

Supported by the Fundamental Research Funds for Central Universities (JB161004).

to their mental health which is fundamentally linked with their physical health. Researchers state that emotion regulation is an essential feature of mental health [1]. Therefore, real-time emotion recognition is very important for people's timely emotional regulation and the clinical diagnosis of mental illness. In recent years, emotion recognition has been studied as a promising technology in a broad range of mental health monitor [2–4].

Emotion recognition has also been an important research topic in the fields of artificial intelligence owing to its significant academic and commercial potential [5]. By 2019, AI-based emotion detection has become a 20 billion dollars industry [6]. Scientists typically identify human emotion based on physiological signals including facial or verbal expressions [7], skin temperature [8], blood volume pulse [9], electromyographic signal [10], etc. However, these signals always need special bio-sensors or equipment to get. This makes trouble to its application. People and their doctors often hope to monitor their emotion changing process to help them do some judgments and decisions. For example, some doctors want to establish a connection between some clinical symptoms and patients' emotion status. To do that, patients always need to rent or buy some special carry-on equipment. This is inconvenient and may cause negative emotions. We know that life is filled with ups and downs. It is common for people to have shifts in emotions. Therefore, it is very important that the emotions of the observed person are unconsciously recorded and recognized in time.

Over recent decades the pace of the Internet has gradually changed from personal computer (**PC**) storm to smart phone competition. Not only are mobile phones more affordable, but mobile phones also are becoming more powerful in functions. People spend more and more time on mobile phones everyday. And most people carry their mobile phones with them all the day. Furthermore, sensor devices are no longer the special configuration of high-end mobile phones. Most mid-end and low-end mobile phones are equipped with dozens of sensor devices, such as gravity, gyroscope, pressure, temperature, lighting, etc. These make unconsciously record and recognize uses' emotion possible [11]. The strength of the finger on the screen, the speed of walking, the volume of the sound, etc. indicate different emotion status of users.

The human emotion status is an important factor in creating good mental or physical health. Therefore, identifying factors that directly influence the emotion of individuals and using them to predict human emotional state in real-time will have enormous societal benefits. In this paper, we develop a novel method to identify the emotion status of mobile users by collecting environmental and physiological features using sensors of their mobile phones. We track daily behavior data of 50 student volunteers using sensors on their smart-phones. Then a machine learning based classifier pool is constructed with considering diversity and complementary. Base classifiers with high inconsistent are combined using a dynamic adaptive fusion strategy. The weights of base classifiers are learned based on their prior probabilities and class-conditional probabilities. Finally, the emotion status of mobile phone users are predicted.

The rest of this paper is organized as follows. The related work is discussed in Sect. 2. Then, the overview of the proposed method is presented in Sect. 3. In Sect. 4, we discuss the methodology. The experiment evaluation is presented in Sect. 5. Finally, we conclude the paper in Sect. 6.

## 2    Related Work

Emotion recognition methods can be easily divided into three categories: text analysis based emotion recognition, facial and verbal expression analysis based emotion recognition, and physiological signal analysis based emotion recognition. In this section, we will introduce this three categories separately.

There are two main types of text analysis based emotion recognition methods: word classification based method and semantic analysis based method. The former requires a special dictionary which stores as many emotional scores of emotionally salient words as possible. With this dictionary, the emotional scores of all emotionally salient words which are contained in a piece of text can be obtained. Then the emotion that this text wants to express can be identified by aggregating these scores. Identifying and collecting as many emotionally salient words as possible is critical to the effectiveness of this method. Therefore, Many studies focus on how to find emotionally salient words and score them [12,13]. The latter performs the emotion recognition of the corresponding text through the semantic network. This method depends heavily on the richness of the semantic knowledge base which it uses. Baccianella presents SentiwordNet which is devised for supporting sentiment classification and opinion mining applications [14]. Cambria combines the largest existing public knowledge classification with a natural language-based common sense knowledge semantic network [15]. The multi-dimensional extension of the generated knowledge base is used for sentiment analysis.

The rapid development and rise of the deep learning neural network make the analysis of images, text, and speech more effective. This makes emotion recognition based on facial and verbal expression a hot topic. Different expressions often reflect different emotions. So we can judge people's emotions by analyzing people's facial expressions and accompanying facial muscle movements. For example, when the corner of the mouth curves up and wrinkles radiate out from the corners of the eyes, we can usually judge the emotion status of someone as happy. When someone's eyes pop out and his brows are puckered in a frown. Obviously, his emotional status at this time is anger. Facial expression based emotion recognition may be either based on local features or based on overall features. The former is based on the fact that people in different situations have different shapes, sizes, and relative positions of facial features. Considering the difference of the overall facial features under different emotions, the latter is based on the overall facial feature. And the range of extracted features is the entire face. Researchers have made many achievements in this field. For example, Affectiva creates a facial expression dataset with a large number of facial expression pictures of different races, ages, and genders [16]. Then Affectiva design artificial intelligence algorithms to identify people's emotions by observing all the changes of facial expression such as changes of textures and wrinkles of the face, changes in the shape of the facial features, etc.

Many people always wear a poker face in their normal life. They like emotionless expressions that give no indication of their thoughts or intentions. This limit the effective of facial and verbal expression based emotion recognition technology. Therefore, researchers try to identify people's emotion by analyzing their physiological signals [17–21]. Liu et al. propose a real-time emotion recognition system based on **EEG** signals. The advantage of this system is that it is combined with the clip database [22]. Wang et al. study various emotional characteristics of EEG signals, track the changes of EEG signals. They establish the connection between EEG characteristics and emotion status and finally identify the classification of emotions [23]. Machine learning has a good effect on the fusion of multimodal information, such as the fusion of various physiological features and EEG signals. This makes machine learning based algorithms have good performance on separating signals with high spatial and frequency dimensions [24, 25].

## 3    Overview of the Method

In this section, we introduce the main idea of the proposed method. Figure 1 shows its flow chart. We choose 50 student volunteers from Xidian University. They install our data acquisition application EAmobile on their Android mobile phones. EAmobile can collect their daily emotion related information through the gravity sensor, pressure sensor, gyroscope sensor, speed sensor and voice sensor. This collection process lasts for half a year. The volunteers help us label their information with their real emotion status. Their labeled information are transferred to our remote server. After preprocessing and feature extraction, the data is stored in Mysql database. This data is used to train our fusion classifier. The trained model is used to predict the real-time emotion status of other mobile phone users based on their instantaneous mobile phone sensing information.

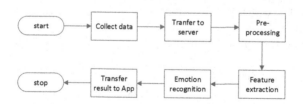

**Fig. 1.** The flow chart of the proposed method

Inspired by [26], we propose an emotional model considering the unsuitability of discrete variables for linear analysis and transformation. It is shown in Fig. 2. This model is consist of two dimensions: activity level and pleasure level. The pleasure level is used to measure the positive or negative degree of the user. The activity level is used to measure the degree of proactiveness or passiveness of the user. The coordinate system is established with the level of pleasure as the x-axis and the level of activity as the y-axis. Then the corresponding emotion status can be positioned in the coordinate system.

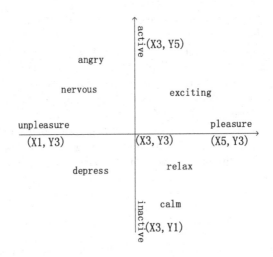

**Fig. 2.** The emotional model

We divide the pleasure level and activity level into five levels separately. We set $(X3, Y3)$ as the origin. Then $(X1, Y1)$ represents very unpleasant and very inactive. $(X5, Y5)$ represents very pleasant and very active. The statistics result based on the collected information shows that most emotion status of the mobile users stay between the third and fourth levels. This means that the value of their emotion status $(Xi, Yi)$ satisfies: $X3 \leq Xi \leq X4$ and $Y3 \leq Yi \leq Y4$. The result indicates that users always are in calm status at most time. And the probability of being in a happy status is greater than the probability of being in a sad status. This is obviously reasonable. Table 1 shows the users' emotion status distribution which is calculated based on our statistics.

**Table 1.** The distribution (%) of users' emotion status.

| Pleasure level \ Activity level | $X1$ | $X2$ | $X3$ | $X4$ | $X5$ |
|---|---|---|---|---|---|
| $Y1$ | 0.25 | 0.23 | 0.21 | 0 | 0 |
| $Y2$ | 0 | 6.79 | 2.18 | 0.32 | 0.41 |
| $Y3$ | 0.57 | 5.62 | 42.06 | 1.32 | 0 |
| $Y4$ | 0.78 | 1.56 | 20.05 | 12.57 | 0.78 |
| $Y5$ | 0 | 0 | 0.58 | 1.95 | 1.77 |

# 4    Methodology

This section focus on the detailed classifier fusion method used in this paper. We first describe how to eliminate the useless data contained in the collected daily behavior data. Then, we explain the extracted features in detail, followed by description to the used classifier fusion method.

## 4.1    Data Pre-processing

Data obtained directly through mobile phone sensors cannot be used directly. Because the collected raw data has many problems such as noise, incomplete, inconsistency, etc. There are many reasons for this useless. For example, the application may not be used correctly. Data collection process may be disrupted or abort by device failures or interference. So we should pre-process the data so that all data can be used. In this paper, the following methods are used for data pre-processing:

- Filtering illegal data. We filter illegal data in the system artificially to avoid importing them into the emotion recognition process and confusing classifiers.
- Eliminating incomplete data. If a large number of data items are missing in a certain sample, the system will reject the sample. If there are only very few items miss in a sample, we just fill it with the default value.
- Medium filtering. We perform median filtering on the acquired waveform data. This method can make the waveform bumpier. Then most key features will be more clear.

## 4.2    Feature Extraction

In this paper, we collect mobile users' daily behavior data through a lot of mobile phone sensors. Data from different sensors have different features which are related the emotion status of users. Table 2 shows the details of these features. Due to the length limitation of this paper, we just give the feature detail of light sensor as an example.

Data which is collected through the light sensor has notable and distinctive values. When we put the phone in a confined space such as a pocket or a bag, the output value of light sensor is small. It is basically in $[0, 100]$. When we put it under normal light, the values are generally concentrated in $[100, 1000]$. When we are outdoors, the values are mostly in $[1000, 2000]$. Therefore, we can easily extract the phone's current status of use by the output value of the light sensor.

## 4.3    Classifier Fusion Method

The classifier fusion method we use can be divided into three phases: generation phase, selection phase, fusion phase. We will describe these three phases separately.

**Table 2.** Features need to be extracted from each types of sensor.

| Sensor | Number of features | Description |
|---|---|---|
| Gravity | 3 | Time of portrait orientation, time of landscape orientation, times of exchanging between these two orientations |
| Gyroscope | 2 | Rotation time, average angular velocity |
| Accelerator | 2 | Total shaking time, severity of shaking, times of shaking |
| Light | 1 | One of following states: no use, outdoor, indoor |
| Pedometer | 2 | Step count, difference between the average speed and the largest speed |
| Global Positioning System (**GPS**) | 1 | Location entropy |
| Network speed | 3 | Fluctuation of speed, speed, strength of signals |

**Generation Phase:** At this stage, a base classifier pool is constructed considering the diversity of base classifiers. We choose most of the classification algorithms based on neural networks and decision trees as the base classifiers. They are trained on different data sets.

**Selection Phase:** At this stage, classifiers which are good for the data set to be classified are selected from the classifier pool for the next phase. Breiman suggested that the more unstable the base classifier the better the combined classifier will achieve [27]. This instability means that even a very subtle change in the training sample may result in a very different final decision classification result. In other words, the more sensitive the base classifier algorithm is to the sample during the learning and training process, the better the final combination result.

Therefore, in order to improve the accuracy and generalization ability of the final decision classifier, it is necessary to select suitable classifiers from the base classifier pool. The selection method generally considers the accuracy and difference of the base classifier. In terms of accuracy, we set a threshold. The base classifier whose accuracy is above the threshold can participate in the selection phase. In terms of difference, we use the inconsistent measurement [28].

For two classifiers $C_i$ and $C_j$, we define their difference as formula 1.

$$\Delta_{i,j} = \frac{S_{\bar{c}_i,c_j} + S_{c_i,\bar{c}_j}}{S_{\bar{c}_i,\bar{c}_j} + S_{\bar{c}_i,c_j} + S_{c_i,\bar{c}_j} + S_{c_i,c_j}} \tag{1}$$

Here, $S_{\bar{c}_i,\bar{c}_j}$ represents the number of the samples that the classifier $C_i$ and $C_j$ both make the wrong prediction. $S_{\bar{c}_i,c_j}$ represents the number of samples that the classifier $C_i$ makes the wrong prediction and $C_j$ makes the right prediction. $S_{c_i,\bar{c}_j}$ represents the number of samples that the classifier $C_i$ makes the wrong prediction and $C_j$ makes the right prediction. $S_{c_i,c_j}$ represents the number of samples that the classifier $C_i$ and $C_j$ both make the right prediction. The difference between the two classifiers $\Delta_{i,j}$ is in the range $[0, 1]$. For all samples, if the two classifiers have the same recognition effect, the difference is 0. A larger value indicates a greater difference between two classifiers.

We assume that there are $n$ classifiers in the base classifier pool. Let $\Gamma_i$ denote the average of the difference between the base classifier $C_i$ and all other base classifiers. It can be defined by formula 2.

$$\Gamma_i = \frac{\sum_{j=1}^{n} \Delta_{i,j}}{n-1} \tag{2}$$

Then the average difference of the base classifier pool can be calculated by formula 3.

$$Avg = \frac{\sum_{i=1}^{n} \Gamma_i}{n} \tag{3}$$

When $\Gamma_i \geq Avg$, it indicates that the base classifier $C_i$ is good. Then $C_i$ will be chosen as a member of the final fusion classifier.

**Fusion Phase:** There are many fusion strategies such as average, majority vote, weight, etc. The average method and majority vote method are not stable in performance and easily affected by extreme values. The weighting method gives different weights to different base classifiers. For example, the reference weight can be given based on the accuracy of the base classifier. The more accurate the base classifier, the greater its weight.

We use a dynamic adaptive weight strategy for classifier fusion. Inspired by Bayesian theory, we use the historical decision of the training data as the prior probability and use the classification confidence of the current input test sample as its class-conditional probability to dynamically compute the weight of each base classifier. In the training phase, we train each base classifier on the training data set and build up the confusion matrix for each classifier. The confusion matrix is be used to measure the apriori behavior of each base classifier. Given $m$ class labels, the confusion matrix of classifier $C_i$ can be calculate as formula 4.

$$M^i = \begin{bmatrix} e_{11}^i & e_{12}^i & \cdots & e_{1m}^i \\ e_{21}^i & e_{22}^i & \cdots & e_{2m}^i \\ \cdots & \cdots & \cdots & \cdots \\ e_{m1}^i & e_{m2}^i & \cdots & e_{mm}^i \end{bmatrix} \tag{4}$$

Here, $e_{lk}^i$ indicates the probability that a sample which belongs to $C_l$ is identified as a member of $C_k$ by the $i$th classifier. We set the confusion matrix which is generated by the training data as the prior probability. In the recognition process, we set the confidence $P_i(D_k|X)$ which is the output of one classifier $C_i$ as the class-conditional probability of class $D_k$. We define the inverse reliability of $C_i$ at $D_k$ according to $D_l$ as formula 5.

$$\phi^i_{kl} = exp(-|P_i(D_k|X) - P_i(D_l|X)|) \tag{5}$$

Then, we combine confusion matrix which is based on the prior probability with the inverse reliability which is based on the class-conditional probability to get the fusion weight of $C_i$. Formula 6 gives the definition of the fusion weight of $C_i$.

$$w_{ki} = \frac{e^i_{kk}}{\sum^m_{l=1,l\neq k} e^i_{kl}\phi^i_{kl} + \sum^m_{l=1,l\neq k} e^i_{lk}\phi^i_{lk}} \tag{6}$$

The confidence of the fusion classifier for class $D_i$ can be got by formula 7.

$$P(D_i|X) = \sum^n_{i=1} w_{ki}P^i(C_k|X) \tag{7}$$

The class with the largest confidence of the fusion classifier is the final classification decision.

## 5   Evaluation

The evaluation results of the proposed system will be described in this section. We choose 8 base classifiers to construct the classifier pool. The original accuracy of these base classifiers on our sample are shown in Table 3.

Table 3. Original accuracy.

| Algorithm | Accuracy (%) | Algorithm | Accuracy (%) |
|-----------|--------------|-----------|--------------|
| LAD | 66.43 | Part Rule | 61.60 |
| CART | 64.87 | MLP | 58.54 |
| REP | 63.76 | SVM | 53.27 |
| DTNB | 62.31 | LR | 51.04 |

We compare three different fusion strategies: majority vote method, accuracy based method and our method. The accuracy of them on our samples are shown in Table 4. All experiments are repeated 1000 times. And the result is the average result of all experiments. We can see that our method can get the largest accuracy. We create two base classifier pools with two different groups of base classifiers. The first one is consist of LAD, REP and CART. These three classifiers have the largest accuracy and average difference. So the accuracy strategy has better performance than the vote strategy. When we change the strategy to the dynamic adaptive strategy with our parameters, the accuracy is significantly improved. In the second group, we replace REP which has the worst accuracy percentage with DTNB. The accuracy of the system is improved under all strategies.

**Table 4.** Experiment results.

| Base classifier1 | Base classifier2 | Base classifier3 | Fusion strategy | accuracy |
|---|---|---|---|---|
| LAD | REP | CART | majority vote | 64.85 |
| LAD | REP | CART | accuracy | 65.49 |
| LAD | REP | CART | dynamic adaptive strategy | 67.53 |
| LAD | CART | DTNB | majority vote | 66.78 |
| LAD | CART | DTNB | accuracy | 68.39 |
| LAD | CART | DTNB | dynamic adaptive strategy | 71.67 |

# 6  Conclusion

In this paper, we present an emotion recognition system for mobile phone users. By collecting behavior data of 50 student volunteers through their mobile phone sensors, we train the dynamic adaptive weight fusion classifier. The system achieve an average accuracy of 71.67%. In the future, we intend to extend our work to combine our system with android smart-phones to manage users' emotions in a real-time environment.

# References

1. Gross, J.J., Muñoz, R.F.: Emotion regulation and mental health. Clin. Psychol.: Sci. Pract. **2**(2), 151–164 (1995)
2. Valstar, M., et al.: Avec 2016: depression, mood, and emotion recognition workshop and challenge. In: Proceedings of the 6th International Workshop on Audio/Visual Emotion Challenge, pp. 3–10. ACM (2016)
3. Trigeorgis, G., et al.: Adieu features? End-to-end speech emotion recognition using a deep convolutional recurrent network. In: 2016 IEEE International Conference on Acoustics, Speech and Signal Processing (ICASSP), pp. 5200–5204. IEEE (2016)
4. Zhao, M., Adib, F., Katabi, D.: Emotion recognition using wireless signals. In: Proceedings of the 22nd Annual International Conference on Mobile Computing and Networking, pp. 95–108. ACM (2016)
5. Ko, B.: A brief review of facial emotion recognition based on visual information. Sensors **18**(2), 401 (2018)
6. Baur, D.: AI-based emotion detection has become a 20B industry (2019). https://www.theguardian.com/technology/2019/mar/06/facial-recognition-software-emotional-science/. Accessed 01 July 2019
7. Li, M., et al.: Facial expression recognition with identity and emotion joint learning. IEEE Trans. Affect. Comput. (2018)
8. Greco, A., et al.: Skin admittance measurement for emotion recognition: a study over frequency sweep. Electronics **5**(3), 46 (2016)
9. Zhao, B., et al.: EmotionSense: emotion recognition based on wearable wristband. In: 2018 IEEE SmartWorld, Ubiquitous Intelligence & Computing, Advanced & Trusted Computing, Scalable Computing & Communications, Cloud & Big Data Computing, Internet of People and Smart City Innovation (SmartWorld/SCALCOM/UIC/ATC/CBDCom/IOP/SCI), pp. 346–355. IEEE (2018)

10. Vijayan, A.E., Sen, D., Sudheer, A.P.: EEG-based emotion recognition using statistical measures and auto-regressive modeling. In: 2015 IEEE International Conference on Computational Intelligence & Communication Technology, pp. 587–591. IEEE (2015)
11. Farhan, A.A.: Modeling Human Behavior using Machine Learning Algorithms (2016)
12. Deng, Z.-H., Luo, K.-H., Yu, H.-L.: A study of supervised term weighting scheme for sentiment analysis. Expert Syst. Appl. **41**(7), 3506–3513 (2014)
13. Khan, F.H., Qamar, U., Bashir, S.: Lexicon based semantic detection of sentiments using expected likelihood estimate smoothed odds ratio. Artif. Intell. Rev. **48**(1), 113–138 (2017)
14. Baccianella, S., Esuli, A., Sebastiani, F.: SentiWordNet 3.0: an enhanced lexical resource for sentiment analysis and opinion mining. In: Lrec, vol. 10, pp. 2200–2204 (2010)
15. Cambria, E., et al.: New avenues in opinion mining and sentiment analysis. IEEE Intell. Syst. **28**(2), 15–21 (2013)
16. McDu, D., et al.: Affectiva-mit facial expression dataset (AM-FED): naturalistic and spontaneous facial expressions collected. In: Proceedings of the IEEE Conference on Computer Vision and Pattern Recognition Workshops, pp. 881–888 (2013)
17. Martini, N., et al.: The dynamics of EEG gamma responses to unpleasant visual stimuli: from local activity to functional connectivity. NeuroImage **60**(2), 922–932 (2012)
18. Frantzidis, C.A., et al.: Toward emotion aware computing: an integrated approach using multichannel neurophysiological recordings and affective visual stimuli. IEEE Trans. Inf. Technol. Biomed. **14**(3), 589–597 (2010)
19. Balconi, M., Mazza, G.: Brain oscillations and BIS/BAS (behavioral inhibition/activation system) effects on processing masked emotional cues: ERS/ERD and coherence measures of alpha band. Int. J. Psychophysiol. **74**(2), 158–165 (2009)
20. Jenke, R., Peer, A., Buss, M.: Feature extraction and selection for emotion recognition from EEG. IEEE Trans. Affect. Comput. **5**(3), 327–339 (2014)
21. Khezri, M., Firoozabadi, M., Sharafat, A.R.: Reliable emotion recognition system based on dynamic adaptive fusion of forehead biopotentials and physiological signals. Comput. Methods Programs Biomed. **122**(2), 149–164 (2015)
22. Liu, Y.-J., et al.: Real-time movie-induced discrete emotion recognition from EEG signals. IEEE Trans. Affect. Comput. **9**(4), 550–562 (2017)
23. Wang, X.-W., Nie, D., Lu, B.-L.: Emotional state classification from EEG data using machine learning approach. Neurocomputing **129**, 94–106 (2014)
24. Balconi, M., Lucchiari, C.: EEG correlates (event-related desynchronization) of emotional face elaboration: a temporal analysis. Neurosci. Lett. **392**(1–2), 118–123 (2006)
25. Iacoviello, D., et al.: A real-time classification algorithm for EEGbased BCI driven by self-induced emotions. Comput. Methods Programs Biomed. **122**(3), 293–303 (2015)
26. Yang, R., Xi, C., Xi, S.: J. Front. Comput. Sci. Technol. **10**(6), 751–760 (2016)
27. Breiman, L.: Bias, variance, and arcing classifiers. Technical report, 460, Statistics Department, University of California, Berkeley (1996)
28. Bowes, D., Randall, D., Hall, T.: The inconsistent measurement of message chains. In: 2013 4th International Workshop on Emerging Trends in Software Metrics (WETSoM), pp. 62–68. IEEE (2013)

# Toward Detection of Driver Drowsiness with Commercial Smartwatch and Smartphone

Liangliang Lin[1,2(✉)], Hongyu Yang[1], Yang Liu[1], Haoyuan Zheng[1],
and Jizhong Zhao[1]

[1] School of Computer Science and Technology,
Department of Telecommunications, Xi'an Jiaotong University,
Xi'an 710004, People's Republic of China
Lin_LL@126.com
[2] Information Department, Xi'an Conservatory of Music, Xi'an 710061,
People's Republic of China

**Abstract.** In the life, there are always many objects that are unable to actively contact with us, such as keychains, glasses and mobile phones. In general, they are referred to non-cooperative targets. Non-cooperative targets are often overlooked by users while being hard to find. It will be convenient if we can localize those non-cooperative targets. We propose a non-cooperative target localization system which based on MEMS. We detect the arm posture changes of the user by using the MEMS sensors which embedded in the smart watch. First distinguish the arm motions, identify the final motion, and then perform the localization. There are two essential models in our system. The first step is arm gesture estimation model which based on MESE sensor in smart watch. we first collect the MEMS sensor data from the watch. And then the arm kinematic model and formulate the mathematical relationship between arm degrees of freedom with and the gestures of watch. We compare the results of the four actions which are important in the later model with the Kinect observations. The errors in the space are less than 0.14 m. The second step is non-cooperative target localization model that based on the first step. We use the 5-degrees data of the arm to train the classification model and identify the key actions in the scene. In this step, we estimate the location of non-cooperative targets through the type of interactive actions. To demonstrate the effectiveness of our system, we implement it on tracking keys and mobile phones in practice. The experiments show that the localization accuracy is >83%.

**Keywords:** Arm gesture · Non-cooperative target · Localization · Smart-watch

## 1 Introduction

There are many small things in life which cannot interact with people actively, such as keychains, glasses and mobile phones. They are collectively referred to as non-cooperative targets. The location of non-cooperative targets is often forgotten by users. If these targets can be located through technology, it will greatly facilitate people's life.

This work is supported by NSFC Grants No.61802299, 61772413, 61672424.

Q. Li et al. (Eds.): BROADNETS 2019, LNICST 303, pp. 227–243, 2019.
https://doi.org/10.1007/978-3-030-36442-7_15

For non-cooperative targets, people often find them by recalling where they were last used (what scene) and then recalling what they were used for (what interactions). However, there are no relevant sensors on the non-cooperative objects, and people cannot live under constant surveillance instead relying on smart devices to record their interactions with the target and guess its location. Today's smart devices are equipped with commercial MEMS sensors, including accelerometers [1–3], gyroscopes [4] and magnetic induction meters [5]. If scenes and interactions could be recorded by using MEMS sensors, it would can help people find non-cooperative targets.

Meanwhile, with the development of science and technology in recent years, people's life has become more and more intelligent. Smart devices emerge in an endless stream. A series of devices begin to enter people's lives, such as smart phones [6], including smart wristbands [7], smart watches and smart glasses [8]. In this paper, we uses smart watch to achieve the localization of non-cooperative targets. The main idea is to record the latest interaction between users and non-cooperative targets through smart watch, and then guess the location of non-cooperative targets. This paper mainly consists of two steps: the arm posture estimation and the non-cooperative target localization.

The rest of this paper is organized as follow:

Section 2 Related work

Related work is mainly about Kinect development. Kinect is an image device with high reliability that can recognize motion. In this paper, we compare the results of skeletal tracking with the results of arm posture estimation model from Kinect.

Section 3 Model Design

This section mainly realizes arm posture estimation model through the data of smartwatch MEMS sensor and non-cooperative target localization model. We take the key tracking in the opening scene and the mobile phone tracking in the calling scene to illustrate the non-cooperative target localization.

Section 4 Simulation and Experiment

This section realizes the visual simulation of the arm posture estimation model and explain the results of our system. Firstly, the arm posture estimation model is tested with common images, and then the scene of non-cooperative target is visualized, and posture changes are also tested. Finally, the localization results of non-cooperative target are tested.

We summarize this paper in Sect. 5.

## 2   Related Work

### 2.1   Kinect

Kinect is released by Microsoft which is a device that interacts with camera sensors [9–11]. It is a 3D device that can recognize objects, perform language operations, and capture objects. Kinect is first announced in 2009, and after years of research and development, Kinect has a richer set of functions and interactive tools.

Kinect relies on several core sensors for its rich functionality. The positions of Kinect infrared (IR) projector, color camera and infrared camera are indicated respectively. Depth information is collected by an infrared projector combined with an infrared camera, both of which are CMOS sensors. An infrared camera is an instrument that emits infrared light using a diffraction grating.

## 2.2   Bone Tracking

Simple arm movement recognition work can be achieved with Kinect. It is done by means of images. When interacting with people, Kinect use bone tracking technology to locate key parts of the body.

### Bone Tracking Technology

In Kinect, human skeleton architecture can be extracted through more than 20 joints [12–14]. When the user enters the Kinect field of vision, the device can represent the user's joint position in space through coordinates $(x, y, z)$. When the computer obtains these coordinates, it can calculate the posture of the body's main limbs in space. Kinect supports simultaneous detection of 6 people, but can only support two skeleton structures at most. In general, the user can track 20 joints while standing and 10 joints while sitting. More specifically, Kinect can provide three kinds of information: (a) the tracking status of relevant bones, only the position of the tester can be detected in passive mode, and the coordinates of the tester's 20 joints can be detected in active mode; (b) give each skeleton a test ID. This ID will be associated with several testers that can be detected to determine which one the skeleton data belongs to. (c) the specific location of the user which is actually the center of mass of the user.

### Kinect Bone Tracking Development

The development of Kinect is described as follows: Firstly, it communicate with Kinect and then take the relevant data of the skeleton, so that skeletal events can be returned and detection functions can be started. Then, the bone tracking function is turned on, which processes the image information from the camera, and reads the bone data.

Secondly, data smoothing is carried out. Mutations in the data may occur while processing the data. For example, in the tracking process, some situations may lead to a large change in the detected position, which is caused by the incoherent actions of the user and the performance problems of the hardware device of Kinect, so jitter should be removed by filtering.

Finally, we transform the frame. Since the depth image data and color image data read by Kinect come from different cameras, and they face different scenes (they are in different positions), the generated images will be different, so the spatial coordinate system needs to be transformed. After that the depth information is removed from the final data and the data are drawn on the plane graph.

# 3   Model Design

## 3.1   Arm Gesture Estimation Model Based on Smart Watch MEMS Sensors

Arm gesture estimation model uses sensors of commercial smartwatches to model and estimate a series of user arm posture changes.

### Kinematics Analysis Model of the Arm

First, we need to introduce the degree of freedom. In the kinematics model of the arm, the degree of freedom is the rotation angle of the joint which affects the arm posture.

There are seven degrees of freedom of the human arm. Since the smart watch is worn on the wrist, We only need to consider the motion state of the upper arm and the lower arm, so it only considers 5 degrees of freedom, including three on the shoulder and two on the elbow. We need to establish three coordinate systems to describe the posture of the arm relative to the human body. The three coordinate systems are: human body, shoulder and elbow.

This paper establishes the right hand system in the human body shoulder, namely human body coordinate system. The origin is the shoulder join. The positive direction of the X axis is from the left shoulder to the right shoulder. The positive direction of the Y axis is the back pointing to the chest. The positive direction of the Z axis is perpendicular to the other two axes. Shoulder coordinate system is the coordinate system established with the upper arm of human body as the reference point. The origin is the shoulder joint. The  axis is parallel to the direction of the zupper arm, with the positive direction pointing from the shoulder to the elbow. In addition, we need to determine another axis for it. Here, when the arm is naturally drooping and the palm is facing the body, the X axis is perpendicular to the direction of the arm pointing to the human body. The elbow coordinate system is similar to the shoulder coordinate system. The origin of the coordinate system is the elbow joint, and the positive direction of the Z axis is from the elbow to the wrist, and the direction of the palm is the X axis.

Now we assume that the length of upper arm is lb, the length of lower arm is ls, the radius of lower arm is rs.

First, deduce the position of the smartwatch in elbow coordinates:

$$P_{we} = (-rs, 0, ls) \tag{1}$$

The subscript $we$ indicates the information of the smart watches in the elbow. In order to determine the watch posture in the space, the X and Z directions of the watch's own coordinate system are introduced here as a description method of the watch posture. Two directions in the elbow coordinate system can be obtained [15]:

$$O_{we} = \begin{pmatrix} 0 & 0 & 1 \\ -1 & 0 & 0 \end{pmatrix} \tag{2}$$

At the same time, the vector $\overrightarrow{S_{we}}$ is introduced to describe the position and state of the smart watch:

$$\overrightarrow{S_{we}} = \begin{pmatrix} P_{we} \\ O_{we} \end{pmatrix} \tag{3}$$

Then, the state of the smartwatch in the shoulder coordinate system is deduced. Consider the rotation first, assuming that there is a transition coordinate system $\overrightarrow{O_t}$ whose origin is at the elbow, but the three axes are parallel to the three axes of the shoulder coordinate system, and the positive direction is the same. Then, according to the rotation matrix R ($\theta_{ef}$, $\theta_{er}$) corresponding to the degrees of freedom $\theta_{ef}$ and $\theta_{er}$, the relationship of the state of the watch in the two coordinate systems is derived [16]:

$$\overrightarrow{S_{wt}} = \overrightarrow{S_{we}} * R(\theta_{ef}, \theta_{er})^{-1} \tag{4}$$

where $\overrightarrow{S_{wt}}$ is the state of the watch in the transition coordinate system. Then consider there is a displacement change between the actual shoulder coordinate system and the coordinate system used for the transition. So we can get:

$$\overrightarrow{S_{ws}} = \overrightarrow{S_{wt}} + \begin{pmatrix} 0 & 0 & lb \\ 0 & 0 & 0 \\ 0 & 0 & 0 \end{pmatrix} \tag{5}$$

Where $\overrightarrow{S_{ws}}$ is the state of the watch in the shoulder coordinate system.

Finally, the state of the smart watch relative to the human coordinate system is calculated. There is three degrees of freedom rotation between the shoulder coordinate system and the human coordinate system, then according to the rotation matrix R ($\theta_{sa}$, $\theta_{sf}$, $\theta_{sr}$) corresponding to the degrees of freedom $\theta_{sa}$, $\theta_{sf}$ and $\theta_{sr}$, the state of the watch in the human coordinate system can be derived:

$$\overrightarrow{S_{wh}} = \overrightarrow{S_{ws}} * R(\theta_{sa}, \theta_{sf}, \theta_{sr})^{-1} \tag{6}$$

Through the above derivation process, the relationship between the position of the watch and the 5 angles can be obtained. Finally, the posture of the smart watch in the human coordinate system can be derived as follows [17]:

$$\overrightarrow{S_{wh}} = \begin{pmatrix} P_{wh} \\ X \\ Z \end{pmatrix} = \left[ \begin{pmatrix} P_{we} \\ O_{we} \end{pmatrix} * R(\theta_{ef}, \theta_{er})^{-1} + \begin{pmatrix} 0 & 0 & lb \\ 0 & 0 & 0 \\ 0 & 0 & 0 \end{pmatrix} \right] \tag{7}$$

The position of the watch almost overlaps with the wrist, so the position of the wrist can be obtained as long as the position of the watch is available. But only one point in the space (the position of the wrist) is not enough to fully describe the posture of the entire arm, but also to obtain the position of the elbow. Here, the position of the elbow is also derived. In the shoulder coordinate system, the position of the elbow $P_{es}$ is as follows:

$$P_{es} = (0, 0, lb) \tag{8}$$

The subscript es indicates the information of the elbow in the shoulder coordinate system. Then, change it into the human coordinate system. The elbow involves two degrees of freedom, $\theta_{sa}$, $\theta_{sf}$, which gives the position of the elbow:

$$\overrightarrow{P_{eh}} = \overrightarrow{P_{es}} * R(\theta_{sa}, \theta_{sf})^{-1} \tag{9}$$

## DOF and Arm Posture Mapping Relationship

In this section, the mapping relationship between the posture of the watch and wrist position, elbow position and 5 degrees of freedom will be established on the basis of last subsection. The mapping here is one to many. That is, given a watch posture, the corresponding other information value is a few. The set of values is called the solution space.

Steps of acquire the mapping relationship between the posture of the watch and other information in body coordinate system are: First, traverse the given 5 degrees of freedom, respectively, and then calculate the values of the corresponding three keys for a given degree of freedom, namely the wristwatch, wrist and elbow position, and finally use the watch posture as the key, and other information as the value to establish a mapping relationship.

**Posture Acquisition**

We obtain the real posture of the smartwatch in the body coordinate system through the data of the smartwatch sensor. It mainly includes two steps: first, obtain the direction of the smartwatch in the world coordinate system, and then obtain the direction of the user in the world coordinate system, so as to obtain the posture of the smartwatch in the user coordinate system.

We get the direction of the smartwatch by combining sensors such as gyroscope, accelerometer and magnetic induction meter. The main steps are: integral calculate the real-time Angle of gyroscope, then when the watch is still, use accelerometer and magnetic induction meter to calibrate Angle obtained by integral calculation.

Next, the connection between the smartwatch and smartphone is used to determine the direction the user is facing. Data from the smartphone's accelerometer, gyroscope and magnetic induction meter are measured and calibrated to determine the user's orientation in the world coordinate system. Use the watch posture in the world coordinate system "minus" user orientation in the world coordinate system, we can obtain the watch posture in the body coordinate system.

**Constraints**

In the mapping relationship, each state of the watch corresponds to the multiple freedom combination of arms and the position of the smartwatch. The data of degree of freedom is relatively abstract, which is not conducive to intuitive display, while the position of the wrist (or elbow) can be clearly visualized. In this paper, the wrist (or elbow) is selected to draw the state cloud.

The state cloud which corresponds to the wrist position in the circle motion, exbtract 20 time points in the whole action, calculate the 20 time points gesture of the watch, and draw the same color graph of multiple wrist positions corresponding to each time point. State clouds of 20 time points are drawn on the same graph to form the state cloud of this action. To distinguish, the state cloud at the same moment becomes a layer of state cloud.

There are multiple wrist positions at each moment, and the estimation results on this basis will be greatly deviated. If the number of corresponding state points at each moment can be reduced, the accuracy of the results will be greatly improved. Therefore, in this subsection, constraints are introduced to compress and further optimize the solution space to make the final "estimation" result more accurate.

The constraint of solution space in this paper mainly includes the following four points:

(1)  Continuity of arm posture;
(2)  Arm posture is limited by human structure;
(3)  Arm posture changes can be captured by sensors;
(4)  The frequency of different arm posture is different.

**Estimation of Arm Posture Change**

In the previous subsection, we obtained the state cloud corresponding to the arm posture change. This subsection will estimate on this basis. Here, we need to select the most likely state from each layer of state cloud as the current state, and then string the most likely state points at different consecutive moments to obtain the change trajectory of wrist position over a period of time. This trajectory is referred to the trajectory line of wrist in this paper. Similarly, there are trajectory lines of elbow.

The method of estimating trajectory lines is mainly divided into two steps. The first step is to use the particle points of the state cloud for estimation directly, and the second step is to smooth the trajectory lines of state changes by a filter.

### 3.2 Non-cooperative Target Localization Model Based on Arm Gesture Estimation Model

This subsection will use the DOF data obtained in the previous subsection to capture the historical interaction between the target object and the user's arm, and then estimate the destination of the non-cooperative target object in the scene.

**Non-cooperative Target Localization Scenes**

In this subsection, the scenes involved in non-cooperative target localization will be described in detail. This includes an introduction to the scenes of the non-cooperative target and a description of the types of actions in the scene.

First of all, it is necessary to define the non-cooperative target localization scene that can be studied in this paper, that is, to explain which non-cooperative targets can be located and in which scene. The scenes in this paper contain two characteristics: 1. There are obvious scene triggering actions when someone enters the scene, so that the smartwatch knows what kind of environment it enters and what target it tracks. Here, the scene is bound to the tracking target. For example, in the opening scene, only the key is tracked; in the smoking scene, the lighter is tracked. 2. A series of interactive actions are carried out between the hand of the user wearing the smartwatch and the non-cooperative target, so that the smartwatch can estimate the destination of the target.

We picks two simple scenes for analysis.

Scene 1: Track the key chain in the opening scene.
Scene 2: Track the phone in the calling scene.

In general, the action is mainly divided into two categories in related scenes: trigger action and historical interactive action.

Trigger action: action indicates entry into a scene. When the trigger scene is detected, it means that the user has entered the scene and the watch can record the following actions. For example, the opening action in the opening scene.

Historical interaction action: after triggering the action, the user enters the scene, interacts with the non-cooperative target, and places it somewhere. The action of placing it somewhere is the historical interactive action studied in this paper. By identifying the historical interaction, we can guess the final destination of the detection target.

There are two scenes in this paper, namely, tracking the key chain in the opening scene and tracking the mobile phone in the calling scene. The trigger scene here is a door opening and a phone call.

(1)  Opening action: it refers to the process of taking out the key, inserting it in the key hole and turning the key to open the door after the user arrives at the door.

(2)  Call action: the user takes the phone from a certain location and holds it to his ear to hold the call.

There are two reasons for dividing actions into trigger actions and interactive actions. First, trigger actions represent a boundary that defines which actions will be related to non-cooperative goals of interest. The action after the trigger action is needed to predict the action of the non-cooperative target position. Then, the sampling rate of the system can be controlled. The strategy we adopt is to collect a small amount of data at the ordinary sampling rate for analysis to determine whether the scene is entered. If the scene is entered, the strategy of high sampling rate will be adopted.

The scenes in this paper are not complicated, and the historical interactive actions mainly include: putting them into coat pockets, putting them into trouser pockets, putting them on the table, and throwing them onto some objects nearby (such as sofa, bed, etc.). In addition to the four types of actions related to where the target is going, there are also some unrecognizable actions. These 5 types of actions are collectively referred to as historical interactive actions. They are: load top A1, load bottom A2, put on top of something A3, throw out A4, and other actions A5.

### Action Segment Segmentation

This subsection will segment the action based on the data of the variation of the freedom of the arm. In the process of moving arms, the movement tends to move from one resting position to another, rather than moving at any given time. If these periods of rest throughout the moving progress can be found, then we can segment the action. The specific approach is as follows: first, track the rest point, extract a short period of sliding window. In this window, if a few degrees of freedom change is detected to be very small, then this segment can be used as a candidate set of rest points. Then the data changes between the rest points are analyzed. Data changes between rest points need to exceed a certain threshold to prevent small changes from affecting the final results. The non-conforming rest points will be excluded. Through the above algorithm, one action segment can be obtained.

### Identification of Key Actions

When the action segment is obtained, each action segment can be identified. We will identify key actions in three steps.

Firstly, feature extraction is carried out to extract the most closely related parameters of posture change, and then dimensionality reduction of redundant features is carried out. Finally, the classifier's parameters are obtained by training the eigenvector.

In order to tell which, action each segment is, features need to be extracted. This paper makes use of the following features, which will play a key role in the later action recognition:

(1) mean

The mean is the mean of the signals. The mean value can reflect the approximate fluctuation position of signal, and its calculation formula is as follow:

$$\mu_x = \frac{1}{N} \sum_{n=0}^{N-1} x_n \tag{10}$$

(2) variance

The variance can represent the fluctuation of the whole data above and below the mean value. Its calculation formula is as follow:

$$\sigma_X^2 = \frac{1}{N} \sum_{n=0}^{N-1} |x_n - \mu_x|^2 \tag{11}$$

(3) kurtosis

Kurtosis represents the peak value of the image at the average value. Intuitively, it represents the height of the image tip. Its calculation formula is as follow:

$$\frac{\sum_{n=1}^{N} (x_n - x)^4}{(N-1)s^4} \tag{12}$$

(4) signal power RMS

Here RMS is actually the square root of signal power. Signals carry energy, and signal power is a measure of the energy carried by a signal. Its calculation formula is $\sqrt{P_x}$.

Where the * calculation is as follow:

$$P_x = \frac{E_x}{N} = \frac{1}{N} \sum_{n=0}^{N-1} |x_n|^2 \tag{13}$$

We select 5 features, which means we need to reduce the dimension of the 25 dimensional vector corresponding to 5 degrees of freedom.

In this paper, principal component analysis (PCA) is used to realize data dimension reduction. PCA is an important algorithm in statistics, which is mainly used to reduce redundancy and extract main features. PCA can reduce the dimension of feature in the last subsection greatly. The next step is to use these features for identification.

The types of actions have been described in the previous subsection. What need to be identified here are the trigger actions such as opening the door, answering the phone, and the actions such as loading top, loading bottom, putting on something, and throwing out. Opening the door and answering the phone are identified as a group, and the last four actions are identified as a group, but the recognition method is universal.

This paper uses support vector machine (SVM) for classification. This is a multi-classification problem. For example, in the first group, three categories of actions, such as opening the door, answering the phone and other actions, need to be distinguished. In this paper, one-to-one support vector machine model is used. It trains the support vector machine model for any two kinds of actions in the same group and finally judges the categories according to the voting results of all the models.

### Non-cooperative Target Localization

In the previous section, several actions that need to be recognized are identified by SVM. In this section, the location/destination of non-cooperative targets is tracked according to the recognition results in the previous subsection.

In this paper, Fig. 1 is used to represent the relationship between motion recognition and the location/destination of non-cooperative targets (i.e. keys and mobile phones). In Fig. 1, Scene judgement identifies which scene it is, and the location is determined by the action, including clothes, table/cabinet, sofa/bed. In this paper, the smartwatch is first used to collect data for action judgment to determine whether a scene is entered. After judging the scene, the following actions can be judged, and these historical interactive actions will be directly linked to the location of the non-cooperative target.

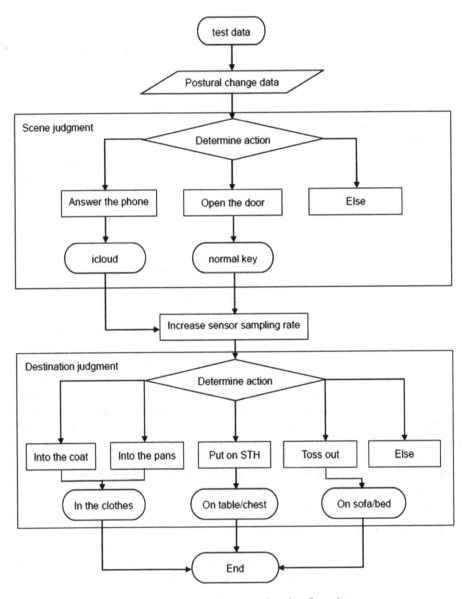

**Fig. 1.** Non-cooperative target location flow chart

So far, this paper has connected the data of arm posture change, classification model and the location of non-cooperative targets together, so as to realize the localization of non-cooperative targets in the scene.

# 4  Simulation and Experiment

In Sect. 3, the arm posture estimation model based on smartwatch MEMS sensor and the non-cooperative target model based on arm posture estimation are respectively established. This section will carry on the simulation, the experiment, and carry on the analysis to the result.

## 4.1  Analysis of Arm Posture Estimation Model Results

This subsection will analyze the reliability and accuracy of the arm posture estimation model. This paper will compare the results of arm posture estimation model based on smartwatch with the results of Kinect.

The whole testing process is carried out in the laboratory environment. A total of 5 participants were invited to test each action multiple times. The testing process mainly includes 3 steps: collecting data of smartwatch, collecting images of Kinect device, and comparing the calculated results of the computing platform with the results of Kinect.

Smartwatch data collection: the tester wears the LG G Watch on the left wrist, with the surface facing the same direction as the back of the hand; The tester stands in a room with few electronic devices (to reduce the effect of magnetic fields) and faced the Kinect camera to perform the relevant action tests.

Calculation results of the platform are compared with the results of Kinect: after collecting all sensor data of the smartwatch, it is transmitted to the computing platform; Calibrate and process sensor data of smartwatch on Matlab, and obtain relevant parameters; The results of relevant parameters were compared with the data collected by Kinect in Matlab.

This paper selects four actions for verification, namely circle, "S" curve, straight line and square. Take a set of test data of drawing circle for display. The tester wears the watch and draws an arbitrary circle in the space. Figure 2 is the trajectory line of the elbow and wrist corresponding to the circle drawing, where the figure on the left is the result of estimation and the figure on the right is the result from Kinect.

In Fig. 2, you can see that the right and left trajectories have the same trend, although they differ in subtle shapes. In the later part of this paper, when locating the non-cooperative target, it is enough to identify the action itself as long as the trend is consistent, without requiring the coordinate points to be exactly identical.

Here, the errors in the circle drawing process were counted, and the cumulative distribution errors of the circle drawing results were randomly counted for 5 times respectively.

The errors are calculated on three axes, and the results are shown in Fig. 3. It can be seen that the wrist error and elbow error are respectively 0.27 m and 0.15 m, the accuracy is well guaranteed. When drawing circles, the wrist error is larger than the elbow error, because when doing the action, the wrist motion range is larger than the

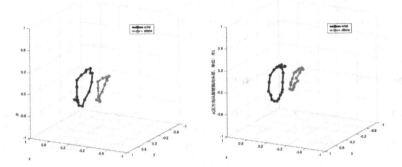

**Fig. 2.** Model estimation results and Kinect observation results of drawing circle trajectory line (unit: meter)

elbow, the error has such a relationship is not surprising. And $Z$ axis error than the other two axes error is large, it is related to the error of Kienct in depth.

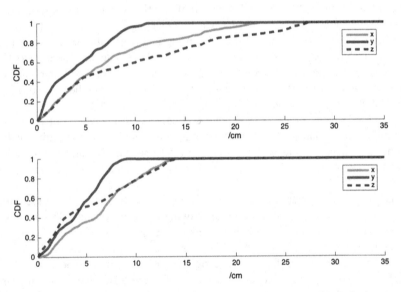

**Fig. 3.** Cumulative distribution of wrist and elbow position errors in circle drawing

## 4.2 Visualization and Recognition Results of Non-cooperative Target Locating Scene

In this subsection, the scene of non-cooperative target locating will be specified to make it visible, and the rationality of the result can be judged directly through the visualization results of the model. This subsection will examine the door opening and phone calling scenarios, as described in Sect. 3.

**Opening the Door Scene**

Compare the estimated results of the smartwatch with the results of Kinect, as shown in Fig. 4. In Fig. 4, blue is the track line of the wrist and red is the track line of the elbow. It can be seen that the trend of the two segments of track lines remains consistent, and the errors of the whole process on the three axes are 0.12 m, 0.14 m and 0.14 m respectively. The estimated results of the smartwatch on the left clearly show the process of turning a key or a door handle, while Kinect's description of the process is not accurate enough. This is because when the wrist and elbow are at the same height, the key points of the wrist and elbow are next to each other in the bone tracking, and the interaction between the two makes it impossible to describe the state in detail.

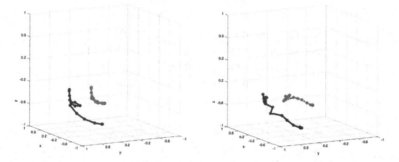

**Fig. 4.** Comparison figure of track lines of two devices (unit: meter) (Color figure online)

**Calling Scene**

From Fig. 5, it can be seen that the track line of the whole phone call is relatively simple, and Significant stagnation can be seen at the top of the trajectory line, which corresponds to the state of the phone at the ear. Compared with the results of Kinect, the errors of the three axes were 0.11 m, 0.13 m and 0.15 m respectively.

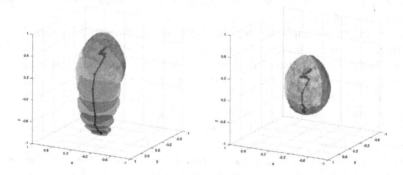

**Fig. 5.** The trajectory of the wrist and elbow during a phone call (unit: meter)

**Historical Interactive Action**

The results of the historical interaction actions involved in the following scenes are described here. Several related historical interactions have been illustrated in the previous section, they respectively are load top, load bottom, put on something and throw. It's also necessary to show the accuracy of their recognition, to prove later non-cooperative target test is conducted on the basis of the accuracy of guarantees. The results are shown in the following table (Table 1):

**Table 1.** The result of historical interactions

| Action | Average wrist distance | | | Average elbow distance | | |
|--------|------|------|------|------|------|------|
|        | X | Y | Z | X | Y | Z |
| Into the coat | 0.10 | 0.08 | 0.14 | 0.14 | 0.09 | 0.12 |
| Into the pants | 0.09 | 0.09 | 0.15 | 0.09 | 0.07 | 0.14 |
| Put on STH | 0.04 | 0.12 | 0.14 | 0.11 | 0.08 | 0.12 |
| Toss out | 0.08 | 0.09 | 0.12 | 0.06 | 0.06 | 0.09 |

It can be seen from the above table that smartwatches have high accuracy in identifying the following actions. Next, this paper will use the data of 5 degrees of freedom changes obtained by the estimation model to locate the non-cooperative targets.

## 4.3    Results Analysis of Non-cooperative Target Locating

**Test Preparation**

This section describes the preparation of a test for a non-cooperative target location. It describes the entire test process and details.

The whole testing process of this paper is carried out in the laboratory environment. The main points include two steps: the smartwatch collects data, and the computing platform trains the data.

Data collection of smartwatch: the experiment invited 5 testers to perform 70 times of 2 trigger actions and 4 historical interactive actions respectively. They also did 70 other random actions. So there are seven types of actions, and each type of action has 350 sets of data. The LG G Watch was worn on the left wrist in the same direction as the back of the hand.

The computing platform trains the data: the collected data are preprocessed, reduced and put into model training on the computing platform, and the parameters are calculated to distinguish the actions. For 7 types of actions, 250 groups are randomly selected from 350 groups of data as the training set and 100 groups as the test set. During the extraction of these 250 groups, the data of each tester should be distributed evenly, so as to facilitate later analysis.

Each of the 5 participants was invited to perform 40 door opening and phone call simulations. That is 200 sets of complete test data for each scene, a total of 400 sets of data.

**Test Results and Analysis**

This subsection will process 350 sets of data for each of 7 types of actions. After dividing the data into training set and test set, the training set is used to train the SVM model, and then test the data and explain the results.

First, 350 groups of data were randomly assigned. Using random Numbers, 250 groups were selected as the training set and 100 groups as the test set. Each group corresponds to 7 actions, namely 7 pieces of data. Here, the tester is required to make a single action each time when collecting data, so as to avoid the workload of dividing the action segments and directly conduct follow-up training.

Then, the arm posture estimation model is used to obtain the changes of the 5 degrees of freedom of these 350 sets of data. The input of non-cooperative target locating is the change data of 5 degrees of freedom. Next, it is necessary to extract features from the data of these degrees of freedom changes and conduct dimensionality reduction processing on these features.

Finally, the training set is put into the support vector machine to train, record the parameters, and use the test set to verify the accuracy and recall rate of the model. Here, 3 support vector machines are trained by using door opening data, phone call data and other movement data, and 10 support vector machines are trained by using 4 movements and other movement data. The final test results are shown in the following table:

**Table 2.** Action test results (unit: %)

| Action | Open the door | Answer the phone | Into the coat | Into the pants | Put on STH | Toss out | Else |
|---|---|---|---|---|---|---|---|
| Open the door | 93 | 2 | | | | | 5 |
| Answer the phone | 2 | 91 | | | | | 7 |
| Into the coat | | | 95 | 1 | 1 | 0 | 3 |
| Into the pants | | | 2 | 91 | 3 | 0 | 4 |
| Put on STH | | | 0 | 3 | 88 | 3 | 6 |
| Toss out | | | 1 | 2 | 2 | 87 | 8 |

Where the columns represent the actual categories of data and the rows represent the categories assigned. Among them, opening the door, answering the phone and other actions are the first group, indicating the distinction of scenes. Putting in coat, putting in trousers, putting it on something, throwing it out and other actions are in the second group to indicate the division of direction. You can see that the classification has good accuracy, which is mainly related to the complexity of the scene. Among them, more than 87% of the samples were correctly judged, and the misjudged movements were mainly other movements, but the proportion was relatively small. Next, the whole process will be tested with the complete scene test data.

It can be seen that the accuracy of Table 3 is lower than that of Table 2, mainly because the overall error is jointly determined by the error of scene judgment and the error of direction judgment. Among them, the main reason for wrong judgment results is that the other actions of the tester are casual, and some actions are similar to the key actions concerned in this paper, which leads to the inaccurate classification results of the classifier. However, the final accuracy rate was more than 76% and the overall accuracy rate was 83%, which enabled the estimation of the destination of non-cooperative targets.

**Table 3.** Scene test results (unit: %)

| Action | Open the door | | | | Answer the phone | | | |
|---|---|---|---|---|---|---|---|---|
| | Into the coat | Into the pants | Put on STH | Toss out | Into the coat | Into the pants | Put on STH | Toss out |
| Accuracy rate | 90 | 86 | 84 | 80 | 88 | 82 | 78 | 76 |

## 5  Conclusion

This paper realizes the locating of non-cooperative targets based on smartwatch, which mainly includes the arm posture estimation model and the non-cooperative target locating model.

This paper verifies and tests the model, and the results are as follows:

(1)  arm attitude estimation model results

In the experimental part of this paper, 5 testers are invited to draw 4 kinds of images in the air, and the triaxial error is no more than 0.30 m. This paper also identifies the scene actions that need to be involved in the locating of non-cooperative targets, and the error is no more than 0.14 m.

(2)  results of non-cooperative target locating

In this paper, 5 testers were invited to perform a total of 350 times of 7 types of actions involved in the scene, 7*250 groups of data were used as the training set, and 7*100 groups of data were used as the test set, with the accuracy of action is more than 87%. In this paper, the testers demonstrated a total of 400 scenes, and the average accuracy of non-cooperative target locating is no less than 83%.

We believe that the model in this paper will be applied to more scenes in the future.

# References

1. Nelson, E.C., Verhagen, T., Noordzij, M.L.: Health empowerment through activity trackers: an empirical smart wristband study. Comput. Hum. Behav. **62**, 364–374 (2016)
2. Li, S.Y., et al.: An exploration of fall-down detection by smart wristband. Appl. Mech. Mater. **687–691**, 805–808 (2014)
3. Cheng, L., Shum, V., Kuntze, G., et al.: A wearable and flexible Wristband computer for on-body sensing. In: 2011 IEEE Consumer Communications and Networking Conference (CCNC), pp. 860–864. IEEE (2011)
4. Al-Nasser, K.: Smart watch: US, US8725842 (2014)
5. Rauschnabel, P.A., Brem, A., Ivens, B.S.: Who will buy smart glasses? Empirical results of two pre market-entry studies on the role of personality in individual awareness and intended adoption of Google Glass wearables. Comput. Hum. Behav. **49**(8), 635–647 (2015)
6. Ji, S., Xu, W., Yang, M., et al.: 3D convolutional neural networks for human action recognition. IEEE Trans. Pattern Anal. Mach. Intell. **35**(1), 221 (2013)
7. Lei, J., Li, G., Zhang, J., Guo, Q., Dan, T.: Continuous action segmentation and recognition using hybrid convolutional neural network-hidden Markov model model. Comput. Vis. IET **10**(6), 537–544 (2016)
8. Martínez, F., Manzanera, A., Romero, E.: Spatio-temporal multi-scale motion descriptor from a spatially-constrained decomposition for online action recognition. Comput. Vis. IET **11**(7), 541–549 (2017)
9. Scovanner, P., Ali, S., Shah, M.: A 3-dimensional sift descriptor and its application to action recognition. In: Proceedings of the 15th ACM International Conference on Multimedia, pp. 357–360 (2007)
10. Blank, M., Gorelick, L., Shechtman, E., Irani, M., Basri, R.: Actions as space-time shapes. In: Proceedings of the Tenth IEEE International Conference on Computer Vision, pp. 1395–1402, 17–20 October 2005 (2005)
11. Csurka, G., et al.: Visual categorization with bags of keypoints. In: ECCV (2004)
12. Ogata, M., Imai, M.: SkinWatch: skin gesture interaction for smart watch. In: Proceedings of the 6th Augmented Human International Conference, pp. 21–24. ACM (2015)
13. Laput, G., Xiao, R., Chen, X.A., Hudson, S.E., Harrison, C.: Skin buttons: cheap, small, low-powered and clickable fixed-icon laser projectors. In: Proceedings of the 27th Annual ACM Symposium on User Interface Software and Technology, Honolulu, Hawaii, USA, 05–08 October 2014 (2014)
14. Weigel, M., Mehta, V., Steimle, J.: More than touch: understanding how people use skin as an input surface for mobile computing. In: Proceedings of the SIGCHI Conference on Human Factors in Computing Systems, Toronto, Ontario, Canada, 26 April–01 May 2014 (2014)
15. Parate, A., Chiu, M.C., Chadowitz, C., et al.: RisQ: recognizing smoking gestures with inertial sensors on a wristband. In: International Conference on Mobile Systems, MobiSys, p. 149 (2014)
16. Xu, C., Pathak, P.H., Mohapatra, P.: Finger-writing with smartwatch: a case for finger and hand gesture recognition using smartwatch. In: International Workshop on Mobile Computing Systems and Applications, pp. 9–14. ACM (2015)
17. Komninos, A., Dunlop, M.: Text input on a smart watch. IEEE Pervasive Comput. **13**(4), 50–58 (2014)

# Security and Privacy

# Fast Algorithm for the Minimum Chebyshev Distance in RNA Secondary Structure

Tiejun Ke, Changwu Wang[✉], Wenyuan Liu, and Jiaomin Liu

Yanshan University, Hebei 066004, China
wcw@ysu.edu.cn

**Abstract.** Minimum Chebyshev distance computation between base-pair and structures cost most time while comparing RNA secondary structures. We present a fast algorithm for speeding up the minimum Chebyshev distance computation. Based on the properties of RNA dot plots and Chebyshev distance, this algorithm uses binary search to reduce the size of base pairs and compute Chebyshev distances rapidly. Compared with $O(n)$ time complexity of the original algorithm, the new one takes nearly $[O(\log_2 n), O(1)]$ time.

**Keywords:** RNA secondary structure · Minimum Chebyshev distance · RNA secondary structure comparison

## 1 Introduction

RNA is a multifunctional molecule that can be used as messenger, enzyme and structural component, which plays an important role in the whole life activity of organisms. Research on RNA not only can improve the understanding of basic biological processes but also facilitates the development of drugs, biotechnology, and genetic engineering. The function of the biological macromolecule mainly depends on its structure [1, 2]. Only when the corresponding structure of the molecule is obtained can the function and biochemical process of the molecule be explored in depth. At present, the most accurate measurement methods of RNA molecular structure, require a large amount of investment in time and economy, such as NMR, x-ray crystallography and cryogenic electron microscopy. It also requires considerable technical expertise. In that way we can see, it has very important realistic significance to use computational algorithms to predict RNA structures.

RNA structure prediction can be divided into secondary structure prediction and tertiary structure prediction. RNA secondary structure is one of the fundamental bases of its tertiary structure prediction, and can effectively improve the efficiency of tertiary structure prediction [3–5].

RNA secondary structure prediction methods includes single sequence prediction based on dynamic programming [6–9] or context-free grammar [10–12], multi-sequences comparison based on similar functions with similar structure, soft computing algorithms [13] such as genetic algorithms [14–17], simulated annealing [18, 19], artificial neural networks [20–23], and fuzzy logic [24]. These methods still have space to promote accuracy and performance [25].

Q. Li et al. (Eds.): BROADNETS 2019, LNICST 303, pp. 247–260, 2019.
https://doi.org/10.1007/978-3-030-36442-7_16

The above prediction methods often involve judging the similarity between different structures. For example, [26] generates many candidate structures and then chooses the representative according to the similarity between these structures by using Boltzmann sampling methods such as minimum free energy folding and sub-optimal free energy folding [27, 28], as well as calculating the probability of base pairing.

The similarity measurement method of the secondary structure is related to its representation. For example, tree edit distance under tree representation, base pair (BP) distance [28], relaxed base pair (RBP) scoring method [29] and Hausdorff distance [30–33] under base pair set representation. These methods are helpful to select one or a few representative structures from large quantities of candidate sets.

At present, the performance of the secondary structure prediction algorithm still needs to be improved. In this paper, we present a fast algorithm for computing the minimum Chebyshev distance between RNA secondary structures. According to the properties of Chebyshev distance calculation and the characteristics of RNA secondary structure, we design an auxiliary computing structure, which reduces the amount of calculation and effectively improves the computing efficiency of the related RNA secondary structure prediction method.

## 2   Minimum Chebyshev Distance Between RNA Secondary Structures

### 2.1   RNA Secondary Structure

**Definition 1.** RNA sequence $R$ is composed of four types of bases A, U, G, and C. Its length is $n$. Its form is defined as follows:

$$R = r_1 r_2 r_3 \ldots r_n, \text{ where } r_i \in \{A, U, G, C\}, i = 1, 2, \ldots n$$

The left end of R represents the starting point of the base sequence, usually represented by 5'; the right end is the endpoint, usually by 3'. This string represents a single directed biological chain consisting of four bases that starts at 5' and ends at 3'. RNA sequences fold themselves through intramolecular base pairs and stabilize their structures by hydrogen bonds that form base pairs.

There are three cases of base pairing in RNA secondary structure: strong Watson Crick pairing for C-G (G-C) and A-U (U-A), and unstable wobble pairing for G-U (U-G).

**Definition 2.** RNA secondary structure is a base pair set $S$ on its sequence $R$. Its elements $(i, j)$ represent the pairing of $r_i$ and $r_j$ in the sequence, where $1 < i < j < n$. At the same time, the following constraints must be satisfied:

1. $(i,j) \in S \Rightarrow j - i \geq 4$;

2. $(i,j) \in S \Rightarrow \begin{cases} (i,j') \notin S, (j' \neq j) \\ \quad or \\ (i',j) \notin S(i \neq i') \end{cases}$ ;

$$3. \left.\begin{array}{l} (i,j) \in S \\ (i',j') \in S \\ i \leq i' \end{array}\right\} \Rightarrow \left\{\begin{array}{l} i = i', j = j' \\ \quad \text{or} \\ i < j < i' < j' \\ \quad \text{or} \\ i < i' < j' < j \end{array}\right.$$

Constraint 1 indicates that the ring formed by base pairing will not be too sharp (the loop area in Fig. 1). Constraint 2 means that the same base participates in at most one base pairing. Constraint 3 denotes that RNA secondary structure does not contain pseudoknots.

**Definition 3.** RNA Sequence $R$'s secondary structure plane graph [34] is a labeled graph with $n$ vertices $G = (V, E)$, where $V = \{r_1, r_2, \ldots, r_n\}$, and its adjacency matrix $A = \{a_{i,j}\}$ satisfies the following conditions:

1. $a_{i,i+1} = 1 \ 0 < i < n$;

2. $(i,j) \in S \Leftrightarrow \begin{cases} a_{i,j} = 1 \\ j - i \geq 4 \end{cases}$.

Take the RNA molecule numbered PDB_00194 in RNA STRAND (the RNA secondary STRucture and statistical ANalysis Database) as an example. According to *Definition 1*, its sequence is:

$$R = \text{UGCUCCUAGUACGAGAGGACCGGAGUG}$$

According to Definition 2, its secondary structure can be expressed as:

$$S = \{(2, 26), (3, 25), (4, 24), (5, 23), (6, 22), (7, 21), (10, 19), (11, 18), (12, 17), (13, 16)\}$$

According to Definition 3, the secondary structure planar graph is shown in Fig. 1.

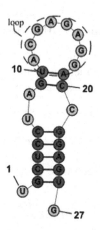

**Fig. 1.** PDB_00194 secondary structure planar graph. Note: This figure is generated by RNAStructure [35]

## 2.2   Chebyshev Distance

**Definition 4.** If two vectors or two points $p$ and $q$ have coordinates of $p_i$ and $q_i$ respectively, then the Chebyshev distance between them is defined as follows:

$$d_{Cheb}(p,q) = \max_i(|p_i - q_i|)$$

In the 2-dimensional plane, the Chebyshev distance between two points $p(i_1, j_1)$ and $q(i_2, j_2)$ is:

$$d_{Cheb}(p,q) = max\{|i_1 - i_2|, |j_1 - j_2|\}$$

Property 1: In a two-dimensional plane, a set of points with equal Chebyshev distance to a certain point constitutes a square [36].

If the fixed point and moving point are satisfied $d_{Cheb}(p,m) = r(r > 0)$ then $\max\{|i - i'|, |j - j'|\} = r$. This formula is equivalent to $\begin{cases} |i - i'| = r \\ |j - j'| \leq r \end{cases}$ or $\begin{cases} |j - j'| = r \\ |i - i'| \leq r \end{cases}$.

Then the trajectory of point $m$ is a square with point $p$ as the center and side length $2r$, as shown in Fig. 2.

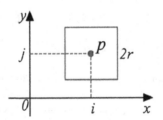

**Fig. 2.** Trajectories of points with the same Chebyshev distance from a certain point $p$.

## 2.3   Minimum Chebyshev Distance of RNA Secondary Structure

Assume that there are two RNA secondary structures $S_1$ and $S_2$ respectively, where $(i,j) \in S_1$, $(i',j') \in S_2$. To calculate the minimum Chebyshev distance between $S_2$ and base pairs $(i, j)$ in $S_1$, we can operate as the following steps:

Step 1: Calculate the distance between two base pairs:

$$d_{Cheb}((i,j), (i',j')) = \max(|i - i'|, |j - j'|)$$

Step 2: Calculate the distance between base pairs $(i, j)$ and $S_2$:

$$d_a((i,j), S_2) = \min_{(i',j') \in S_2} \{d_{Cheb}((i,j), (i',j'))\}$$

The specific calculation process is shown in Algorithm 1.

**Algorithm 1.** original minimum Chebyshev distance between base pair and secondary Structure

---
**Input:** Secondary structure $S = \{(i_1, j_1)(i_2, j_2), \dots ,(i_m, j_m)\}$; base pair $p = (i', j')$
**Output:** minimum Chebyshev distance *mindist* between $p$ and $S$
**processing:**
    $mindist = \text{chebyshev}((i', j'),(i_1, j_1))$
    **for** $t = 2, 3, \dots, m$ **do**
        $dist = \text{chebyshev}((i', j'),(i_t, j_t))$
        **if** *mindist* > *dist* **then**
            *mindist* = *dist*
        **end if**
    **end for**
Note: chebyshev$((i, j), (i', j'))$ is used to calculate the Chebyshev distance between point (i, j) and point (i', j').

---

# 3 Fast Algorithm for Computing Minimum Chebyshev Distance of RNA Secondary Structure

## 3.1 Details of the Algorithm Idea

**Definition 5.** RNA secondary structure dot plot [37] is shown in Fig. 3. In the $n*n$ square grid, the intersection points of row $i$ and column $j$ are used to represent base pairs $(i, j)$, which represent the entire secondary structure $S$, where $n$ represents the length of the RNA sequence.

The grid in Fig. 3 and its black square point are shown as a dot plot representation of PDB_00194, where the numbers in the x-axis and y-axis directions represent the position of the base in the RNA respectively, and the base pairs are represented by black squares. All black square points in the ellipse represent the secondary structure of PDB_00194. The dot plot can be viewed as part of a plane rectangular coordinate system.

The minimum chebyshev distance from one base pair in the RNA secondary structure to another secondary structure is the minimum chebyshev distance calculated from a point to a point set. Let a base pair in the RNA secondary structure be $p(i', j')$ and another secondary structure $S$ where and the minimum chebyshev distance between $p$ and $S$ is $r_{min}$. If the base pair $p$ is the center and $r_{min}$ is the radius of square $G$, then $G$ and $S$ will obtain the following properties:

Property 2: $G$ does not contain any base pairs in $S$
Property 3: At least one of the base pairs in $S$ is on the edge of $G$.

Based on the above two properties, the algorithm uses the binary search method to quickly screen out the square $G$ that meets the condition, and the radius is the minimum Chebyshev distance.

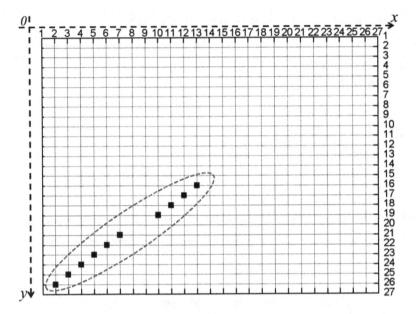

**Fig. 3.** PDB_00194 dot plot

To facilitate the description of the algorithm, $i$ in the base pair $(i, j)$ is referred to as the *left base*, and $j$ is referred to as the *right base*. Set the secondary structure $S = \{(i_1, j_1), (i_2, j_2), \ldots, (i_m, j_m)\}$, and the other two sets of structures $S_x$ and $S_y$ are

$$p_x = (i, j, leftNN, rightNN) \in S_x$$

$$p_y = (i', j', downNN, upNN) \in S_y,$$

$p_x$, $p_y$ has the following characteristics:

1. The elements in $S_x$ and $S_y$ are arranged in ascending order according to left base $i$ and right base $j$ respectively;
2. $1 \le i \le i_m, 1 \le j' \le \max\{j_1, j_2, \ldots, j_m\}$;
3. $(i, j)$ and $(i', j')$ represent the pairing of bases. If the i-number bases are not paired, then $j = 0$; if the j'-number bases are not paired, then $i' = 0$;
4. *LeftNN* and *rightNN* in $p_x$ represent the nearest left and right nearest neighbor points in the x-axis direction of the secondary structure dot plot that are closest to the straight line $x = i$. Similarly, when $i' = 0$, *downNN* and *upNN* in $p_y$ refer to the nearest lower nearest neighbor and upper nearest neighbor in y-axis direction from $y = j'$.

$S_x$ and $S_y$ are used to quickly locate the nearest point to a fixed point in the direction of the x-axis and y-axis respectively. In this paper, they are called NNI (Nearest Neighbor Indexes), represented by arrays in the algorithm.

The calculation process mainly consists of the following two steps:

Step 1: Generate the NNI

Take PDB_00194 as an example, the base pair set which represents its secondary structure is sorted by $i$ by default, while the set sorted by $j$ is $\{(13,16), (12,17), (11,18), (10,19), (7,21), (6,22), (5,23), (4,24), (3,25), (2,26)\}$, and the generated NNI is shown in Fig. 4.

Step 2: Calculate the minimum Chebyshev distance between the base pair $p(i', j')$ and S

Quickly locate the closest point in the $x$-axis direction to $i'$ based on the generated NNI:

Case 1: If there is a base pair in $S$ that coincides with $p(i', j')$, we can determine that the minimum Chebyshev distance is 0, and end the calculation;

Case 2: If $i$ or $j$ has a paired base in $S$, return the base pair;

Case 3: If neither $i$ nor $j$ in $S$ has no paired bases, return two base pairs in the $x$-axis direction from $x = i$ and the $y$-axis direction closest to $y = j$ respectively.

Calculate the minimum Chebyshev distance $r_0$ according to the base pair returned in case 2 or case 3. Let the square area with $p(I', j')$ as the center and $r_0$ as the initial radius be G. According to property 2, first determine whether there is a base pair in G and if so, reduce the radius of G to $r/2$; if it does not exist, and there are no base pairs on the four sides of G, then the radius is enlarged to $3r/4$. Thus, using the binary search to

| No | $i$ | $j$ | LeftNN | RightNN |
|----|-----|-----|--------|---------|
| 1 | 1 | 0 | 0 | 2 |
| 2 | 2 | 26 | 0 | 3 |
| 3 | 3 | 25 | 2 | 4 |
| 4 | 4 | 24 | 3 | 5 |
| 5 | 5 | 23 | 4 | 6 |
| 6 | 6 | 22 | 5 | 7 |
| 7 | 7 | 21 | 6 | 10 |
| 8 | 8 | 0 | 7 | 10 |
| 9 | 9 | 0 | 7 | 10 |
| 10 | 10 | 19 | 7 | 11 |
| 11 | 11 | 18 | 10 | 12 |
| 12 | 12 | 17 | 11 | 13 |
| 13 | 13 | 16 | 12 | 0 |

a)NNI sorted by $i$

| No | $i$ | $j$ | DownNN | UpNN |
|----|-----|-----|--------|------|
| 1 | 0 | 1 | 0 | 16 |
| 2 | 0 | 2 | 0 | 16 |
| 3 | 0 | 3 | 0 | 16 |
| 4 | 0 | 4 | 0 | 16 |
| 5 | 0 | 5 | 0 | 16 |
| 6 | 0 | 6 | 0 | 16 |
| 7 | 0 | 7 | 0 | 16 |
| 8 | 0 | 8 | 0 | 16 |
| 9 | 0 | 9 | 0 | 16 |
| 10 | 0 | 10 | 0 | 16 |
| 11 | 0 | 11 | 0 | 16 |
| 12 | 0 | 12 | 0 | 16 |
| 13 | 0 | 13 | 0 | 16 |
| 14 | 0 | 14 | 0 | 16 |
| 15 | 0 | 15 | 0 | 16 |
| 16 | 13 | 16 | 0 | 17 |
| 17 | 12 | 17 | 16 | 18 |
| 18 | 11 | 18 | 17 | 19 |
| 19 | 10 | 19 | 18 | 21 |
| 20 | 0 | 20 | 19 | 21 |
| 21 | 7 | 21 | 19 | 22 |
| 22 | 6 | 22 | 21 | 23 |
| 23 | 5 | 23 | 22 | 24 |
| 24 | 4 | 24 | 23 | 25 |
| 25 | 3 | 25 | 24 | 26 |
| 26 | 2 | 26 | 25 | 0 |

b)NNI sorted by $j$

**Fig. 4.** NNI of PDB 00194

scale the radius of G until the property 3 is satisfied, that is, G is empty, and at least one base pair is on the four sides, the radius of the current G is the minimum Chebyshev distance sought. The specific process is shown in Algorithm 2.

Figure 5 shows the minimum Chebyshev distance calculation process between the points $p = (12, 25)$ and the PDB_00194 structure.

**Algorithm 2.** fast calculation for the minimum Chebyshev distance between base pair and secondary structure

---

**Input:** point $p(i', j')$, secondary structure $S = \{(i_1, j_1), (i_2, j_2), ..., (i_m, j_m)\}$

**Output:** minimum Chebyshev distance $r$

**Process:**

According to $S$ to generate $S_x$, $S_y$, and

$$S_x = \{(g_k, h_k, leftNN_k, rightNN_k)\}_{k=1}^{i_m},$$
$$S_y = \{(u_k, v_k, downNN_k, upNN_k)\}_{k=1}^{\max(j_1, j_2, ..., j_m)}$$

///To get the initial fixed-length Chebyshev distance square radius

**if** $i > i_m$ **then**

   $Xchebydist = chebyshev(p, S_x.leftNN_{i_m})$

**else**

   $Xchebydist$=min(chebyshev($p,S_x.leftNN_{i'}$),

                     chebyshev($p,S_x.rightNN_{i'}$))

**end if**

**if** $Xchebydist$==0 **then**

   **return** 0

**end if**

Similarly, the minimum Chebyshev distance in the y-axis direction from the nearest two points to the point $p$ can be obtained as $Ychebydist$

$SquareRadius$=min($Xchebydist$, $Ychebydist$)

//Find the minimum Chebyshev distance using the //binary search method

$radiusStart = 0$

$radiusEnd = SquareRadius$

**while** $pointnumber$>0 **do**

   $r$=($radiusStart$+$radiusEnd$)/2

   $pointnumber$=pointinsquare($p,r,S_x,S_y$)

   **if** $pointnumber > 0$ **then**

      $radiusEnd = r - 1$

   **else if** sidehaspoint($p,r,S_x,S_y$) == False **then**

      $radiusStart = r + 1$

      $pointnumber = 1$// Let the loop continue

   **end if**

**end while**

Note: pointinsquare($p$, $r$, $S_x$, $S_y$) is used to calculate the number of base pairs of the secondary structure $S$ in the square with the radius $r$ and the center point $p$; sidehaspoint($p$, $r$, $S_x$, $S_y$) is used to determine whether the secondary structure $S$ has a base pair just on or in the square with the radius $r$ and the center point $p$.

---

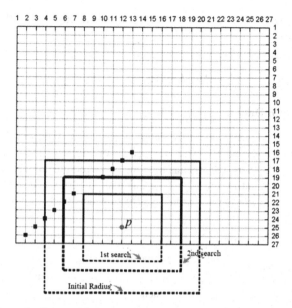

**Fig. 5.** Schematic diagram of the minimum Chebyshev distance calculation process

It can be seen intuitively from Fig. 5 that after the algorithm zooms through the search radius of 3 times, the minimum Chebyshev distance from the point $p$ is 5.

### 3.2   Algorithm Complexity Analysis

Assuming that the length of the RNA sequence $R$ is $n$, and the set of secondary structures generated by folding is $\{S_1, S_2, ..., S_m\}$. We want to calculate the minimum Chebyshev distance from a base pair of $S_1$ to $S_2$. According to Definition 3, it can be seen that the number of base pairs in $S_1$ and $S_2$ does not exceed $n/2$.

When solving the problem with Algorithm 1, a base pair in $S_1$ needs to traverse all base pairs in $S_2$ to calculate Chebyshev distance to find the minimum value. The number of traversals is at most $n/2$, and the time complexity is $O(n)$.

NNI generation and traversal calculation analysis are needed when solving with Algorithm 2. According to the characteristics of RNA secondary structure in Definition 2, the NNI could be generated in $O(n)$ time complexity by the counting sorting algorithm. Since the NNI only needs to be initialized before comparison, if each secondary structure in the set calculates the minimum Chebyshev distance from other secondary structures, the average time complexity is $O(n/m^2)$. According to Table 1 in the following experiment, this time is negligible.

When traversing the computational analysis, the best case traversal is 1, so the time complexity is $O(1)$; in the worst case, the time complexity $O(\log_2 n)$ of the binary search. Therefore, the time complexity of Algorithm 2 is nearly $[O(\log_2 n), O(1)]$.

Compared to the original algorithm, the improved algorithm requires two arrays sorted by the left base and right base respectively, and the length of the array is determined by the maximum number of base numbers participating in the pairing. In the worst case, the lengths of the two arrays are all Sequence length, so its spatial complexity is $O(n)$.

## 4    Experiment and Results Analysis

### 4.1    Experimental Data

The experimental data (see Table 1 for details) are derived from mRNA sequences of different lengths from the NCBI (National Center for Biotechnology). The data details are shown in Table 1 where TSC means the times for structure comparison and then the default parameters are set by mfold (v2.3). Calculate the fold, and finally compare the CT secondary data of the RNA secondary structure generated by the same sequence folding.

**Table 1.** Experimental data details

| Locus | Gene | Length | Number of generated structures | TSC | Base pair repetition rate |
|---|---|---|---|---|---|
| NM_130776 | XAGE3 | 493 | 26 | 650 | 48.29% |
| NM_002988 | CCL18 | 793 | 24 | 552 | 37.52% |
| NM_002475 | MYL6B | 869 | 27 | 702 | 36.09% |
| NM_033142 | CGB7 | 880 | 22 | 462 | 41.05% |
| NM_016815 | GYpC | 1019 | 41 | 1640 | 45.48% |
| NM_004019 | DMD | 1634 | 33 | 1056 | 45.42% |
| NM_002231 | CD82 | 1715 | 49 | 2352 | 42.81% |
| NM_013958 | NRG1 | 1740 | 38 | 1406 | 48.60% |
| NM_000781 | CYp11A1 | 1821 | 65 | 4160 | 39.85% |
| NM_021785 | RAI2 | 2338 | 35 | 1190 | 40.00% |
| NM_004381 | ATF6B | 2622 | 35 | 1190 | 44.71% |
| NM_033172 | B3GALT5 | 2711 | 43 | 1806 | 45.70% |
| NM_005656 | TMpRSS2 | 3226 | 42 | 1722 | 60.89% |

### 4.2    Experimental Environment

The experimental environment is as follows:

Operating system: ubuntu 17.10 64-bit version; development language: python 2.7.14; CPU: Core i7 3630QM; memory: 16 GB.

### 4.3    Experimental Steps

Experiments were carried out by taking the common Hausdorff distance and RBP scoring methods as examples to calculate the distance between RNA secondary structures. Each method uses the minimum Chebyshev distance original algorithm and the improved algorithm proposed in this paper to compare the running time overhead.

Because there are more identical base pairs in different secondary structures generated by the same sequence, the base pair repetition in the structure comparison is considered in the experiment, and the Algorithm 1 is slightly improved, as shown in the Algorithm 3.

| Algorithm 3. Improved original algorithm |
|---|
| **Input:**secondary structure $S = \{(i_1, j_1),(i_2, j_2), \ldots ,(i_m, j_m)\}$;base pair $p = (i', j')$<br>**Output:**Minimum Chebyshev distance $mindist$ between $p$ and $S$<br>**Process:**<br>    $mindist = $ chebyshev$((i', j'),(i_1, j_1))$<br>    $t = 2$<br>    **while** $(mindist > 0$ **and** $t < m + 1)$ **do**<br>        $dist = $ chebyshev$((i', j'),(i_t, j_t))$<br>        **if** $mindist > dist$ **then**<br>            $mindist = dist$<br>        **end if**<br>        $t = t + 1$<br>    **end while** |

## 4.4  Experimental Results

The experiment uses the running time of the average structure comparison in the same sequence to measure the pros and cons of the algorithm. For example, the sequence NM_130776 with length 493 is folded to generate 26 candidate structures. The comparison times of the two structures are 26 * (26 − 1) = 650 times. The total time of the improved hausdorff distance calculation algorithm is 7.291402102 s compared with the original 130.0833249 s, and the average time of the improved algorithm is 11.2175 417 ms compared with the original 200.1281922 ms.

The experimental results are as follows:

## 4.5  Analysis of Results

It can be seen from Table 1 that the secondary sequence generated by the same sequence has a higher rate of coincidence of base pairs, and further exerts the advantage of rapid calculation of the improved algorithm. The experimental results in Table 2 show that the improved algorithm significantly improves the calculation speed.

**Table 2.** Experimental results

| Locus | TSC | Average running time per structure (unit: milliseconds) | | | |
|---|---|---|---|---|---|
| | | Improved Hausdorff | Original Hausdorff | Improved RBP | Original RBP |
| NM_130776 | 650 | 11.2175417 | 200.1281922 | 11.49805692 | 187.2433091 |
| NM_002988 | 552 | 29.89899076 | 504.4981216 | 29.81576668 | 471.2811341 |
| NM_002475 | 702 | 50.83871499 | 985.6605442 | 53.64886037 | 925.020497 |
| NM_033142 | 462 | 32.92536788 | 670.0163032 | 34.59124028 | 622.7872052 |
| NM_016815 | 1640 | 53.65096767 | 1292.021342 | 56.01231529 | 1204.217655 |
| NM_004019 | 1056 | 82.25197819 | 1951.261859 | 85.54889781 | 1800.510011 |
| NM_002231 | 2352 | 112.8718082 | 2922.761558 | 120.2368014 | 2744.817355 |
| NM_013958 | 1406 | 92.43744808 | 2507.75789 | 97.35820982 | 2313.373188 |
| NM_000781 | 4160 | 154.4170339 | 3445.17768 | 162.1970053 | 3162.275498 |
| NM_021785 | 1190 | 174.5501597 | 4165.3774 | 187.9682958 | 3904.11921 |
| NM_004381 | 1190 | 270.955447 | 6404.746261 | 293.4771656 | 5991.996467 |
| NM_033172 | 1806 | 244.5065211 | 7262.653743 | 261.8198942 | 6713.290925 |
| NM_005656 | 1722 | 201.0514477 | 9254.406022 | 213.9117422 | 8557.3886 |

## 5  Conclusion

The distance calculation optimization method proposed in this paper is a great improvement in the computational efficiency of the original RNA secondary structure minimum Chebyshev distance algorithm. This method locates the target calculation points quickly by using the characteristics and regularity of RNA secondary structure. It also reduces the calculation range. Although only different structures generated by the same sequence are tested in this paper. This method can be generalized to compare the structures generated by different sequences.

The minimum Chebyshev distance also has applications in image processing, such as digital image forensics [38], cell tracking [39], hyperspectral imaging [40], and neural network research [41, 42]. The algorithm can also be used to calculate the minimum Chebyshev distance between other points and point sets.

## References

1. National Natural Science Foundation of China, Chinese Academy of Sciences. Major Scientific Issues in the Study of RNA in China's Discipline Development Strategy. Science Press, China (2017)
2. Krieger, E., Sander, B., Vriend, G.: Homology modeling. In: Structural Bioinformatics, vol. 44 (2003)
3. Thiel, B.C., Flamm, C., Hofacker, I.L.: RNA structure prediction: from 2D to 3D. Emerg. Top. Life Sci. **1**(3), 275–285 (2017)

4. Galvanek, R., Hoksza, D.: Template-based prediction of RNA tertiary structure using its predicted secondary structure. In: 2017 IEEE International Conference on Bioinformatics and Biomedicine (BIBM), pp. 2238–2240. IEEE (2017)
5. Zhao, Y., Huang, Y., Gong, Z., Wang, Y., Man, J., Xiao, Y.: Automated and fast building of three-dimensional RNA structures. Sci. Rep. **2**, 1–6 (2012)
6. Waterman, M.S., Smith, T.F.: RNA secondary structure: a complete mathematical analysis. Math. Biosci. **42**(3–4), 257–266 (1978)
7. Nussinov, R., Jacobson, A.B.: Fast algorithm for predicting the secondary structure of single-stranded RNA. Proc. Natl. Acad. Sci. U.S.A. **77**(11), 6309–6313 (1980)
8. Waterman, M.S., Smith, T.F.: Rapid dynamic programming algorithms for RNA secondary structure. Adv. Appl. Math. **7**(4), 455–464 (1986)
9. Akutsu, T.: Dynamic programming algorithms for RNA secondary structure prediction with pseudoknots. Discret. Appl. Math. **104**(1–3), 45–62 (2000)
10. García, R.: Prediction of RNA pseudoknotted secondary structure using stochastic context free grammars (SCFG). CLEI Electron. J. **9**(2) (2006)
11. Sakakibara, Y., Brown, M., Hughey, R., Mian, I.S., Sjölander, K., Underwood, R.C., Haussler, D.: Stochastic context-free grammars for tRNA modeling. Nucleic Acids Res. **22** (23), 5112–5120 (1994)
12. Anderson, J.W.J., Tataru, P., Staines, J., Hein, J., Lyngsø, R.: Evolving stochastic context-free grammars for RNA secondary structure prediction. BMC Bioinform. **13**(1), 78 (2012)
13. Pal, S.K., Ray, S.S., Ganivada, A.: RNA secondary structure prediction: soft computing perspective. Granular Neural Networks, Pattern Recognition and Bioinformatics. SCI, vol. 712, pp. 195–222. Springer, Cham (2017). https://doi.org/10.1007/978-3-319-57115-7_7
14. Batenburg, F.H.V., Gultyaev, A.P., Pleij, C.W.: An APL-programmed genetic algorithm for the prediction of RNA secondary structure. J. Theor. Biol. **174**(3), 269–280 (1995)
15. Shapiro, B.A., Bengali, D., Kasprzak, W., Wu, J.C.: RNA folding pathway functional intermediates: their prediction and analysis. J. Mol. Biol. **312**, 27–44 (2001)
16. Wiese, K.C., Deschenes, A., Glen, E.: Permutation based RNA secondary structure prediction via a genetic algorithm. In: Proceedings of the 2003 Congress on Evolutionary Computation, pp. 335–342 (2003)
17. Wiese, K.C., Deschênes, A.A., Hendriks, A.G.: RnaPredict - an evolutionary algorithm for RNA secondary structure prediction. IEEE/ACM Trans. Comput. Biol. Bioinf. **5**(1), 25–41 (2008)
18. Schmitz, M., Steger, G.: Description of RNA folding by simulated annealing. J. Mol. Biol. **255**(1), 254–266 (1996)
19. Tsang, H.H., Wiese, K.C.: SARNA-predict: accuracy improvement of RNA secondary structure prediction using permutation based simulated annealing. IEEE/ACM Trans. Comput. Biol. Bioinf. **7**(4), 727–740 (2010)
20. Liu, Q., Ye, X., Zhang, Y.: A Hopfield neural network based algorithm for RNA secondary structure prediction. In: Proceedings of the 1st International Conference on Multi-symposiums on Computer and Computational Sciences, pp. 1–7 (2006)
21. Haynes, T., Knisley, D., Knisley, J.: Using a neural network to identify secondary RNA structures quantified by graphical invariants. MATCH Commun. Math. Comput. Chem. **60**, 277–290 (2008)
22. Zou, Q., Zhao, T., Liu, Y., Guo, M.: Predicting RNA secondary structure based on the class information and Hopfield network. Comput. Biol. Med. **39**(3), 206–214 (2009)
23. Koessler, D.R., Knisley, D.J., Knisley, J., Haynes, T.: A predictive model for secondary RNA structure using graph theory and a neural network. BMC Bioinform. **11**, S6–S21 (2010)

24. Song, D., Deng, Z.: A fuzzy dynamic programming approach to predict RNA secondary structure. In: Bücher, P., Moret, B.M.E. (eds.) WABI 2006. LNCS, vol. 4175, pp. 242–251. Springer, Heidelberg (2006). https://doi.org/10.1007/11851561_23

25. Oluoch, I.K., Akalin, A., Vural, Y., Canbay, Y.: A review on RNA secondary structure prediction algorithms. In: 2018 International Congress on Big Data, Deep Learning and Fighting Cyber Terrorism (IBIGDELFT), Ankara, Turkey, pp. 18–23. IEEE (2018)

26. Zuker, M.: Mfold web server for nucleic acid folding and hybridization prediction. Nucleic Acids Res. **31**, 3406–3415 (2003)

27. Markham, N.R., Zuker, M.: UNAFold: software for nucleic acid folding and hybridization. In: Keith, J.M. (ed.) Bioinformatics: Structure, Functions and Applications, vol. 453, pp. 3–31. Humana Press, Totowa (2008)

28. Ding, Y., Chan, C.Y., Lawrence, C.E.: Clustering of RNA secondary structures with application to messenger RNAs. J. Mol. Biol. **359**, 554–571 (2006)

29. Agius, P., Bennett, K.P., Zuker, M.: Comparing RNA secondary structures using a relaxed base-pair score. RNA **16**(5), 865–878 (2010)

30. Schirmer, S., Ponty, Y., Giegerich, R.: Introduction to RNA secondary structure comparison. Methods Mol. Biol. **1097**(1097), 247–273 (2014)

31. Lopez, M.A., Reisner, S.: Hausdorff approximation of convex polygons. Comput. Geom. **2**(32), 139–158 (2005)

32. Moulton, V., Zuker, M., Steel, M., Pointon, R., Penny, D.: Metrics on RNA secondary structures. J. Comput. Biol.: J. Comput. Mol. Cell Biol. **7**(1–2), 277–292 (2000)

33. Chen, Q., Chen, B., Zhang, C.: Interval based similarity for function classification of RNA pseudoknots. In: Chen, Q., Chen, B., Zhang, C. (eds.) Intelligent Strategies for Pathway Mining. LNCS (LNAI), vol. 8335, pp. 175–192. Springer, Cham (2014). https://doi.org/10.1007/978-3-319-04172-8_8

34. Fu, W., Huang, J., Xu, L.: RNA secondary structure representation and conversion algorithms. Comput. Eng. Appl. (14), 43–45, 85 (2004)

35. Reuter, J.S., Mathews, D.H.: RNAstructure: software for RNA secondary structure prediction and analysis. BMC Bioinform. **11**(1), 129 (2010)

36. Zhang, Y.: Track exploration under Chebyshev distance. Math. Teach. **8**, 19–22 (2015)

37. Tsang, H.H., Jacob, C.: RNADPCompare: an algorithm for comparing RNA secondary structures based on image processing techniques, pp. 1288–1295. IEEE (2011)

38. Kang, X., Wei, S.: Identifying tampered regions using singular value decomposition in digital image forensics. In: 2008 International Conference on Computer Science and Software Engineering, Wuhan, Hubei, pp. 926–930 (2008)

39. Chowdhury, A.S., Chatterjee, R., Ghosh, M., Ray, N.: Cell tracking in video microscopy using bipartite graph matching. In: 2010 20th International Conference on pattern Recognition, Istanbul, pp. 2456–2459 (2010)

40. Demirci, S., Erer, I., Ersoy, O.: Weighted Chebyshev distance classification method for hyperspectral imaging. In: Proceedings of SPIE 9482, Next-Generation Spectroscopic Technologies VIII, p. 948218 (2015)

41. Ritter, G.X., Urcid-Serrano, G., Schmalz, M.S.: Lattice associative memories that are robust in the presence of noise. In: Proceedings of SPIE 5916, Mathematical Methods in Pattern and Image Analysis, p. 59160Q (2005)

42. Ritter, G.X., Urcid, G.: Learning in lattice neural networks that employ dendritic computing. In: Kaburlasos, V.G., Ritter, G.X. (eds.) Computational Intelligence Based on Lattice Theory, vol. 67, pp. 25–44. Springer, Heidelberg (2007). https://doi.org/10.1007/978-3-540-72687-6_2

# Fine-Grained Access Control in mHealth with Hidden Policy and Traceability

Qi Li[1,2(✉)], Yinghui Zhang[3], and Tao Zhang[4]

[1] School of Computer Science, Nanjing University of Posts and Telecommunications,
Nanjing 210023, China
liqics@njupt.edu.cn
[2] Jiangsu Key laboratory of Big Data Security and Intelligent Processing,
Nanjing University of Posts and Telecommunications, Nanjing, China
[3] National Engineering Laboratory for Wireless Security,
Xi'an University of Posts and Telecommunications, Xi'an 710121, China
[4] School of Computer Science and Technology, Xidian University,
Xi'an 710071, China

**Abstract.** Ciphertext-Policy Attribute-Based Encryption (CP-ABE) is a well-received cryptographic primitive to securely share personal health records (PHRs) in mobile healthcare (mHealth). Nevertheless, traditional CP-ABE can not be directly deployed in mHealth. First, the attribute universe scale is bounded to the system security parameter and lack of scalability. Second, the sensitive data is encrypted, but the access policy is in the plaintext form. Last but not least, it is difficult to catch the malicious user who intentionally leaks his access privilege since that the same attributes mean the same access privilege. In this paper, we propose HTAC, a fine-grained access control scheme with partially hidden policy and white-box traceability. In HTAC, the system attribute universe is larger universe without any redundant restriction. Each attribute is described by an attribute name and an attribute value. The attribute value is embedded in the PHR ciphertext and the plaintext attribute name is clear in the access policy. Moreover, the malicious user who illegally leaks his (partial or modified) private key could be precisely traced. The security analysis and performance comparison demonstrate that HTAC is secure and practical for mHealth applications.

**Keywords:** CP-ABE · Partially hidden policy · Traceability · Large universe · Adaptive security

## 1 Introduction

Mobile Healthcare (mHealth) provides remote, on-demand, accurate health service for patients. Unlike traditional medical service, mHealth enables a patient to collect his comprehensively physical information by various wearable sensors, integrate it into the personal health record (PHR) via smart devices, and share his PHR to request health services over cloud platform. Despite its convenience,

Q. Li et al. (Eds.): BROADNETS 2019, LNICST 303, pp. 261–274, 2019.
https://doi.org/10.1007/978-3-030-36442-7_17

mHealth service raises high security risks. The PHR contains a vast amount of private and sensitive information, including glycemic index, infectious disease, genetic history, etc. The patient expects his PHR can only be read by authorized entities. However, when the PHR data is stored on the cloud server, which may lack of effective security mechanism. Even worse, the cloud server may sell the PHR files for benefit. In such situation, the unauthorized access is unavoidable.

As a promising technique, Attribute-based encryption (ABE) [19] was designed to provide data confidentiality and fine-grained access control. Due to the fact that the ciphertext is computed from an access policy, Ciphertext-Policy ABE (CP-ABE) [2,5,22] is more suitable for the PHR owner to set the access policy for his PHR data. Although there are various ABE schemes proposed for policy expressiveness [20,23], multiple authorities [4,9,21], attribute revocation [10,24] and adaptively security [7,11], it is still worth considering the concerns of *large universe, policy hiding* and *traceability*.

*Large Universe.* In terms of the scale of attribute universe, ABE can be divided into two types: small universe and the large one. In the former ABE, the attributes should be stated in the initialization phase and the size of public parameters is usually linear with the scale of attribute universe which is polynomially bounded. If the bound is set to be too small, the system might be rebuilt as the amount of attributes exceeds such bound. If the bound is set to be too large, it will cause redundant performance. On the contrary, the scale of attribute universe in large universe schemes can be exponentially large. In addition, the size of public parameters is constant. Lewko *et al.* [8] and Rouselakis *et al.* [18] proposed a large universe ABE scheme on the composite and prime order groups, respectively. In [13], Ning *et al.* proposed an efficient large universe ABE scheme with verifiable outsourced decryption.

*Policy Hiding.* In traditional ABE schemes, the access policy is sent along with the ciphertext in the plaintext form. That is, any user who gets the ciphertext can obtain the access policy even if he is not authorized. Suppose that a patient defines the access policy ('Diabetes Mellitus' AND 'Doctor'), then anyone can learn the policy and infer that the patient is suffering from diabetes mellitus. The patient's privacy is entirely violated. Takashi *et al.* [15] addressed this issue and presented the first hidden policy CP-ABE scheme, where the access policy was not directly sent along with the ciphertext. Lai *et al.* [6] introduced an adaptively secure and partially hidden CP-ABE scheme, where each attribute is denoted by an attribute name and an attribute value. In [6], only the access policy with attribute names is sent along with the ciphertext, while the attribute values are embedded in the ciphertext. Compared with fully-hidden policy ABE schemes, the partially-hidden schemes can achieve a tradeoff between efficiency and fully-hidden policies. In [26], Zhang *et al.* also constructed a partially-hidden access control scheme with large universe in smart health.

*Traceability.* If a user abuses or leaks his private key for benefit, such malicious user must be catched to enhance the data confidentiality. However, in CP-ABE, the user's private key is described by his attributes and multiple users may have the same attributes. It is difficult to identify who leaks his private key. To solve this issue, Liu *et al.* [12] constructed a traceable CP-ABE scheme with adaptive security and expressive access policy. Li *et al.* [17] presented a traceable and large universe CP-ABE scheme with efficient user decryption in eHealth cloud.

In this paper, we simultaneously addressed the above three main properties and proposed HTAC, a traceable access control scheme with partially-hidden access policy in mHealth. The PHR owner can define any monotonic access policies himself. In summary, our contributions are as follows:

1. *Large Universe.* There is no extra bound on the attribute universe and the size of system public parameter is constant.
2. *Policy Hiding.* Only the access policy with generic attribute names is transmitted along with the ciphertext. The sensitive attribute values are embedded in the ciphertext and unknowable to any unauthorized users.
3. *Traceability.* The source of the illegally leaked key could be precisely traced. The private PHR is protected from being abused and exposed to unauthorized access.
4. *Adaptive Security and Efficiency.* We construct HTAC on composite order groups and prove the adaptive security in the standard model. The performance analysis show that HTAC is almost as efficient as the underline scheme [26].

## 2   Preliminaries

### 2.1   Linear Secret Sharing Schemes (LSSS)

**Definition 1 (LSSS [1]).** *Let $\mathcal{U}$ denote the system attribute universe: $\mathcal{U} = (At_1, At_2, \ldots, At_n)$, where each attribute $x$ is composed of two parts: the attribute name $At_x$ and multiple attribute values. $\mathcal{VU}_x = \{J_{x,1}, J_{x,2}, \ldots, J_{x,n_x}\}$ refers to the set of all possible values of attribute $x$.*

*$\mathbf{A} \in \mathbb{Z}_p^{\ell \times n}$ is a share-generating matrix and $\rho$ is a function which maps the $i - th$ row of $\mathbf{A}$ to an attribute name $At_x \in \mathcal{U}$. An LSSS is composed of two following algorithms:*

- *Secret Share: It takes in a secret value $s \in \mathbb{Z}_p$ and $\mathbf{A}$. It computes the secret share $\lambda_x = A_x \cdot v$ for each row $A_x$ of $\mathbf{A}$, where $v = (s, y_2, \ldots, y_n)^T$ and $y_2, \ldots, y_n$ are randomly chosen from $\mathbb{Z}_p$.*
- *Secret Reconstruction: It takes in the secret shares $\{\lambda_x\}$ and any authorized set $\mathcal{P}$. It sets $\mathcal{I} = \{i | \rho(i) \in \mathcal{P}\} \subseteq \{1, 2, \ldots, \ell\}$ and calculates the coefficients $\{\omega_i \in \mathbb{Z}_p\}_{i \in \mathcal{I}}$ such that $\sum_{i \in \mathcal{I}} \omega_i A_i = (1, 0, \ldots, 0)$. Then the secret $s$ is reconstructed by $s = \sum_{i \in \mathcal{I}} \omega_i \lambda_i$.*

Similar as [7,25,26], the LSSS matrices in our scheme is constructed over $\mathbb{Z}_N$, where $N$ is a product of multiple primes. In HTAC, the user's attribute set is denoted by $\mathcal{S} = (\mathcal{NAM}_S, \mathcal{VU}_S)$, where $\mathcal{NAM}_S \subseteq \mathbb{Z}_N$ is the attribute name index and $\mathcal{VU}_S = \{J_{x,i}\}_{x \in \mathcal{NAM}_S}$ is the attribute value set.

In the employed access structure $\mathbb{A} = (\mathbf{A}, \rho, \mathcal{T})$, $\mathcal{T} = (t_{\rho(1)}, t_{\rho(2)}, \ldots, t_{\rho(\ell)})$ refers to the set of attribute value for each row of $\mathbf{A}$. $\mathcal{S}$ satisfies $\mathbb{A}$ means that there exists $\mathcal{I} \subseteq \{1, 2, \ldots, \ell\}$ satisfying $(\mathbf{A}, \rho)$, $\{\rho(i)|i \in \mathcal{I}\} \subseteq \mathcal{NAM}_\mathcal{S}$ and $\mathcal{J}_{\rho(i)} = t_{\rho(i)} \forall i \in \mathcal{I}$.

## 2.2  Composite Order Bilinear Group

A group generator $\mathcal{G}$ takes in a security parameter $\lambda$ and outputs the terms $(\mathbb{G}, \mathbb{G}_1, p_1, p_2, p_3, p_4, e)$, where $p_1$, $p_2$, $p_3$ and $p_4$ are 4 different primes, the order of cyclic groups $\mathbb{G}$ and $\mathbb{G}_1$ is $N = p_1 p_2 p_3 p_4$, and $e : \mathbb{G} \times \mathbb{G} \to \mathbb{G}_1$ is a bilinear map with such properties:

1. Bilinearity: $\forall \varrho, \varpi \in \mathbb{G}$ and $a, b \in \mathbb{Z}_N$, we have $e(\varrho^a, \varpi^b) = e(\varrho, \varpi)^{ab}$.
2. Non-degeneracy: $\exists \varrho \in \mathbb{G}$ such that the order of $e(\varrho, \varrho)$ is $N$.

Denote $\mathbb{G}_{p_x}$ as the subgroup of order $p_x$ in $\mathbb{G}$. If $\varrho_x \in \mathbb{G}_{p_x}$ and $\varrho_y \in \mathbb{G}_{p_y}$, for $x \neq y$, we have $e(\varrho_x, \varrho_y) = 1$.

## 2.3  Complexity Assumptions

$\mathbb{GD}$ represents the terms $(\mathbb{G}, \mathbb{G}_1, N = p_1 p_2 p_3 p_4, e)$ generated by a group generator $\mathcal{G}$.

**Assumption 1.** Given $\mathcal{G}$ and the following distribution:

$$g \xleftarrow{R} \mathbb{G}_{p_1}, P_3 \xleftarrow{R} \mathbb{G}_{p_3}, P_4 \xleftarrow{R} \mathbb{G}_{p_4},$$
$$\Phi = (\mathbb{GD}, g, P_3, P_4), \partial_1 \xleftarrow{R} \mathbb{G}_{p_1} \times \mathbb{G}_{p_2}, \partial_2 \xleftarrow{R} \mathbb{G}_{p_1}.$$

The algorithm $\mathcal{A}$'s advantage in breaking this assumption is $\mathsf{Adv1}_{\mathcal{G}, \mathcal{A}}(\lambda) = |\Pr[\mathcal{A}(\Phi, \partial_1) = 1] - \Pr[\mathcal{A}(\Phi, \partial_2) = 1]|$.

**Definition 2.** $\mathcal{G}$ satisfies Assumption 1 if $\mathsf{Adv1}_{\mathcal{G}, \mathcal{A}}(\lambda)$ is negligible for any probabilistic polynomial time (PPT) algorithm $\mathcal{A}$.

**Assumption 2.** Given $\mathcal{G}$ and the following distribution:

$$g, P_1 \xleftarrow{R} \mathbb{G}_{p_1}, P_2, Q_2 \xleftarrow{R} \mathbb{G}_{p_2}, P_3, Q_3 \xleftarrow{R} \mathbb{G}_{p_3}, P_4 \xleftarrow{R} \mathbb{G}_{p_4},$$
$$\Phi = (\mathbb{GD}, g, P_1 P_2, Q_2 Q_3, P_3, P_4),$$
$$\partial_1 \xleftarrow{R} \mathbb{G}_{p_1} \times \mathbb{G}_{p_2} \times \mathbb{G}_{p_3}, \partial_2 \xleftarrow{R} \mathbb{G}_{p_1} \times \mathbb{G}_{p_3}.$$

The algorithm $\mathcal{A}$'s advantage in breaking this assumption is $\mathsf{Adv2}_{\mathcal{G}, \mathcal{A}}(\lambda) = |\Pr[\mathcal{A}(\Phi, \partial_1) = 1] - \Pr[\mathcal{A}(\Phi, \partial_2) = 1]|$.

**Definition 3.** $\mathcal{G}$ satisfies Assumption 2 if $\mathsf{Adv2}_{\mathcal{G}, \mathcal{A}}(\lambda)$ is negligible for any PPT algorithm $\mathcal{A}$.

**Assumption 3.** Given $\mathcal{G}$ and the following distribution:

$$g \xleftarrow{R} \mathbb{G}_{p_1}, g_2, P_2, Q_2 \xleftarrow{R} \mathbb{G}_{p_2}, P_3 \xleftarrow{R} \mathbb{G}_{p_3}, P_4 \xleftarrow{R} \mathbb{G}_{p_4},$$

$$\Phi = (\mathbb{GD}, g, g_2, g^\alpha P_2, g^s Q_2, P_3, P_4),$$

$$\partial_1 = \hat{e}(g, g)^{\alpha s}, \partial_2 \xleftarrow{R} \mathbb{G}_1.$$

The algorithm $\mathcal{A}$'s advantage in breaking this assumption is $\mathsf{Adv3}_{\mathcal{G},\mathcal{A}}(\lambda) = |\Pr[\mathcal{A}(\Phi, \partial_1) = 1] - \Pr[\mathcal{A}(\Phi, \partial_2) = 1]|$.

**Definition 4.** $\mathcal{G}$ satisfies Assumption 3 if $\mathsf{Adv3}_{\mathcal{G},\mathcal{A}}(\lambda)$ is negligible for any PPT algorithm $\mathcal{A}$.

**Assumption 4.** Given $\mathcal{G}$ and the following distribution:

$$g, h \xleftarrow{R} \mathbb{G}_{p_1}, g_2, P_2, A_2, B_2, D_2 \xleftarrow{R} \mathbb{G}_{p_2}, t', r' \xleftarrow{R} \mathbb{Z}_N$$

$$P_3 \xleftarrow{R} \mathbb{G}_{p_3}, P_4, Z, A_4, D_4 \xleftarrow{R} \mathbb{G}_{p_4},$$

$$\Phi = (\mathbb{GD}, g, g_2, g^{t'} B_2, h^{t'} Q_2, P_3, P_4, hZ, g^{r'} D_2 D_4),$$

$$\partial_1 = h^{r'} A_2 A_4, \partial_2 \xleftarrow{R} \mathbb{G}_{p_1} \times \mathbb{G}_{p_2} \times \mathbb{G}_{p_4}.$$

The algorithm $\mathcal{A}$'s advantage in breaking this assumption is $\mathsf{Adv4}_{\mathcal{G},\mathcal{A}}(\lambda) = |\Pr[\mathcal{A}(\Phi, \partial_1) = 1] - \Pr[\mathcal{A}(\Phi, \partial_2) = 1]|$.

**Definition 5.** $\mathcal{G}$ satisfies Assumption 4 if $\mathsf{Adv4}_{\mathcal{G},\mathcal{A}}(\lambda)$ is negligible any PPT algorithm $\mathcal{A}$.

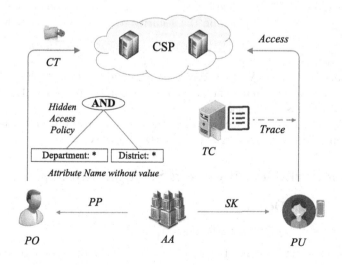

**Fig. 1.** System architecture of HTAC

# 3  System Model and Security Goals

## 3.1  System Model

As depicted in Fig. 1, the system architecture of HTAC consists of 5 types of entities: (1) Attribute Authority (AA), (2) Cloud Service Provider (CSP), (3) PHR Owner (PO), (4) PHR User (PU), (5) Trace Center (TC).

- AA governs the attribute universe, sets the public parameters and grants the private keys for PU according to his attributes.
- CSP stores the encrypted PHRs along with the corresponding hidden access structures. It can also delete the PHRs if necessary.
- PO integrates his own PHRs via smart devices and some wearable or implantable sensors. PO can define appropriate access structures to encrypt his PHRs before outsourcing it to CSP.
- PU is a PHR user who needs to access the encrypted PHRs to offer health care service, such as a physician. Every PU possesses some attributes and the corresponding private keys. A PU can successfully recover the encrypted PHR only if his attribute set matches the attached hidden access policy.
- TC is responsible for tracing the malicious PU who leaks his private key for some illegal purpose.

## 3.2  Security Goals

In HTAC, AA and TC are trustworthy. CSP is assumed to be honest-but-curious as in [24]. That is, it honestly executes the specified procedures but tries to gain secret information from encrypted PHRs. The adversary could be a malicious PU or a group of multiple PUs and CSP. Moreover, the adversary also aims to learn the attribute values of the hidden access policies from encrypted PHRs.

Concretely, we mostly focus on the security requirements as follows.

- *PHR Confidentiality.* The PHRs contains sensitive information and should be kept secrecy from any unauthorized user.
- *Collusion-Resistance.* Various malicious PUs and CSP may collude to recover the PHR ciphertext that none of them is authorized to access. HTAC should resist such collusion attacks.
- *Attribute Privacy.* In an access policy, the concrete attribute value is sensitive and should be hidden to preserve the attribute privacy.

# 4  Framework Definition and Security Models

## 4.1  Framework Definition

HTAC consists of the following five algorithms.

- $\mathsf{Setup}(\kappa, U) \to (\mathsf{PP}, \mathsf{MSK})$: By taking in a security parameter $\kappa$ and the system attribute universe description $U$, this algorithm returns the system public parameter $\mathsf{PP}$ and the master key $\mathsf{MSK}$. Additionally, it initializes an identity table $IT = \varnothing$.

- KeyGen(PP, MSK, $id, S$) $\rightarrow$ SK$_{id,S}$: This algorithm takes in PP, MSK, an identity $id$, and a set of attributes $S$. It then returns a private key SK$_{id,S}$.
- Encrypt(PP, $\mathbb{A}, M$) $\rightarrow$ CT$_{\mathbb{A}}$: This algorithm takes in PP, an access structure $\mathbb{A} = (\mathbf{A}, \rho, \mathcal{T})$, a plaintext message $M$. It then returns a ciphertext CT$_{\mathbb{A}}$.
- Decrypt(PP, CT$_{\mathbb{A}}$, SK$_{id,S}$) $\rightarrow$ $M$ or $\perp$: This algorithm takes in PP, CT$_{\mathbb{A}}$ and SK$_{id,S}$. It returns $M$ only if $S$ matches $\mathbb{A}$. Otherwise, it returns $\perp$. Concretely, the decryption algorithm contains two subroutines, Matching Test and Final Decryption. If $S$ does not match $\mathbb{A}$, Matching Test outputs $\perp$ to terminate the decryption process. Otherwise, it invokes the Final Decryption to return $M$.
- Trace($PP$, SK$_{id,S}$, $IT$) $\rightarrow$ $id$ or $\top$. Given $PP$, SK$_{id,S}$ and $IT$. This algorithm first checks whether SK$_{id,S}$ is well-formed. If so, it returns the $id$ associated with SK$_{id,S}$. Otherwise, it returns $\top$ to claim that SK$_{id,S}$ is not required to be traced. SK$_{id,S}$ is called well-formed if it can pass a 'key sanity check' [12].

## 4.2 CPA Security Model

The security model is described as a security game between a simulator $\mathcal{B}$ and an adversary $\mathcal{A}$.

- **Setup.** $\mathcal{B}$ runs Setup to create PP and MSK. Only PP is sent to $\mathcal{A}$.
- **Phase 1.** $\mathcal{A}$ can request the private keys of the following attribute sets $(id_1, S_1), \ldots, (id_{q_1}, S_{q_1})$.
- **Challenge.** $\mathcal{A}$ gives $\mathcal{B}$ two equal-length messages $M_0$, $M_1$ and two challenge access structures $\mathbb{A}_1 = (\mathbf{A}, \rho, \mathcal{T}_0)$, $\mathbb{A}_2 = (\mathbf{A}, \rho, \mathcal{T}_1)$. $\mathcal{B}$ randomly picks $\beta \in \{0, 1\}$, computes CT$_{\mathbb{A}_\beta}$ $\leftarrow$ Encrypt(PP, $M_\beta, \mathbb{A}_\beta$) and gives CT$_{\mathbb{A}_\beta}$ to $\mathcal{A}$.
- **Phase 2.** $\mathcal{A}$ requests the private keys of the following attribute sets $(id_{q_1+1}, S_{q_1+1}), \ldots, (id_q, S_q)$.
- **Guess:** $\mathcal{A}$ outputs its guess $\beta' \in \{0, 1\}$.

$\mathcal{A}$ wins if $\beta' = \beta$ under such restriction that neither $\mathbb{A}_1$ nor $\mathbb{A}_2$ can be matched by any queried set in Phase 1 and Phase 2. The advantage of $\mathcal{A}$ in the above game can be defined as $\left| \Pr[\beta' = \beta] - \frac{1}{2} \right|$.

**Definition 6.** *HTAC is adaptively secure if none polynomial time adversary can win the security game with a non-negligible advantage.*

## 4.3 Traceability Model

- **Setup.** $\mathcal{B}$ runs Setup to create PP and MSK. PP is sent to $\mathcal{A}$.
- **Key Query.** $\mathcal{A}$ queries the private keys of the tuples $(id_1, S_1)$, $(id_2, S_1), \ldots, (id_q, S_q)$ from $\mathcal{B}$.
- **Key Forgery.** $\mathcal{A}$ outputs $SK_\star$. $\mathcal{A}$ wins if **Trace**($IT, PP, SK_\star$) $\neq \top$ and **Trace**($IT, PP, SK_\star$) $\notin \{id_1, id_2, ..., id_q\}$. $\mathcal{A}$'s advantage is defined as $Pr[\mathbf{Trace}(IT, PP, SK_\star) \neq \{\top\} \cup \{id_1, id_2, ..., id_q\}]$.

**Definition 7.** *HTAC is fully traceable if all PPT attackers have at most negligible advantage in the above game.*

# 5   Our Construction

## 5.1   Initialization

AA generates the system parameters by running the Setup algorithm.

– Setup: Firstly, AA runs the group generator $\mathcal{G}$ by taking in the system security
  parameter $\kappa$ and obtain $\mathbb{GD} = (\mathbb{G}, \mathbb{G}_1, N = p_1 p_2 p_3 p_4, e)$. Secondly, AA sets
  $\mathbb{Z}_N$ as the system attribute universe and randomly picks $\alpha, a, b \in \mathbb{Z}_N$, $g, \vartheta \in$
  $\mathbb{G}_{p_1}$, $X_3 \in \mathbb{G}_{p_3}$, $X_4, \Upsilon \in \mathbb{G}_{p_4}$ and calculates $Y = e(g, g)^\alpha$, $H = \vartheta \Upsilon$. Finally,
  $\mathsf{PP} = (g, g^a, g^b, H, Y, X_4, N)$ is published and $\mathsf{MSK} = (\vartheta, \Upsilon, \alpha)$ is kept as
  secret.

  It also initializes the table $IT$ to be empty.

## 5.2   PU Authorization

When a PU joins in the system, he will be labeled by an $id$ and issued an attribute
set $\mathcal{S} = (\mathcal{NAM}_S, \mathcal{VU}_S)$, where $\mathcal{NAM}_S \subseteq \mathbb{Z}_N$ and $\mathcal{VU}_S = \{s_i\}_{i \in \mathcal{NAM}_S}$. AA
then generates the private keys for PU by running KeyGen.

– KeyGen: AA randomly picks $t \in \mathbb{Z}_N$, $c \in \mathbb{Z}_N^*$ and $R, R', R'', R_i \in \mathbb{G}_{p_3}$ for $i \in$
  $\mathcal{NAM}_S$. The private key is set as $\mathsf{SK}_{id, \mathcal{S}} = (\mathcal{S}, K, K', L, L', \{K_i\}_{i \in \mathcal{NAM}_S})$,
  where $K = g^{\frac{\alpha}{(b+c)}} g^{at} R, K' = g^t R', L = c, L' = g^{bt} R'', K_i = (g^{s_i} \vartheta)^{(b+c)t} R_i$.

  Finally, AA records $(id, c)$ in $IT$.

## 5.3   PHR Outsourcing

As in the KEM scheme [13], PO first employs a symmetric encryption scheme
and selects a symmetric key SYK to encrypt his PHRs. Then, PO defines a
hidden access policy $\mathbb{A} = (\mathbf{A}, \rho, \mathcal{T})$ and encrypts SYK by running Encrypt. SYK
is imply set as an element in $\mathbb{G}_1$.

– Encrypt: PO randomly picks two vectors $v, \mu \in \mathbb{Z}_N^n$ where $v = (s, v_2, \ldots, v_n)$
  and $\mu = (s_1, \mu_2, \ldots, \mu_n)$. Based on $X_4$, It also randomly chooses
  $Q, Q_1, Q_{\Delta,x}, Q_{c,x}, Q_{d,x} \in \mathbb{G}_{p_4}$ and $r_x \in \mathbb{Z}_N$ for $1 \leq x \leq \ell$. It then
  set $\mathsf{CT}_\mathbb{A} = ((\mathbf{A}, \rho), CT_T, CT_D)$, where $CT_T$ is used for decryption test
  and $CT_D$ is the real ciphertext of SYK. More specifically, $CT_T$ is com-
  puted as $CT_T = (C_\Delta, C_{\Delta,0}, C'_{\Delta,0}, \{C_{\Delta,x}\}_{1 \leq x \leq \ell})$ and $CT_D$ is calculated as
  $(C, C_0, C'_0, \{C_x, D_x\}_{1 \leq x \leq \ell})$ where $C_\Delta = Y^{s_1}, C_{\Delta,0} = g^{s_1} Q, C'_{\Delta,0} = g^{bs_1} Q_1$,
  $C_{\Delta,x} = g^{aA_x \cdot \mu} (g^{t_{\rho(x)}} H)^{-s_1} Q_{\Delta,x}, C = \mathsf{SYK} \cdot Y^s, C_0 = g^s, C'_0 = g^{bs}$ and
  $C_x = g^{aA_x \cdot v} (g^{t_{\rho(x)}} H)^{-r_x} Q_{c,x}, D_x = g^{r_x} Q_{d,x}$.

  Finally, PO uploads the PHR ciphertext data and $\mathsf{CT}_\mathbb{A}$ to CSP.

### 5.4 PHR Access

If PU owns the appropriate attributes that match the access policy, SYK can be recovered by the running Decrypt.

- Decrypt: PU first computes $\mathbf{I_{A,\rho}}$, which refers to the set of minimum subsets of $\{1, 2, \ldots, \ell\}$ that satisfies $(\mathbf{A}, \rho)$. Then PU works as follows.
  - Matching Test: This algorithm checks if there exists $\mathcal{I} \in \mathbf{I_{A,\rho}}$ that satisfies $\{\rho(i)|i \in \mathcal{I}\} \subseteq \mathcal{NAM}_S$ and $C_\Lambda^{-1} = \Lambda\Theta\Xi$, where $\Lambda = e\left(\prod_{i \in \mathcal{I}} C_{\Lambda,i}^{\omega_i}, (K')^L L'\right)$, $\Theta = e\left(C_{\Lambda,0}, \prod_{i \in \mathcal{I}} K_{\rho(i)}^{\omega_i}\right)$, $\Xi = e\left(((C_{\Lambda,0})^L C_{\Lambda,0}', K^{-1}\right)$ and $\sum_{i \in \mathcal{I}} \omega_i A_i = (1, 0, \ldots, 0)$ for some coefficients $\{\omega_i\}_{i \in \mathcal{I}}$. If no such $\mathcal{I}$ can be found, it returns $\perp$ to point out that $\mathcal{S}$ does not match $\mathbb{A}$. Otherwise, it runs Final Decryption by invoking the qualified $\mathcal{I}$ and $\{\omega_i\}_{i \in \mathcal{I}}$.
  - Final Decryption: This algorithm recovers SYK by computing $\mathsf{SYK} = C/B$, where

$$B = \frac{e\left((C_0)^L C_0', K\right)}{\prod_{i \in \mathcal{I}}\left((e(C_i, (K')^L L')e(D_i, K_{\rho(i)})\right)^{\omega_i}}.$$

Finally, PU can use SYK to decrypt the PHR ciphertext.

### 5.5 Trace

To catch the malicious PU who leaks the key $\mathsf{SK}_{id,\mathcal{S}}$ in the form of $(\mathcal{S}, K, K', L, L', \{K_i\}_{i \in \mathcal{NAM}_S})$. TC can runs Trace by taking in $\mathsf{SK}_{id,\mathcal{S}}$.

- Trace: TC first calls Key Sanity Check to check if $\mathsf{SK}_{id,\mathcal{S}}$ is well-formed. If $\mathsf{SK}_{id,\mathcal{S}}$ can pass the following checks, it could be called a well-formed key. TC then searches $L$ in $IT$: If $L$ can be found, TC returns the corresponding $id$. Otherwise, it returns $\top$.
  - Key Sanity Check:
    (1) $L \in \mathbb{Z}_N, K, K', L', K_i \in \mathbb{G}$;
    (2) $e(g^b, K') = e(g, L') \neq 1$;
    (3) $e(g^b \cdot g^L, K) = e((K')^L \cdot L', g^a) \cdot e(g, g)^\alpha \neq 1$;
    (4) $\exists i \in \mathcal{S}, s.t. e(g^{s_i} H, (K')^L \cdot L') = e(g, K_i) \neq 1$.

## 6 Security Analysis

### 6.1 CPA Security

For simplicity, we reduce the CPA security of HTAC to that of [26]. In the following security proof, we denote the scheme [26] and HTAC be $\sum_{HAC}$ and $\sum_{HTAC}$, respectively.

**Lemma 1.** $\sum_{HAC}$ *is adaptively secure if Assumptions 1, 2, 3, and 4 hold.*

PROOF. For the detailed security proof, please refer to [26].

**Lemma 2.** $\sum_{HTAC}$ *is adaptively secure in the security game of Sect. 4.2, if* $\sum_{HAC}$ *is adaptively secure.*

PROOF. Suppose there exists a PPT adversary $\mathcal{A}$ which can break our scheme $\sum_{HTAC}$ with advantage $ADV_{\sum_{HTAC}}$, then we can build a PPT simulator $\mathcal{B}$ to break the underline $\sum_{HAC}$ with advantage $ADV_{\sum_{HAC}}$, which is identical to $ADV_{\sum_{HAC}}$.

**Setup.** After receiving the public parameters $\mathsf{PP}_{\sum_{HAC}} = (g, g^a, H, Y, X_4, N)$ from $\sum_{HAC}$, $\mathcal{B}$ randomly chooses $b \in \mathbb{Z}_N$ and gives $\mathsf{PP}_{\sum_{HTAC}} = (g, g^a, g^b, H, Y, X_4, N)$ to $\mathcal{A}$. Besides, $\mathcal{B}$ initializes $IT = \varnothing$

**Phase 1.** When $\mathcal{A}$ queries the private key of $(id, \mathcal{S})$, $\mathcal{B}$ submits $\mathcal{S}$ to $\sum_{HAC}$ and obtains $\mathsf{SK}_{\mathcal{S}} = (\mathcal{S}, \hat{K}, \hat{K}', \{\hat{K}_i\}_{i \in \mathcal{NAM}_{\mathcal{S}}})$, where $\hat{K} = g^\alpha g^{a\hat{t}} R, \hat{K}' = g^{\hat{t}} R', \hat{K}_i = (g^{s_i} \vartheta)^{\hat{t}} R_i$. $\mathcal{B}$ then randomly picks $c \in \mathbb{Z}_N^*$, $R'' \in \mathbb{G}_{p_3}$ and computes

$$K = (\hat{K})^{\frac{1}{b+c}} = (g^\alpha g^{a\hat{t}} R)^{\frac{1}{b+c}} = g^{\frac{\alpha}{(b+c)}} g^{at} R^{\frac{1}{b+c}}$$
$$K' = (\hat{K}')^{\frac{1}{b+c}} = (g^{\hat{t}} R')^{\frac{1}{b+c}} = g^t R'^{\frac{1}{b+c}}$$
$$L' = (\hat{K}')^{\frac{b}{b+c}} R'' = g^{bt} R'^{\frac{1}{b+c}} R''$$
$$K_i = \hat{K}_i = (g^{s_i} \vartheta)^{\hat{t}} R_i = (g^{s_i} \vartheta)^{(b+c)t} R_i$$

$\mathcal{B}$ then gives $\mathsf{SK}_{id,\mathcal{S}} = (\mathcal{S}, K, K', L = c, L', \{K_i\}_{i \in \mathcal{NAM}_{\mathcal{S}}})$ to $\mathcal{A}$ and adds $(id, c)$ to $IT$.

Remark that, $\mathcal{B}$ implicitly sets $t = \frac{\hat{t}}{b+c}$ in generating $\mathsf{SK}_{id,\mathcal{S}}$. Moreover, if $gcd(b+c, N) \neq 1$ or $c$ has been put into $IT$, $\mathcal{B}$ has to randomly pick a new $c \in \mathbb{Z}_N^*$ and rebuild the private key.

**Challenge.** $\mathcal{A}$ gives $\mathcal{B}$ two access structures $\mathbb{A}_1 = (\mathbf{A}, \rho, \mathcal{T}_0)$, $\mathbb{A}_2 = (\mathbf{A}, \rho, \mathcal{T}_1)$ and two messages $M_0, M_1$ of equal length. $\mathcal{B}$ submits them to $\sum_{HAC}$ and gets the corresponding challenge ciphertext $\mathsf{CT}_{\sum_{HAC}} = (\mathbf{A}*, \rho), \hat{CT}_T = (\hat{C}_\Delta, \hat{C}_{\Delta,0}, \hat{C}'_{\Delta,0}, \{\hat{C}_{\Delta,x}\}_{1 \leq x \leq \ell}), \hat{CT}_D = (\hat{C}, \hat{C}_0, \hat{C}'_0, \{\hat{C}_x, \hat{D}_x\}_{1 \leq x \leq \ell}))$, where

$$\hat{C}_\Delta = Y^{s_1}, \hat{C}_{\Delta,0} = g^{s_1} Q, \hat{C}_{\Delta,x} = g^{aA_x \cdot \mu} (g^{t_{\rho(x)}} H)^{-s_1} Q_{\Delta,x}, \hat{C} = \mathsf{SYK} \cdot Y^s,$$
$$\hat{C}_0 = g^s \text{ and } \hat{C}_x = g^{aA_x \cdot v} (g^{t_{\rho(x)}} H)^{-r_x} Q_{c,x}, \hat{D}_x = g^{r_x} Q_{d,x}.$$

$\mathcal{B}$ then sets

$$C_\Delta = \hat{C}_\Delta = Y^{s_1}, C_{\Delta,0} = \hat{C}_{\Delta,0} = g^{s_1} Q, C_{\Delta,x} = \hat{C}_{\Delta,x} = g^{aA_x \cdot \mu} (g^{t_{\rho(x)}} H)^{-s_1} Q_{\Delta,x},$$
$$C = \hat{C} = \mathsf{SYK} \cdot Y^s, C_0 = \hat{C}_0 = g^s \text{ and } C_x = \hat{C}_x = g^{aA_x \cdot v} (g^{t_{\rho(x)}} H)^{-r_x} Q_{c,x},$$
$$D_x = \hat{D}_x = g^{r_x} Q_{d,x}.$$

Additionally, $\mathcal{B}$ computes $C'_{\Delta,0} = (\hat{C}_{\Delta,0})^b$ and $C'_0 = (\hat{C}_0)^b$.

Finally, $\mathcal{B}$ gives $\mathsf{CT}_{\mathbb{A}*} = (\mathbf{A}, \rho), CT_T = (C_\Delta, C_{\Delta,0}, C'_{\Delta,0}, \{C_{\Delta,x}\}_{1 \leq x \leq \ell})$, $CT_D = (C, C_0, C'_0, \{C_x, D_x\}_{1 \leq x \leq \ell}))$ to $\mathcal{A}$.

**Phase 2.** Same as in Phase 1.

**Guess.** $\mathcal{A}$ returns its guess $\beta'$. $\mathcal{B}$ gives $\beta'$ to $\sum_{HAC}$.

**Theorem 1.** *The proposed HTAC is adaptively secure if Assumptions 1, 2, 3 and 4 hold.*

PROOF. The theorem follows Lemmas 1 and 2.

**Table 1.** Characteristic comparison with related work

| Schemes | Policy hidden | Standard model | Large universe | Adaptive security | Expressiveness | Group order | Traceability |
|---------|---------------|----------------|----------------|-------------------|----------------|-------------|--------------|
| [18] | ✗ | ✓ | ✓ | ✗ | LSSS | Prime | ✗ |
| [14] | ✗ | ✓ | ✓ | ✗ | LSSS | Prime | ✓ |
| [12] | ✗ | ✓ | ✗ | ✓ | LSSS | Composite | ✓ |
| [15] | ✓ | ✗ | ✗ | ✗ | AND | Prime | ✗ |
| [16] | ✓ | ✓ | ✗ | ✗ | AND | Prime | ✗ |
| [6] | ✓ | ✓ | ✗ | ✓ | LSSS | Composite | ✗ |
| [26] | ✓ | ✓ | ✓ | ✓ | LSSS | Composite | ✗ |
| Ours | ✓ | ✓ | ✓ | ✓ | LSSS | Composite | ✓ |

**Table 2.** Parameter length comparison

| Scheme | Public parameter | Private key | Ciphertext |
|--------|------------------|-------------|------------|
| [26] | $4\|\mathbb{G}\| + 1\|\mathbb{G}_1\|$ | $(\|S_K\| + 2)\|\mathbb{G}\|$ | $(3\|S_C\| + 2)\|\mathbb{G}\| + 2\|\mathbb{G}_1\|$ |
| Ours | $5\|\mathbb{G}\| + 1\|\mathbb{G}_1\|$ | $(\|S_K\| + 1)\|\mathbb{G}\| + 1\|\mathbb{Z}_N\|$ | $(3\|S_C\| + 4)\|\mathbb{G}\| + 2\|\mathbb{G}_1\|$ |

### 6.2 Traceability

**Theorem 2.** *If $\ell$-SDH assumption [3, 12] and Assumption 2 hold, the proposed scheme is fully traceable provided that $q < \ell$.*

PROOF. We briefly introduce the proof of traceability. The simulator $\mathcal{B}$ is given two independent instances from $\ell$-SDH assumption and Assumption 2. Then $\mathcal{B}$ can interact with an adversary $\mathcal{A}$ as in [12,14]. If $\mathcal{A}$ has non-negligible advantage in the traceability game, then $\mathcal{B}$'s advantage is non-negligible in breaking Assumption 2 and $\ell$-SDH assumption.

### 6.3 Performance Comparison

Table 1 demonstrates the characteristic comparison between related works [6, 12, 14–16, 18, 26] and ours, including access policy privacy, security model, universe scale, security level, policy expressiveness, group order and traceability. From Table 1, we can learn that our HTAC simultaneously achieves policy hidden, large universe and traceability, while the others can only realize one or two of them.

Tables 2 and 3 give the numeric performance comparison between our work and [26]. $P$ represents a bilinear pairing. $|S_C|$, $|S_K|$ and $|I|$ represent the number of attributes associated with $\mathsf{CT_A}$, $\mathsf{SK}_{id,\mathcal{S}}$ and $\mathcal{I}$, respectively. $E$ and $E_1$ refer to an exponential operation in $\mathbb{G}$ and $\mathbb{G}_1$, respectively. From Tables 2 and 3, we learn that the size of $\mathsf{PP}$, $\mathsf{SK}_{id,\mathcal{S}}$ and $\mathsf{CT_A}$ is slightly longer than that of [26]. In the phase of encryption, matching test and final decryption, the computation cost is a little bit more than that in [26]. However, the additional parameters

**Table 3.** Computing cost comparison

| Scheme | Encryption cost | Matching cost | Final decryption cost |
|--------|-----------------|---------------|----------------------|
| [26] | $(6|S_C|+2)E+2E_1$ | $2|I|E+2P$ | $|I|E_1+(2|I|+1)P$ |
| Ours | $(6|S_C|+4)E+2E_1$ | $(2|I|+2)E+3P$ | $2E+|I|E_1+(2|I|+1)P$ |

and incurred computation overhead are employed to support the traceability of HTAC, regardless of the number of involved attributes. In summary, our scheme only sacrifices tiny parameter elements and computation cost to realize traceability in comparison to the scheme [26].

## 7    Conclusion

We have constructed HTAC which simultaneously supported large universe, partially hidden policy and white-box traceability. The size of the system public parameters is constant. The PHR owner can define any LSSS access policy and hide the specific attribute value. None of the unauthorized users can obtain any sensitive information of attributes from the ciphertext. We proved the adaptive security in the standard model. To support the property of traceability, only a few group elements and tiny computation overhead are incurred and have no concern with the attributes.

Our future work is to alleviate the user decryption overhead by designing appreciate outsourced technique to offload the heavy matching test and decryption operations to the cloud.

**Acknowledgment.** This research is sponsored by The National Natural Science Foundation of China under grant No. 61602365, No. 61502248, and the Key Research and Development Program of Shaanxi [2019KW-053]. Yinghui Zhang is supported by New Star Team of Xi'an University of Posts and Telecommunications [2016-02]. We thank the anonymous reviewers for invaluable comments.

## References

1. Beimel, A.: Secure schemes for secret sharing and key distribution. DSc dissertation (1996)
2. Bethencourt, J., Sahai, A., Waters, B.: Ciphertext-policy attribute-based encryption. In: IEEE Symposium on Security and Privacy, SP 2007, pp. 321–334, May 2007. https://doi.org/10.1109/SP.2007.11
3. Boneh, D., Boyen, X.: Short signatures without random oracles. In: Cachin, C., Camenisch, J.L. (eds.) EUROCRYPT 2004. LNCS, vol. 3027, pp. 56–73. Springer, Heidelberg (2004). https://doi.org/10.1007/978-3-540-24676-3_4
4. Chase, M.: Multi-authority attribute based encryption. In: Vadhan, S.P. (ed.) TCC 2007. LNCS, vol. 4392, pp. 515–534. Springer, Heidelberg (2007). https://doi.org/10.1007/978-3-540-70936-7_28

5. Cheung, L., Newport, C.: Provably secure ciphertext policy ABE. In: Proceedings of the 14th ACM Conference on Computer and Communications Security, CCS 2007, New York, NY, USA, pp. 456–465. ACM (2007). https://doi.org/10.1145/1315245.1315302

6. Lai, J., Deng, R.H., Li, Y.: Expressive CP-ABE with partially hidden access structures. In: Proceedings of the 7th ACM Symposium on Information, Computer and Communications Security, ASIACCS 2012, New York, NY, USA, pp. 18–19. ACM (2012). https://doi.org/10.1145/2414456.2414465

7. Lewko, A., Okamoto, T., Sahai, A., Takashima, K., Waters, B.: Fully secure functional encryption: attribute-based encryption and (hierarchical) inner product encryption. In: Gilbert, H. (ed.) EUROCRYPT 2010. LNCS, vol. 6110, pp. 62–91. Springer, Heidelberg (2010). https://doi.org/10.1007/978-3-642-13190-5_4

8. Lewko, A., Waters, B.: Unbounded HIBE and attribute-based encryption. In: Paterson, K.G. (ed.) EUROCRYPT 2011. LNCS, vol. 6632, pp. 547–567. Springer, Heidelberg (2011). https://doi.org/10.1007/978-3-642-20465-4_30

9. Li, J., Chen, X., Chow, S.S., Huang, Q., Wong, D.S., Liu, Z.: Multi-authority fine-grained access control with accountability and its application in cloud. J. Netw. Comput. Appl. 112, 89–96 (2018). https://doi.org/10.1016/j.jnca.2018.03.006. http://www.sciencedirect.com/science/article/pii/S1084804518300870

10. Li, Q., Ma, J., Li, R., Liu, X., Xiong, J., Chen, D.: Secure, efficient and revocable multi-authority access control system in cloud storage. Comput. Secur. 59, 45–59 (2016). https://doi.org/10.1016/j.cose.2016.02.002. http://www.sciencedirect.com/science/article/pii/S0167404816300050

11. Li, Q., Zhu, H., Xiong, J., Mo, R., Wang, H.: Fine-grained multi-authority access control in IoT-enabled mhealth. Ann. Telecommun. 4, 1–12 (2019)

12. Liu, Z., Cao, Z., Wong, D.S.: White-box traceable ciphertext-policy attribute-based encryption supporting any monotone access structures. IEEE Trans. Inf. Forensics Secur. 8(1), 76–88 (2013). https://doi.org/10.1109/TIFS.2012.2223683

13. Ning, J., Cao, Z., Dong, X., Liang, K., Ma, H., Wei, L.: Auditable *sigma*-time outsourced attribute-based encryption for access control in cloud computing. IEEE Trans. Inf. Forensics Secur. 13(1), 94–105 (2018). https://doi.org/10.1109/TIFS.2017.2738601

14. Ning, J., Dong, X., Cao, Z., Wei, L., Lin, X.: White-box traceable ciphertext-policy attribute-based encryption supporting flexible attributes. IEEE Trans. Inf. Forensics Secur. 10(6), 1274–1288 (2015). https://doi.org/10.1109/TIFS.2015.2405905

15. Nishide, T., Yoneyama, K., Ohta, K.: Attribute-based encryption with partially hidden encryptor-specified access structures. In: Bellovin, S.M., Gennaro, R., Keromytis, A., Yung, M. (eds.) ACNS 2008. LNCS, vol. 5037, pp. 111–129. Springer, Heidelberg (2008). https://doi.org/10.1007/978-3-540-68914-0_7

16. Phuong, T.V.X., Yang, G., Susilo, W.: Hidden ciphertext policy attribute-based encryption under standard assumptions. IEEE Trans. Inf. Forensics Secur. 11(1), 35–45 (2016). https://doi.org/10.1109/TIFS.2015.2475723

17. Qi, L., Zhu, H., Ying, Z., Tao, Z.: Traceable ciphertext-policy attribute-based encryption with verifiable outsourced decryption in ehealth cloud. Wirel. Commun. Mob. Comput. 2018, 1–12 (2018)

18. Rouselakis, Y., Waters, B.: Practical constructions and new proof methods for large universe attribute-based encryption. In: Proceedings of the 2013 ACM SIGSAC Conference on Computer and Communications Security, CCS 2013, New York, NY, USA, pp. 463–474. ACM (2013). https://doi.org/10.1145/2508859.2516672

19. Sahai, A., Waters, B.: Fuzzy identity-based encryption. In: Cramer, R. (ed.) EURO-CRYPT 2005. LNCS, vol. 3494, pp. 457–473. Springer, Heidelberg (2005). https://doi.org/10.1007/11426639_27

20. Waters, B.: Ciphertext-policy attribute-based encryption: an expressive, efficient, and provably secure realization. In: Catalano, D., Fazio, N., Gennaro, R., Nicolosi, A. (eds.) PKC 2011. LNCS, vol. 6571, pp. 53–70. Springer, Heidelberg (2011). https://doi.org/10.1007/978-3-642-19379-8_4

21. Xue, K., Xue, Y., Hong, J., Li, W., Yue, H., Wei, D.S.L., Hong, P.: RAAC: robust and auditable access control with multiple attribute authorities for public cloud storage. IEEE Trans. Inf. Forensics Secur. **12**(4), 953–967 (2017). https://doi.org/10.1109/TIFS.2016.2647222

22. Yang, K., Jia, X., Ren, K.: Secure and verifiable policy update outsourcing for big data access control in the cloud. IEEE Trans. Parallel Distrib. Syst. **26**(12), 3461–3470 (2015). https://doi.org/10.1109/TPDS.2014.2380373

23. Yang, Y., Liu, X., Deng, R.H.: Lightweight break-glass access control system for healthcare internet-of-things. IEEE Trans. Ind. Inform. **14**, 3610–3617 (2017). https://doi.org/10.1109/TII.2017.2751640

24. Yu, S., Wang, C., Ren, K., Lou, W.: Achieving secure, scalable, and fine-grained data access control in cloud computing. In: IEEE INFOCOM 2010 Proceedings, pp. 1–9, March 2010. https://doi.org/10.1109/INFCOM.2010.5462174

25. Zhang, L., Hu, G., Mu, Y., Rezaeibagha, F.: Hidden ciphertext policy attribute-based encryption with fast decryption for personal health record system. IEEE Access **7**, 33202–33213 (2019). https://doi.org/10.1109/ACCESS.2019.2902040

26. Zhang, Y., Zheng, D., Deng, R.H.: Security and privacy in smart health: efficient policy-hiding attribute-based access control. IEEE Internet Things J. **5**(3), 2130–2145 (2018)

# Construction of Laboratory Refined Management in Local Applied University

Lan Liu[1,2(✉)], Xiankun Sun[1], Chengfan Li[3], and Yongmei Lei[3]

[1] School of Electronic and Electrical Engineering,
Shanghai University of Engineering Science, Shanghai 201620, China
lanny718@sina.com, xksun@sues.edu.cn
[2] Shanghai Key Laboratory of Computer Software Evaluating and Testing,
Shanghai 201112, China
[3] School of Computer Engineering and Science, Shanghai University,
Shanghai 200444, China
{lchf,lei}@shu.edu.cn

**Abstract.** The refined management of laboratory in local applied university is conducive to the formation of mode to innovative talent training and the improvement the ability to innovate. To ensure the efficient operations of laboratory in local applied university, in this paper we put forward to s new construction method of laboratory refined management in local applied university on the basis of analyzing the current actual situation of laboratory managements. Finally, taking the Computer laboratory of School of Electrical and Electronic Engineering in Shanghai University of Engineering and Technology as an example, the refined management case of laboratory in local applied university is constructed in terms of developing and designing of the refined laboratory management information system (MIS), and future works on the following refined managements of laboratory in local applied university was presented. It has a certain impact on promoting the management efficiency of local applied university and the service ability of the society.

**Keywords:** Local applied university · Construction · Refined management · Laboratory · Management information system (MIS)

## 1 Introduction

Local applied university is an important part of China's higher education. In accordance with the economic and social development and education reform under the new situation, the major decisions that guiding and promoting some undergraduate universities transformed into applied functions have made and implemented [1]. The State and local governments have continuous deepen the education and teaching reform and improve the running conditions. Meanwhile, to strengthen and improve the practical teaching ability of local applied university, the national development and reform commission (NDRC) and the Ministry of Education and other departments have proposed continuously the implementation of education modernization projects for local applied university [2]. The State will take the lead in supporting 100 universities in

© ICST Institute for Computer Sciences, Social Informatics and Telecommunications Engineering 2019
Published by Springer Nature Switzerland AG 2019. All Rights Reserved
Q. Li et al. (Eds.): BROADNETS 2019, LNICST 303, pp. 275–284, 2019.
https://doi.org/10.1007/978-3-030-36442-7_18

strengthening the construction of experimental training platforms and experimental bases so as to improve the quality of higher education by project demonstration of new modes of personnel training.

Correspondingly, with the rapid increase of State and local governments' economic investment, the construction of Shanghai science and technology innovation center has intensive demand for refined management for instruments and equipment and open sharing. However, due to various factors in the current laboratory construction, the laboratory management cannot serve the economic and social development and personnel training of local applied university [3, 4]. As an important institution of scientific research and teaching in university, laboratory plays an important role in the daily operation of universities and social services [5]. How to carry on refined management has become the key of improving our laboratory management level and the use efficiency of instrument and equipment, and further realizing its open sharing to the society and win-win situation of teaching and research.

Based on analyzing for the laboratory management in local applied university, we first summary the existing problems of laboratory management and point out the suggestions and countermeasures. And then, taking the Computer laboratory of School of Electronic and Electrical Engineering in Shanghai University of Engineering Science as an example, the refined laboratory management has been constructed and implemented by developing and designing the laboratory management information system (MIS) in this paper. It has great significance to improve the use efficiency of laboratory equipment and teaching and scientific research level, and to serve the laboratory information construction of local applied university. In this paper we present a practical laboratory MIS of Computer laboratory of School of Electronic and Electrical Engineering in Shanghai University of Engineering Science, it is possible to improve the laboratory management level and the use efficiency and open sharing of instrument and equipment to some extent.

The rest of the paper is constructed as follows: Sect. 2 explains the related work. Section 3 presents the refined management approach in university laboratory. Section 4 devotes the refined management experiments of laboratory MIS in local applied universities. Section 5 discusses the shortcomings and deficiencies of this work. Finally, our conclusions and future work are drawn in Sect. 6.

## 2    Related Work

### 2.1    Current Status of Laboratory Management

Limited by the financial input in the early stage, the funds available for laboratory construction in local applied university are relatively less, especially in laboratory management hardware, which is obviously different from those universities affiliated to the Ministry of Education [6–8]. Traditionally, the management of laboratory is mainly registered and recorded by manual operation, and then generates reports and other output results. These traditional manual methods are not only time-consuming and laborious, easy to produce errors, and the effect is not ideal in practical use. In recent years, the rapid popularity of the Internet has become the important grasp of information construction and refined management for laboratory, and currently it basically covers almost all universities [9, 10]. In addition, on the basis of hardware resources, the refined management of the laboratory is mainly accompanied by corresponding software. At present, many local applied universities have initially established corresponding laboratory MIS by various ways [11]. However, subject to the conditions of local applied university, it has different management functions and development levels, and there are various differences in specific use. According to the statistical analysis for the current mainstream university laboratory, most of those have routine operation, maintenance and laboratory safety management, and the users include student, teacher, management stewardship, departments in charge of the leadership, etc.

At present, a large number of hardware and software products compatible with laboratory management on the market are usually expensive. Laboratory construction and management have long been of little contribution to teaching and scientific research [12]. Faced with limited funds, many universities prefer to purchase scientific research-related instruments and equipment, rather than spend money on laboratory management. As a result, the laboratory refined management is mostly in words, and the actual implementation is not really promoted and implemented. And it seriously restricts the improvement of laboratory management level and the effect of instrument open sharing.

As an important part of laboratory refined management in local applied university, experimental teaching and talent cultivation are far from keeping pace with the times. In recent years, with the advancement of education informatization, a series of new teaching forms have been emerging, such as Mooc, Micro class, Boutique class, Flip class and Virtual reality simulation, etc. [13, 14]. However, these new technologies and means have almost never been involved in fact, and most experimental teachers are limited by age and scope of knowledge, it is difficult to effectively and skillfully apply these new media means [15, 16].

## 2.2    Problems of Laboratory Management in Local Applied University

The function is incomplete and missing overall planning. At present, the laboratory management of the most universities is still stays in the primary stage, and generally lacks the laboratory website, and the content of laboratory management is relatively simple. And there is no effective integration for different functions. On the one hand, for the network construction in the laboratory informatization, the laboratory adopts network and security-related equipment eliminated by other scientific research departments. There is relatively little newly purchased equipment, and many functions cannot be effectively developed and utilized. On the other hand, for laboratory management software, many universities design primary laboratory management software by themselves to save funds. These management software not only function is simple, the system is not perfect, comprehensive, and the security is not effectively guaranteed, easy to cause laboratory information leakage and security risks. It is not conducive to the implementation and rapid progress of laboratory refined management.

The management mode is unreasonable. The management mechanism is backward, and the management mode is unreasonable is one of the most important factors. At present, many universities still adopt the traditional university-college-department three-level management mode. Since the management system was initially set up and divided in terms of majors, the rise of cross-disciplines was not considered. To some extent, there are some problems in the practical use, such as repeated acquisition of equipment, overlapping of laboratory management functions and low efficiency, etc. And what's more, the laboratory usually adopts the single-machine management mode, which is difficult to adapt to the needs of laboratory personnel training in local applied university, and it is not conducive to the construction of comprehensive information MIS platform by cloud computing and other new technologies.

Shortage of openness and uniform standards. At present, the teaching and management of laboratory in many colleges and universities are mainly closed, the experiments are usually carried out in classes, and it is rarely open to students all day. On the one hand, the lack of effective communication and sharing between different types of laboratories and reduces the overall management and use efficiency of laboratories, which is not conducive to serve the society by local applied university. On the other hand, there is a lack of unified standards for the construction of laboratory MIS among different laboratories, which makes the management system established difficult to connect with each other. The laboratory management of local applied university did not consider the unified standards of openness under the background of information construction.

## 3    Methods

### 3.1    Updating the Concept of Laboratory Management

In the practical use, we should change the old traditional concept of emphasizing scientific research and neglecting experiment, strengthen the new concept of opening and refined management of laboratory construction, and further establish a brand-new open experimental environment. And then it significantly improves the management and use efficiency of the laboratory in local applied university. In an open and centralized experimental environment, by the centralized management and open use of valuable laboratory instruments and equipment, it breaks the shackles of existing equipment sharing service mode and improves the degree of open sharing and service of local application-oriented university laboratories. In the newly management mode, it is necessary to break the traditional three-level horizontal management mode between colleges and departments, learn from the current laboratory management mode of foreign universities and research institutes, design laboratory complex with specific application scenarios as the unit, and establish the experimental platform environment for centralized management and use. In addition, for the work of laboratory management, attention should be paid to the connection and coordination among different applications so as to avoid repeated setting of projects and repeated procurement of instruments and equipment, and to finally improve the level and efficiency of laboratory management.

### 3.2    Promoting the Construction of Experimental Technical Team

An experiment team with complete structure and reasonable collocation is the fundamental guarantee to realize refined management and efficient operation of university laboratories. It is necessary to increase the professional training of experimental technical and managerial personnel to improve the professional level and overall quality of the experimental technical team. Detailed and feasible rules and regulations for laboratory operation and assessment should be established and improved, and the post responsibilities of new experimental technical team are revised and formulated according to the actual situation, meanwhile, laboratory managers are encouraged to actively participate in specific scientific research projects, and stimulated the innovation ability of experimental technology. In addition, the related department should take effective measures, actively try to hire experimental series leading talents from well-known universities or research institutes, and attract high-quality experimental talents to participate in the experimental technical team of universities by setting up a number of specific post plans, so as to drive the rapid improvement of the overall level of laboratory technology and management. At the same time, local applied university laboratories should be good at breaking the routine according to their own character-istics and practical needs, innovating the regulations on the professional title evaluation of experimental series, setting up the reward and evaluation measures for experimental series, improving the enthusiasm and enthusiasm of experimental personnel, and fur-ther broadening the professional title promotion channel for experimental personnel. It has great significance to the long-term stable development of the laboratory and to improve the level of laboratory fine management.

### 3.3   Establishing the Portal Website and Construction Standard

The local applied universities should strengthen vigorously the investment in laboratory construction, actively promote laboratory informatization, realize fine management of laboratory, change the previous manual registration management mode of laboratory, and effectively overcome the current overlapping functions, unclear division of labor and horizontal management mode of laboratory. By the establishment of the laboratory portal website, the laboratory teaching, scientific research, management, maintenance and other work can be integrated into one. It can not only clearly understand and complete the refined management of the overall laboratory daily operation and online performance assessment, but also realize the functions of experimental teaching and open sharing of valuable instruments, which greatly saves the development cost and improves the ability of local applied universities to cultivate talents and serve the society innovation. In addition, for portal construction based on a multimedia laboratory of refined management, it needs the related department planning design and functions in advance, establish function module that both suitable for local applied college labora-tory itself characteristic and has a certain standard of interoperability standards, and finally improve the efficiency and scalability of the laboratory of refined management.

## 4   Experimentation

In this section we take the development and design of Computer laboratory MIS of School of Electronic and Electrical Engineering in Shanghai University of Engineering Science as an example, and try to carry out preliminary construction and implemen-tation of laboratory refined management software.

## 4.1   Design Objective

The design objective is to establish a laboratory refined and comprehensive management platform, including experimental teaching, laboratory inquiry and appointment, equipment inquiry and appointment, laboratory management, data export and summary, report generation and other functions. It can reduce the complexity and labor intensity of daily laboratory management, improve the level and efficiency of laboratory management, and meet the needs of local application-oriented university laboratory innovation personnel training mode and serving the society by means of multimedia means.

## 4.2   Implementation

Platform model architecture. The laboratory refined MIS platform is composed of graphical interface and developed by using SQL Server database, B/S three-layer architecture and object-oriented technology in the Visual Studio environment. It basically realizes the overall refined management process of laboratory or instrument reservation, experimental teaching and daily management, maintenance, data export, report generation and so on. Users include administrators, students, and teachers. Software platform features complete, clear and simple, friendly interface, with strong portability, security and data statistics summary ability.

According to the demand analysis and the overall architecture model of the system platform, the constructed MIS software platform consists of three main functional modules, including administrator, teacher and student. Each kind of management module has corresponding function authority and usage instruction, respectively. Figure 1 shows the detailed entity relation (ER) diagram designed for the database of laboratory management system platform.

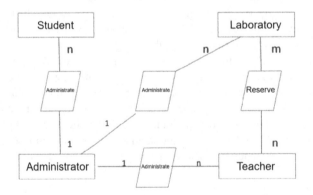

**Fig. 1.**  ER diagram of laboratory refined MIS platform.

As shown in Fig. 1, administrators and students, administrators and teachers, and administrators and laboratories are one-to-many relations, as well as teachers and laboratories are many-to-many relations.

Platform function realization and description. This system adopts the current mainstream framework and divides the interface into upper and lower areas. Meanwhile, the following area is divided into left menu and right content. The left menu shows the permissions and functions of the current user's role, and the right content area shows the contents during operation.

The main interface of refined MIS is shown in Fig. 2. When the user submits the login information, the systems first check whether the user's username and password are correct. Teachers are not able to access the administrator interface. On the opposite, students are not able to access the administrator interface, too. If the user enters the information incorrectly, the system will pop up an error message warning the user.

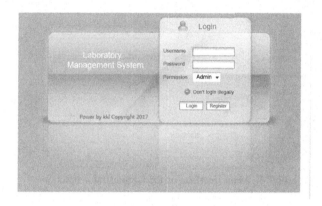

**Fig. 2.** Main interface laboratory refined MIS platform.

The main interface of the student and teacher function is shown in Fig. 3. As shown in Fig. 3, students can modify and improve their personal information, and make an appointment for the use of the required laboratory and equipment and make an appointment record inquiry.

(a)                                                        (b)

**Fig. 3.** Main interface of the student (a) and teacher (b).

Although the functions of teachers are similar to those of students, the differences in authority have been taken into account at the beginning of the design. Students can only reserve some instruments and equipment and special laboratories for undergraduate experimental teaching, while teachers can reserve all instruments and equipment and laboratories.

The main interface of the administrator function is shown in Fig. 4. The administrator has the full authority of this system platform.

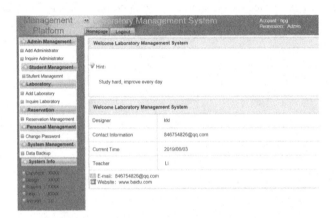

**Fig. 4.** Main interface of the administrator function.

System administrators are divided into general administrators and super administrators. Super administrator can add roles to realize the allocation and management of permissions of different users (ordinary administrators, teachers and students), as well as the summary, statistics and output report generation of each type of reservation and registration. It can provide data support for annual laboratory management and equipment usage.

## 5   Discussions

The Computer laboratory of the School of Electronic and Electrical Engineering in Shanghai University of Engineering Science has made a preliminary exploration on the refined management of the laboratory, and some consensus and achievements have been obtained on the relationship between laboratory fine management and information construction. However, considering the current mode of scientific innovation construction and talent cultivation of local applied universities, the following four problems are worth further discussions:

Laboratory refined management should fully consider the open and shared inquiry, appointment and performance assessment of existing valuable instruments and equipment in universities. The constructed MIS experimental management system is only preliminary about the sharing of laboratory instruments and equipment, without too much reference to specific sharing service mechanism and functions.

The laboratory refined management needs to be the university "a board of chess", which should be included in the laboratory management of the whole university and planned, coordinated and promoted the refined management of the laboratory from top to bottom. It should further refinement and integration of the same or similar laboratory management functions.

With the emerging of the multimedia technology means, the laboratory refined management should not only rely on Internet technology, but also rely more on new multimedia technology. For example, two-dimension code, WeChat, microblog, etc. Experimental teaching and laboratory refined management can be achieved on PC and mobile terminals by cloud platform and Internet of things (IoT) technology; it greatly enriched the technical means and content of laboratory refined management.

Laboratory refined management not only involves the contents of experimental teaching, laboratory management, instrument and equipment management and maintenance, but also involves how to evaluate the quality of these services. By linking it with department and individual performance appraisal, keeping crisis consciousness at all times, and fully mobilizes the enthusiasm of laboratory staff.

## 6  Conclusions and Future Work

As an important part of local applied university, the personnel allocation, management and service level of laboratory have become one of the key factors to cultivate innovative talents in colleges and universities and serve the social innovation. The popularity of the Internet and the rise of new media have provided the possibility of laboratory refined management. In view of many factors affecting the construction of laboratory information and refined management, local applied university should change the current stereotypes emphasizing scientific research and teaching and neglecting laboratory, update the concept of laboratory management, vigorously promote the construction of experimental technical team, and explore the unified standard of refined management of laboratory in local applied universities. It will accelerate the process of laboratory fine management and construction, and finally realizing the win-win situation of personnel training and laboratory refined management via the construction of newly laboratory refined MIS platform. The result shows that it is useful to improve the use efficiency of management level of laboratory in local applied university. In addition, laboratory refined management is a systematic engineering, involving not only network, multimedia and other technical means, but also laboratory management and technical personnel. Future work includes compare and evaluate different construction approaches and platforms of refined management with other management modes.

**Acknowledgement.** The work was supported by the Science and Technology Development Foundation of Shanghai in China under Grant No. 19142201600.

# References

1. Zhou, Y., Yao, Q.G.: Research on the construction of applied laboratory information in universities. China Manag. Informationization **20**(19), 225–226 (2017)
2. Zuo, T.Y.: The role and thinking of university laboratory construction. Res. Explor. Lab. **30**(4), 1–5 (2011)
3. Li, C.F., Zhou, S.Q., Liu, L., Song, Y., Zhao, J.J.: Exploration and practice of shared service and demonstration application for precious equipment in universities. Res. Explor. Lab. **37**(7), 296–300 (2018)
4. Zhang, H.H., Yao, J.B.: Practice and reflection on laboratory construction and management informatization of pharmacy specialty. Health Vocat. Educ. **37**(1), 87–88 (2019)
5. Song, X.H., Zhou, X.F.: Innovation and practice on construction of sharing platform for large-scale instruments and equipment of colleges and universities: taking a public platform of School of Medicine of Zhejiang University as an example. Exp. Technol. Manag. **30**(12), 37–40 (2013)
6. Georghiou, L., Halfpenny, P., Flanagan, K.: Benchmarking the provision of scientific equipment. Sci. Public Policy **28**(4), 303–311 (2001)
7. Bi, W.M.: Learning form top universities to promote the reform and development of large-scale instrument sharing platforms. Res. Explor. Lab. **29**(7), 258–261 (2011)
8. Fang, W.M., Jiang, X.Q., Meng, Y., Wang, B.Z., Wu, B., Fang, L.H.: Discussion on laboratory management and construction of university based on creative personnel cultivation. Res. Explor. Lab. **30**(3), 330–333 (2011)
9. Zhao, Y.E., He, J., Le, Y.: Constructing information management platform and strengthening laboratory safety education. Res. Explor. Lab. **34**(6), 290–293 (2015)
10. Liu, Z.M., Zhou, L., Guan, S.G., Liu, Z.Q., Min, Q.Y., Ye, P.H.: Information construction exploration on university physics laboratory. China Educ. Technol. Equip. **20**(10), 13–16 (2017)
11. Wang, H., Wang, J.: Construction of university laboratory safety management platform based on "Intelligent campus". Exp. Technol. Manag. **36**(2), 49–52 (2019)
12. Li, C.F., Liu, L., Zhao, J.J., Zhou, S.Q., Song, Y.: Exploration on promotion of popular science by laboratories in universities. Res. Explor. Lab. **38**(3), 214–217 (2019)
13. Meng, L.J., Liu, Y., Li, C.L., Jiang, D., Meng, Q.F.: Construction of comprehensive management platform with informatization of university laboratory. China Med. Equip. **16**(2), 117–120 (2019)
14. Huang, T.C., Lou, J.A., Lang, B., Li, N.: Design and implementation of informationization platform on open experimental teaching. Tech. Autom. Appl. **37**(12), 43–47 (2018)
15. Georghiou, L., Halfpenny, P.: Equipping researchers for the future. Nature **383**, 663–664 (1996)
16. Yan, W., Yuan, Y.S.: Promoting laboratory opening, supporting creative talents cultivating. Res. Explor. Lab. **28**(5), 16–17 (2009)

# Design of VNF-Mapping with Node Protection in WDM Metro Networks

Lidia Ruiz(✉) ⓘ, Ramón J. Durán ⓘ, Ignacio de Miguel ⓘ,
Noemí Merayo ⓘ, Juan Carlos Aguado ⓘ, Patricia Fernández ⓘ,
Rubén M. Lorenzo ⓘ, and Evaristo J. Abril ⓘ

Optical Communications Group, Universidad de Valladolid, Paseo de Belén, 15,
47011 Valladolid, Spain
lruiper@ribera.tel.uva.es, rduran@tel.uva.es

**Abstract.** Network Function Virtualization (NFV) is considered to be one of the enabling technologies for 5G. NFV poses several challenges, like deciding the virtual network function (VNF) placement and chaining, and adding backup resources to guarantee the survivability of service chains. In this paper, we propose a genetic algorithm that jointly solves the VNF-placement, chaining and virtual topology design problem in WDM metro ring network, with the additional capacity of providing node protection. The simulation results show how important is to solve all of these subproblems jointly, as well as the benefits of using shared VNF and network resources between backup instances in order to reduce both the service blocking ratio and the number of active CPUs.

**Keywords:** NFV · Resilience · VNF-Provisioning · VNF-Chaining · Virtual topology design · VNF-Protection

## 1 Introduction

The advent of 5G promises to increase the bandwidth to support the ever-increasing number of connected devices, and to drastically reduce latency, among other advantages. That evolution is the key factor for the development of machine-to-machine services like IoT, industry 4.0, e-healthcare services or connected vehicles. Nevertheless, 5G poses great challenges to network operators, who must face deep architectural changes in their networks. This architectural transformation, however, can be addressed thanks to the appearance of new architectural paradigms and technologies like Network Function Virtualization (NFV), Software Defined Networking (SDN) and Multi-access Edge Computing (MEC).

NFV is an architectural paradigm that deploys network functions such as firewalls or packet inspectors as virtual appliances, called Virtual Network Functions (VNF), in Commercial-Off-The-Shelf (COTS) servers, instead of using traditional proprietary hardware. This technology can increase network flexibility, reduce capital and operational expenditures, and decrease the time required for the instantiation of new services, since operators can easily deploy new network functions without the need of installing new equipment.

Q. Li et al. (Eds.): BROADNETS 2019, LNICST 303, pp. 285–298, 2019.
https://doi.org/10.1007/978-3-030-36442-7_19

Although COTS are commonly hosted at datacenters, they can also be located at the nodes of the network or even at the edge nodes, closer to the end-user, thanks to the MEC paradigm [1]. This technology provides cloud computing capabilities to edge nodes by adding IT resources to them. Therefore, it is able to bring data processing closer to the end-user, hence reducing latency [2], one of the key performance indicators of 5G.

When network functions are deployed as software using NFV, operators must solve the VNF-Placement problem, i.e., they must decide the number of instances of each VNF to run on each COTS (located in both datacenters and MEC) according to an estimation of the required services. Then, for each service request, the traffic must traverse a sequence of VNFs in a fixed order. This concatenated set of VNFs is referred to as Service Chain (SC), and operators must decide which of the instances should be employed to set up the chain (known as the VNF-chaining problem), according to the SC requirements, VNF availability and also to the available network resources. The combination of both VNF placement and chaining problems are usually known as service mapping.

In [3], we proposed a genetic algorithm for effective service mapping with virtual topology design (GASM-VTD). That algorithm solves the VNF-placement problem in WDM ring 5G-access network topologies and, after allocating the VNFs to hosting MEC nodes of the network, it also solves the virtual topology design, i.e., it decides which lightpaths must be established between MEC/datacenter nodes, the routing and wavelength assignment problem (RWA) for each lightpath and performs traffic grooming over the lightpaths to support the establishment of the SCs required by the service requests. The simulation results presented in that paper show that thanks to the use of GASM-VTD, the service blocking ratio (i.e., the probability that a service cannot be established due to the lack of resources) can be reduced while optimizing the number of active CPU cores and, therefore, the energy consumption in COTS.

However, the VNF placement and chaining problems also present extra challenges in terms of survivability against failures, and GASM-VTD does not take that problem into account. In a distributed scenario, e.g., one with datacenters and MEC resources, the failure of one node will cause the failure of multiples SCs and the overlying services [4, 5]. Moreover, if that problem is not solved properly, the failure can spread across the network. This challenge can be addressed by instantiating (but not using) backup VNFs when solving both the VNF placement and the chaining problems for each SC, using both primary and backup VNFs in case of the failure of any network node. The reason behind instantiating instead of simply reserving is avoiding the set-up time of backup VNFs in case of failure.

In this paper, we focus on the survivability problem, extending the work proposed in [3] to solve the service mapping problem including resilience against simple node failure, and therefore, protecting the SCs. For that aim, we explore two kinds of protection schemes: dedicated, in which each backup VNF protects only one primary VNF or shared, in which a backup VNF can protect multiple primary VNFs without causing any collision problem (i.e., when a backup VNF must deal with more traffic than its capacity) in case of failure. To avoid that collision problem in single-node protection scenario, a backup VNF cannot protect two or more primary VNFs hosted at the same MEC node. Furthermore, the proposed algorithm also solves the virtual topology design problem, including the virtual links required to build the backup SCs also considering dedicated and shared protection schemes for the use of these backup VNFs. In many previous studies the protection of VNFs does not take into account the network connecting the nodes. However, in this study, we present the results of a simulation study comparing different types of protection schemes and showing the importance of considering the network when solving the survivable VNF placement and chaining problems.

## 2  Related Work

Resilience in NFV placement and chaining have raised great interest in the scientific community. Hmaity et al. [6] propose three SC end-to-end protection schemes: the first one provisions a backup SC ensuring that both the physical links and hosting nodes are completely disjoint from those of the primary SC; the second only provides link protection to the primary SC; and in the third one backup nodes are disjoint, but the virtual links can share the same physical path than the primary ones. Ye et al. [7] propose a heuristic for SC mapping but not for solving the virtual topology design of the optical network. In that paper, the authors further enhance their method with a second step to provide either dedicated or shared end-to-end protection. In contrast to the previous works, which consider end-to-end protection, Beck et al. [8] propose a heuristic method for online SC provisioning which is enhanced with individual link/node protection. However, in that work, they do not consider the configuration (planning) of the network, i.e., the virtual topology design of the WDM network in case of using that technology. Casazza et al. [9] solve the reliable VNF-placement problem in geo-distributed datacentres with the objective of maximizing the VNF availability but they do not consider the chaining problem. Tomassilli et al. [10] focus only on network resilience (not VNF protection) and propose two optimization models to provide dedicated and shared protection against single-link failure. Similarly, Gao et al. [11] propose an ILP formulation and a heuristic to determine a multipath transmission scheme as a way to protect essential traffic between the VNFs hosted in data centres and connected through an elastic optical network. However, they concentrate on the SC problem but do not solve the VNF placement problem and do not implement VNF protection. Finally, Qing et al. [12] propose a method to identify the VNF hosting nodes to be protected in a network to ensure that, in case of failure, the number of surviving SCs is superior to a given threshold but they do not take into account the network design.

In contrast to previous studies, we address the VNF mapping problem with node protection in a WDM metro network equipped with MEC resources solving: (i) the VNF-placement, (ii) the VNF-chaining, (iii) the virtual topology design problems and (iv) determining the set of backup resources (VNFs and links) to protect the system against a node failure.

There are two ways of protecting a SC:

- **End-to-end SC protection**: in case of a node failure, the full SC is replaced by a complete backup SC [6, 7].

- **Individual node protection:** in case of node failure, only the affected VNFs are preplaced by its backup VNFs and the not-affected VNFs still continue being part of the SC [8, 9].

In this paper, we use the second approach, as it allows a reduction of the number of resources required (mainly when shared protection techniques are applied). In the following section we formally describe the problem statement.

## 3    Problem Statement

We assume 5G nodes connected between them and to a Central Office via a WDM ring network, as shown in Fig. 1. These nodes are MEC-enabled and host COTS to instantiate VNFs (besides offering other services). In that scenario, the algorithm must solve the VNF-placement and chaining problems for both primary and backup SCs, and design the virtual topology to be established in the WDM network.

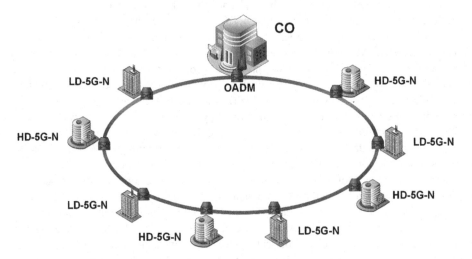

**Fig. 1.** WDM metro ring network.

In order to provide VNF protection against a node failure, a backup VNF must be assigned for each primary VNF. The backup VNF must be instantiated in a different node than the primary one. The corresponding virtual links to establish the alternative SCs must be also reserved. When a node fails, all the SCs using VNFs from that node must start using the backup VNFs. The SCs will continue using those not-affected VNFs. Suitable backup virtual links must also be used. Figure 2 shows an example of the protection-enabled SC problem for a service with three VNFs. For example, if node A fails, the SC will use backup $VNF_1$ and the rest of primary VNFs of the SC. If node B fails, the SC will start with primary $VNF_1$ and will continue with backup $VNF_2$ (node A) and backup $VNF_3$ (node D) and primary $VNF_4$ (node C). Finally, if node C fails, the SC will use primary $VNF_{1-3}$ and backup $VNF_4$ (node D).

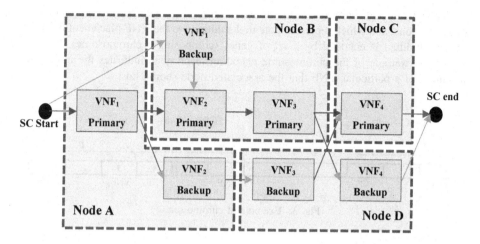

**Fig. 2.** Network resource allocation between primary and backup VNF hosting nodes.

Like in other problems related to survivability, there are two ways of delivering protection: using dedicated resources or sharing resources between those SCs that are not affected by the same node failure. Therefore, we have the following schemes regarding VNF protection:

- **Dedicated VNF Protection**: A backup VNF instance can protect only one primary VNF instance.
- **Shared VNF Protection**: A backup VNF instance can protect multiple primary VNF instances, provided the primary instances are not located in the same node.

Moreover, the network resources assigned for backup connections can be also assigned in dedicated or shared protection schemes:

- **Dedicated network resources**: connections between primary to backup resources (red arrows in Fig. 2) are dedicated for a SC.
- **Shared network resources**: different SCs can share the same network resources (wavelengths, transceivers) between primary to backup resources provided that they do not have to make use of the same backup resources at the same time in case of an individual node failure.

## 4   A Genetic Algorithm to Provide VNF Protection Against VNF/Node Failure

In this paper, we propose a new VNF-Protected Genetic Algorithm for Service Mapping and Virtual Topology Design (VNF-P-GASM-VTD) that enhances our previous proposal, GASM-VTD [3]. In contrast, VNF-P-GASM-VTD provisions backup resources for VNF protection and solves the VNF-chaining problem including the VNF chaining using backup resources as explained in the previous section.

The genetic algorithm operates with individuals, which are evolved along a number of generations, and which represent potential solutions to the VNF-placement problem. Each individual is encoded by a set of genes (composing a chromosome). Figure 3 shows an example of the chromosome encoding: each gene indicates the number of instances of a particular VNF that the associated node should host.

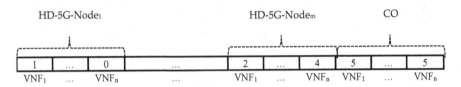

**Fig. 3.** Example of chromosome.

VNF-P-GASM-VTD follows the classical genetic loop [13]: we create an initial population of individuals which undergo classical genetic operations as crossover and mutation. In crossover, the algorithm randomly chooses two individuals of the population and a crossover point, and the genes of the second part of the chromosome, i.e., from the crossover point to the end of the chromosome, are interchanged to produce two new individuals. The offspring undergoes then a mutation operation, in which the algorithm randomly changes the values of the genes with a user-defined probability. If the algorithm cannot create the number of instances of each VNF at every hosting node as the chromosome of the resulting individual indicates, due to lack of computing resources as CPU cores, hard disk or RAM, then it is discarded and a new one is created. The algorithm repeats this process until completing a user-defined population size.

Valid individuals go then through a translation stage. The algorithm translates each individual, i.e., it creates the instances of the VNFs at the nodes as the chromosome indicates. When all instances are created, the algorithm sorts the incoming service connection requests according to the operator's chosen priority and begins the chaining process. To establish the primary chain of each SC, the algorithm uses the chaining strategy proposed in [3], referred to as GASM-VTD-Collaborative. This strategy starts looking for available VNFs, i.e., VNFs with enough free processing capacity to set up the primary chain at the 5G node to which the user is connected, also known as the local node. If the algorithm cannot find available VNFs at the local node, it looks for available ones at the CO. If it is not possible to chain VNFs hosted at the CO, then the algorithm looks for the VNFs at the other nodes of the network, starting with those of higher allocated IT capacity and then those equipped with less IT resources.

When the primary SC search finishes, the reservation of the backup VNFs starts. The VNFs already in use for any primary SC cannot be selected for backup. Then, the version with dedicated VNF protection can only use the VNFs hosted by a different node from the primary VNF to be protected and which do not protect any other VNF. The shared version can use those VNFs and also those backup VNFs that do not protect a primary VNF hosted in the same node than the primary VNF in evaluation. Like in the primary SC reservation the order of search is: CO and VNFs at the other nodes of the network, starting with those of higher allocated IT capacity and then those equipped with less IT resources. If the algorithm can create the primary SC and reserve backup VNFs, VNF-P-GASM-VTD allocates network resources to the SC.

When two consecutive VNFs of the SC are located at different nodes, then the algorithm must establish a virtual link with enough capacity to connect the nodes and establish the service. Consequently, if a lightpath with enough available bandwidth exists between the nodes, the algorithm will use it to create the virtual link, i.e., traffic grooming is allowed. Otherwise the algorithm will create a new lightpath, if there are enough network resources. Lightpaths are created using the Shortest-Path and First-Fit methods [14]. That mechanism is also used to establish the connections required by the backup SC (red arrows in Fig. 2), together with dedicated or shared protection schemes, and also employing traffic grooming. It is important to note that the lightpaths used for primary and backup SC are completely independent. When resources (VNF and network) are found for both primary and backup SC, the resources (primary and backup) are reserved and the connection is established. Otherwise, the connection is blocked.

The algorithm determines the fitness of the solution through three parameters: the service blocking ratio, the percentage of active CPU cores and the number of employed wavelengths. Then, the best individuals are selected to repeat the genetic loop, and this procedure is repeated for a number of times, or generations, defined by the user. If there are ties between two individuals in terms of service blocking ratio, the algorithm chooses the individual with less active CPU cores (and therefore, with lower energy consumption). If the individuals are also tied in this parameter, then it selects the solution which uses less wavelengths. At the end of the process, the algorithm provides the best solution, composed by the VNF-Placement, the SCs for each connection request and its corresponding backup VNF resources, and the virtual topology with primary and backup connections.

## 5   Simulation Set up and Results

In order to compare the different protection alternatives, a simulation study has been conducted using the OMNeT++ simulator. The simulation scenario has been a WDM-ring topology network (like the one of Fig. 1) with a Central Office (CO) and ten 5G nodes equipped with MEC resources with two different levels of equipment and demands: five High Demand (HD) 5G nodes and five Low Demand (LD) 5G nodes. Table 1 shows the computational resources allocated in each one of those nodes [3, 15, 16]. All the nodes have 10 Gbps optical transceivers and a ROADM, and support the switching of different number of wavelengths.

**Table 1.** IT resources for CO and XD-5G-Nodes

| Location | Computational resources |
| --- | --- |
| CO | 100 CPU cores, 480 GB RAM and 27 TB HDD |
| HD-5G-Ns | 16 CPU cores, 64 GB RAM and 10 TB HDD |
| LD-5G-Ns | 8 CPU cores, 32 GB RAM and 7 TB HDD |

Regarding traffic, we assume an operator that offers three kind of services, VoIP, Video and Web services, and users may request them with a probability of 30%, 20% and 50% respectively [3]. Each service has an associated SC and bandwidth requirements, which are shown in Table 2 [3, 15–19].

**Table 2.** Service Chain requirements.

| Service | Chained VNFs* | Bandwidth |
|---|---|---|
| VoIP | NAT-FW-TM-FW-NAT | 64 kbps |
| Video | NAT-FW-TM-VOC-IDPS | 4 Mbps |
| Web Services | NAT-FW-TM-WOC-IDPS | 100 kbps |

*\* NAT:Network Address Translator, FW: Firewall, TM: Traffic Monitor, WOC: WAN Optimization Controller, VOC: Video Optimization Controller, IDPS: Intrusion Detection Prevention System.*

Furthermore, each VNF has associated hardware requirements and processing capacity, which are shown in Table 3.

**Table 3.** VNF HW requirements and processing capacity.

| Service | HW requirements | Throughput |
|---|---|---|
| NAT | CPU: 2 cores, RAM: 4 GB, HDD: 16 GB | 2 Gbps [20] |
| FW | CPU: 2 cores, RAM: 4 GB, HDD: 16 GB | 2 Gbps [20] |
| TM | CPU: 1 core, RAM: 2 GB, HDD: 16 GB | 1 Gbps [21] |
| VOC | CPU: 2 cores, RAM: 4 GB, HDD: 2 GB | 2 Gbps* |
| WOC | CPU: 1 core, RAM: 2 GB, HDD: 40 GB | 0.5 Gbps [22] |
| IDPS | CPU: 2 cores, RAM: 4 GB, HDD: 8 GB | 1 Gbps [23] |

*\* Values shown in table are derived from the figures of the other VNFs.*

We defined the parameter $\bar{u}$ as the number of average users per HD-5G-node. The connected users to each HD-5G node is randomly generated in each simulation using a uniform distribution between $[0, 2\bar{u}]$. The connected users to LD-5G-nodes are also randomly generated using a uniform distribution between $[0, 2\bar{u}/10]$. For each number of $\bar{u}$, we repeated the simulation 500 times and all the graphs are shown with 95% confidence intervals.

VNF-P-GASM-VTD has been configured to create 10 individuals per generation and the finish criterion was set to the evolution during 50 generations. Different versions of the algorithm have been compared to analyze the influence of selecting the different types of protections:

- **No protection**: no backup resources are applied. This algorithms is GASM-VTD [3].
- **Dedicated-VNF protection with no network restrictions (Ideal D-VNF** in figures): ideal case where the VNF-P-GASM-VTD is designed to provide

node-protection but without imposing any kind of restriction to the WDM network, i.e., considering unlimited number of optical transceivers and wavelength channels.

- **Shared-VNF protection with no network restrictions** (**Ideal S-VNF** in figures) similar as the previous one (no network restrictions) but using shared-VNF protection.
- VNF-P-GASM-VTD implementing **Dedicated-VNF protection with Dedicated network resources** for the backup (**D-VNF & D-Net** in figures).
- VNF-P-GASM-VTD implementing **Dedicated-VNF protection with Shared network resources** for the backup (**D-VNF & S-Net** in figures).
- VNF-P-GASM-VTD implementing **Shared-VNF protection with Dedicated network resources** for the backup (**S-VNF & D-Net** in figures).
- VNF-P-GASM-VTD implementing **Shared-VNF protection with Shared network resources** for the backup (**S-VNF & S-Net** in figures).

Initially, we have assumed that the network is equipped with optical equipment that allows the use of 10 wavelengths. Figure 4 shows the Service Blocking Ratio (SBR) as a function of the number of users and with the different protection schemes. The corresponding values of the percentage of actives CPU cores are shown in Fig. 5. In that figure it is shown the total number of CPUs in use considering both the ones dedicated for primary VNFs and the ones reserved for backup.

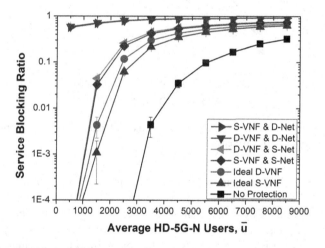

**Fig. 4.** Service Blocking Ratio for the different protection schemes when the WDM network can use 10 wavelengths.

Figure 4 shows that implementing protection leads to an increment in the service blocking ratio even in the case of not considering network restrictions (like other studies do). This is due to the fact that the computing resources in nodes (CO and MEC) are limited. Comparing the versions with protection and without considering network restrictions (ideal) with the ones that consider those restrictions, it is possible to see the huge difference in performance. Therefore, when implementing protection (even when only nodes or VNFs are protected) is essential to take into account the network connecting the nodes as it will be a very restricting factor. It should be noticed that, in the case of using end-to-end protection, the results are even worse as more resources have to be employed for establishing the backup SC.

Then, comparing the different versions of the algorithm with protection, it can be seen that those versions using dedicated network resources for protection are the ones that show a higher SBR. In fact, the values obtained make unfeasible to use these options in a real network. Again, the impact of network communication resources is evident. Finally, the use of shared VNF protection instead of dedicated protection allows a reduction in the SBR for both ideal and the version using shared network resources.

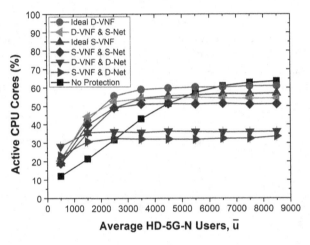

**Fig. 5.** Percentage of total active CPU cores for the different protection schemes when the WDM network can use 10 wavelengths.

Regarding Fig. 5, the scheme without protection is the one that uses the smallest number of CPU cores as it does not reserve backup VNFs. Then, shared VNF policies make a better use of the CPU cores than their dedicated counterparts. Therefore, the use of shared VNF schemes is not only better than using dedicated ones only in terms of resources in use (what is obvious due to the fact of sharing) but also in terms of SBR. Finally, the versions that use dedicated network resources for backup network require less CPU cores than the other alternatives when the number of users grows. This happens because, network resources become the limiting factor before the computing resources, so that there are still CPU resources available to instantiate more VNFs, but there is not enough network capacity to support communication between VNFs.

Figure 6 shows a comparison of the percentage of active CPU cores (out of the total number of CPU cores) allocated for primary and backup VNFs with the different protection schemes with three number of users: $\bar{u} = 500$, $\bar{u} = 2{,}500$ and $\bar{u} = 4{,}500$.

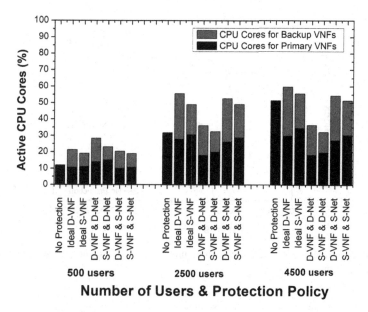

**Fig. 6.** Percentage active CPU cores (out of the total CPU cores of the network) for the different protection schemes when the WDM network can use 10 wavelengths distinguishing the ones used for primary and backup.

Figure 6 shows that, as the number of users increases, higher is the number of active CPUs. However, the difference between the $\bar{u} = 500$ case and the $\bar{u} = 2,500$ case is higher than the difference between $\bar{u} = 2,500$ case and the $\bar{u} = 4,500$ (except for the case of no protection) because in the second case the unavailability of network resources is the limiting factor when establishing new SCs. When no protection is used, the number of used resources (CPUs and wavelengths/transceivers) is lower as none of them has to be used for backup. When comparing CPU utilization for the policies with dedicated network resources (D-Net policies), it is lower than the CPU utilization for shared-network (S-Net) policies. This behavior is again due to the fact that network resources become the limiting factor before the computing resources (there are still CPU resources available to instantiate more VNFs, but there is not enough network capacity to support communication between VNFs). Comparing the percentage of CPUs for primary and backup VNFs, the percentages of CPUs for backup with shared-VNF policies is lower than those of the dedicated-VNF policies (which, obviously, use the same number of active CPUs for backup and primary VNFs, since there is a one-to-one relationship). Finally, shared-VNF policies use a lower number of active CPUs than the dedicated versions (D-VNF) since the formers aim at reutilizing backup resources among different primary VNFs hosted at different nodes.

The corresponding values of SBR and percentage of CPUs in use when the number of network resources increases (the number of wavelength grows up to 20) are shown in Fig. 7 and Fig. 8 respectively.

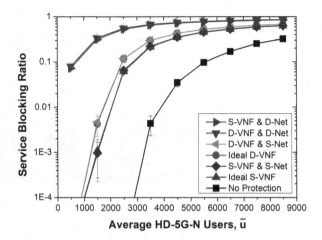

**Fig. 7.** Service Blocking Ratio for the different protection schemes when the WDM network can use 20 wavelengths.

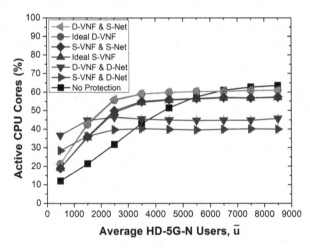

**Fig. 8.** Percentage of total active CPU cores for the different protection schemes when the WDM network can use 20 wavelengths.

When the number of network resources increases (Figs. 7 and 8), the results in terms of both SBR and active CPU cores improve and the versions that implement the protections using shared network resources for the backup get the same performance of the ideal ones (when no network restrictions is taking into account). However, this is done at the expense of increasing the CAPEX (note that 20 wavelengths are required for a metro network with only 10 nodes). The results of the versions using dedicated network resources for backup also improves its results, although high SBR ($>10^{-1}$) are still obtained.

In conclusion, the results show that when solving the VNF placement and chaining problems with node protection is essential to take into account the network topology and design it at the same time when solving the problem. Moreover, for an efficient use of resources, the reutilization of network resources for backup connections is almost mandatory and sharing backup VNFs can also reduce both the service blocking ratio and the number of active CPUs (and, therefore, the energy consumption).

## 6   Conclusions

In this paper, we have addressed the VNF-provisioning and chaining problems including node protection in a WDM metro network. The method also solves the virtual network design of the WDM network. Different methods for protection have been implemented and compared using shared or dedicated resources for the backup in both the VNF and network levels.

The results of a simulation study show that, in contrast with other proposal, it is essential to solve all those problems (including node protection) taking into consideration the network restrictions and its design. Among the different schemes of protection, the one that shares backup resources at both NFV and network levels is the one that achieves the best results in terms of both service blocking ratio and number of active resources (and therefore, in energy consumption).

**Acknowledgements.** This work has been supported by Spanish Ministry of Economy and Competitiveness (TEC2017-84423-C3-1-P), the fellowship program of the Spanish Ministry of Industry, Trade and Tourism (BES 2015-074514) and the INTERREG V-A España-Portugal (POCTEP) program (0677_DISRUPTIVE_2_E).

## References

1. Savi, M., Tornatore, M., Verticale, G.: Impact of processing costs on service chain placement in network functions virtualization. In: Proceedings of the 2015 IEEE Conference on Network Function Virtualization and Software Defined Network (NFV-SDN), pp. 191–197. IEEE, San Francisco (2015)
2. Patel, M., Naughton, B., Chan, C., Sprecher, N., Abeta, S., Neal, A.: Mobile-edge computing introductory technical white paper. Mob.-Edge Comput. (MEC) Ind. Initiative (2014)
3. Ruiz, L., et al.: Joint VNF-provisioning and virtual topology design in 5G optical metro networks. In: Proceedings of the 2019 21st International Conference of Transparent Optical Networks (ICTON), pp. 1–4. IEEE, Angers (2019)
4. ETSI: GS NFV-REL 003 V1.1.1 Network Functions Virtualisation (NFV); Reliability; Report on Models and Features for End-to-End Reliability
5. Casazza, M., Bouet, M., Secci, S.: Availability-driven NFV orchestration. Comput. Netw. **155**, 47–61 (2019). https://doi.org/10.1016/j.comnet.2019.02.017
6. Hmaity, A., Savi, M., Musumeci, F., Tornatore, M., Pattavina, A.: Virtual network function placement for resilient service chain provisioning. In: 2016 8th International Workshop on Resilient Networks Design and Modeling (RNDM), pp. 245–252. IEEE, Halmstad (2016). https://doi.org/10.1109/RNDM.2016.7608294

7. Ye, Z., Cao, X., Wang, J., Yu, H., Qiao, C.: Joint topology design and mapping of service function chains for efficient, scalable, and reliable network functions virtualization. IEEE Netw. **30**, 81–87 (2016). https://doi.org/10.1109/MNET.2016.7474348

8. Beck, M.T., Botero, J.F., Samelin, K.: Resilient allocation of service Function chains. In: 2016 IEEE Conference on Network Function Virtualization and Software Defined Networks (NFV-SDN), pp. 128–133 (2016). https://doi.org/10.1109/NFV-SDN.2016.7919487

9. Casazza, M., Fouilhoux, P., Bouet, M., Secci, S.: Securing virtual network function placement with high availability guarantees. In: 2017 IFIP Networking Conference (IFIP Networking) and Workshops, pp. 1–9 (2017). https://doi.org/10.23919/IFIPNetworking. 2017.8264850

10. Tomassilli, A., Huin, N., Giroire, F., Jaumard, B.: Resource requirements for reliable service function chaining. In: 2018 IEEE International Conference on Communications (ICC), pp. 1–7 (2018). https://doi.org/10.1109/ICC.2018.8422774

11. Gao, T., Li, X., Zou, W., Huang, S.: Survivable VNF placement and scheduling with multipath protection in elastic optical datacenter networks. In: 2019 Optical Fiber Communications Conference and Exhibition (OFC), pp. 1–3 (2019)

12. Qing, H., Weifei, Z., Julong, L.: Virtual network protection strategy to ensure the reliability of SFC in NFV. In: Proceedings of the 6th International Conference on Information Engineering - ICIE 2017, pp. 1–5. ACM Press, Dalian Liaoning (2017). https://doi.org/10. 1145/3078564.3078583

13. Goldberg, D.: Genetic Algorithms in Optimization Search and Machine Learning. Addison-Wesley, Reading (1989)

14. Zang, H., Jue, J.P., Mukherjee, B., et al.: A review of routing and wavelength assignment approaches for wavelength-routed optical WDM networks. Opt. Netw. Mag. **1**, 47–60 (2000)

15. Pedreno-Manresa, J.-J., Khodashenas, P.S., Siddiqui, M.S., Pavon-Marino, P.: Dynamic QoS/QoE assurance in realistic NFV-enabled 5G access networks. In: 2017 19th International Conference on Transparent Optical Networks (ICTON), pp. 1–4. IEEE, Girona (2017). https://doi.org/10.1109/ICTON.2017.8025149

16. Pedreno-Manresa, J.-J., Khodashenas, P.S., Siddiqui, M.S., Pavon-Marino, P.: On the need of joint bandwidth and NFV resource orchestration: a realistic 5G access network use case. IEEE Commun. Lett. **22**, 145–148 (2018). https://doi.org/10.1109/LCOMM.2017.2760826

17. Savi, M., Tornatore, M., Verticale, G.: Impact of processing costs on service chain placement in network functions virtualization. In: Proceedings of the 2015 IEEE Conference on Network Function Virtualization and Software Defined Network (NFV-SDN), pp. 191–197. IEEE, San Francisco (2015). https://doi.org/10.1109/NFV-SDN.2015.7387426

18. Savi, M., Hmaity, A., Verticale, G., Höst, S., Tornatore, M.: To distribute or not to distribute? Impact of latency on Virtual Network Function distribution at the edge of FMC networks. In: 2016 18th International Conference on Transparent Optical Networks (ICTON), pp. 1–4. IEEE (2016)

19. Ruiz, L., et al.: A genetic algorithm for VNF provisioning in NFV-Enabled Cloud/MEC RAN architectures. Appl. Sci. **8**, 2614 (2018). https://doi.org/10.3390/app8122614

20. Juniper Networks: vSRX Virtual Firewall (2018)

21. Brocade Communications System: Virtual Traffic Manager (2015). https://www.accyotta. com/assets/uploads/docs/Brocade_-_Virtual_Traffic_Manager.pdf

22. Talari Networks: Talari SD-WAN Solutions (2018)

23. Cisco: Cisco Adaptive Security Virtual Appliance (ASAv) (2018)

# Author Index

Printed in the United States
By Bookmasters